AF365432

Remote Sensing Technology Applications in Forestry and REDD+

Remote Sensing Technology Applications in Forestry and REDD+

Special Issue Editors

Kim Calders
Inge Jonckheere
Joanne Nightingale
Mikko Vastaranta

MDPI • Basel • Beijing • Wuhan • Barcelona • Belgrade • Manchester • Tokyo • Cluj • Tianjin

Special Issue Editors

Kim Calders
Ghent University
Belgium

Inge Jonckheere
FAO of the United Nations
Italy

Joanne Nightingale
National Physical Laboratory
UK

Mikko Vastaranta
University of Eastern Finland
Finland

Editorial Office
MDPI
St. Alban-Anlage 66
4052 Basel, Switzerland

This is a reprint of articles from the Special Issue published online in the open access journal *Forests* (ISSN 1999-4907) (available at: https://www.mdpi.com/journal/forests/special_issues/RS_REDD).

For citation purposes, cite each article independently as indicated on the article page online and as indicated below:

LastName, A.A.; LastName, B.B.; LastName, C.C. Article Title. *Journal Name* **Year**, *Article Number*, Page Range.

ISBN 978-3-03928-470-2 (Pbk)
ISBN 978-3-03928-471-9 (PDF)

© 2020 by the authors. Articles in this book are Open Access and distributed under the Creative Commons Attribution (CC BY) license, which allows users to download, copy and build upon published articles, as long as the author and publisher are properly credited, which ensures maximum dissemination and a wider impact of our publications.

The book as a whole is distributed by MDPI under the terms and conditions of the Creative Commons license CC BY-NC-ND.

Contents

About the Special Issue Editors

Kim Calders received B.S. and M.S. degrees in bioscience engineering from KULeuven, Leuven, Belgium, in 2006 and 2008, respectively, a M.S. degree in remote sensing from University College London (UCL), London, U.K., in 2010, and a Ph.D. degree in LiDAR remote sensing from Wageningen University, Wageningen, The Netherlands, in 2015. He was a post-doctoral researcher with the National Physical Laboratory and the Department of Geography, UCL, in the U.K. from 2015 to 2017. Since 2017, he has been a post-doctoral researcher with CAVElab, Ghent University, Ghent, Belgium and an honorary fellow at the School of Earth and Environmental Sciences, University of Queensland, Brisbane, Australia. His research interests include measurements of full 3D vegetation structure and the relation to airborne or spaceborne remote sensing data and forest functioning.

Inge Jonckheere was born in Ghent, she has a MS. in Environmental Engineering and obtained her Ph.D. in applied bioscience and engineering in 2005, at the Katholieke Universiteit Leuven. She worked at the Università degli di Milano, Italy, and has been a team leader and Guest Lecturer of the Geomatics Group at the Department of Land Management at the KUL, combined with a mandate as a science program manager for the environment at the European Science Foundation, Strasbourg, France, in the Life, Earth and Environmental Sciences (LESC) unit. She was granted a postdoctoral fellowship from the Research Foundation Flanders, and started working at FAO as a remote sensing expert in the Forestry Department in 2009. Since 2010, she has been working as a remote sensing expert and team leader of Remote Sensing and Geoportals in the UN-REDD program, supporting developing countries to set up their national monitoring reporting and verification (MRV) systems in the framework of the reducing emissions from deforestation and forest degradation (REDD+) mechanism under the UNFCCC. She has been an official reviewer for Belgium for LULUCF at the UNFCCC since 2012 and is a lead author for IPCC, including the upcoming Assessment Report 6. Her main fields of interest and expertise are vegetation monitoring, (digital) image processing, remote sensing applications, earth observation, and environmental modeling for climate impact.

Joanne Nightingale obtained her Ph.D. in geography and remote sensing from the University of Queensland, Australia and completed two post-doctoral research positions at universities within Canada and the United States. Dr. Nightingale joined the Earth Observation, Climate and Optical Group at the National Physical Laboratory (London, U.K.) in 2013 from NASA's Goddard Space Flight Center, USA, and has over 12 years' experience in coordinating Earth Observation System global land product validation activities. Joanne chaired the CEOS Working Group on Calibration and Validation, sub-group for Land Product Validation 2010–2013, and continues to be involved in international activities involving assessing the quality of information about forests derived from in situ, near surface, and Earth Observation Satellites; improving global satellite-derived biophysical product validation strategies; and contributing to good practice guidance for the evaluation of essential climate data records.

Mikko Vastaranta is currently an associate professor in digitalization and knowledge leadership in forest-based bioeconomy at the School of Forest Sciences at the University of Eastern Finland. He received his M.S. (2007) and Ph.D. (2012) in forest resource science and technology from the University of Helsinki, where he also holds an adjunct professorship of remote sensing of forests.

Over the last five years, he has worked as a research scientist at the Centre of Excellence in Laser Scanning Research (CoE-LaSR), university lecturer in forest planning (University of Helsinki), and as a visiting research scientist at the Pacific Forestry Centre (Canadian Forestry Service, Natural Resources Canada). His current research interests include detailed remote sensing of forests for fostering ecological understanding and supporting sustainable climate-smart forestry. He has published more than 100 peer-reviewed scientific articles.

Editorial

Remote Sensing Technology Applications in Forestry and REDD+

Kim Calders [1,*], Inge Jonckheere [2], Joanne Nightingale [3] and Mikko Vastaranta [4]

[1] CAVElab—Computational & Applied Vegetation Ecology, Faculty of Bioscience Engineering, Ghent University, 9000 Ghent, Belgium

[2] FAO of the United Nations, Forestry Department, viale delle Terme di Caracalla, 00153 Roma, Italy; inge.jonckheere@fao.org

[3] Earth Observation, Climate and Optical Group, National Physical Laboratory, Teddington TW11 0LW, UK; joanne.nightingale@npl.co.uk

[4] School of Forest Sciences, University of Eastern Finland, Yliopistokatu 7, P.O.Box 111, FI-80101 Joensuu, Finland; mikko.vastaranta@uef.fi

* Correspondence: kim.calders@ugent.be

Received: 4 February 2020; Accepted: 5 February 2020; Published: 7 February 2020

Abstract: Advances in close-range and remote sensing technologies drive innovations in forest resource assessments and monitoring at varying scales. Data acquired with airborne and spaceborne platforms provide us with higher spatial resolution, more frequent coverage and increased spectral information. Recent developments in ground-based sensors have advanced three dimensional (3D) measurements, low-cost permanent systems and community-based monitoring of forests. The REDD+ mechanism has moved the remote sensing community in advancing and developing forest geospatial products which can be used by countries for the international reporting and national forest monitoring. However, there still is an urgent need to better understand the options and limitations of remote and close-range sensing techniques in the field of degradation and forest change assessment. This Special Issue contains 12 studies that provided insight into new advances in the field of remote sensing for forest management and REDD+. This includes developments into algorithm development using satellite data; synthetic aperture radar (SAR); airborne and terrestrial LiDAR; as well as forest reference emissions level (FREL) frameworks.

Keywords: airborne laser scanning; terrestrial laser scanning; remote sensing; REDD+; forestry

---

Forest ecosystems cover approximately 31% of the world's land area, with a total forested area of approximately 4 billion hectares [1]. Forests play an important role in today's society and serve as a source for the production of paper products, lumber and fuel wood. In addition, forests produce freshwater from mountain watersheds, purify the air, offer habitat to wildlife and offer recreational opportunities among many other ecosystem services. To keep forests productive, and ecological and recreational functions balanced, accurate and precise information about forest structure and its biophysical parameters are needed for supporting informed decision making and sustainable management [2].

Many decisions made by natural resource managers and policymakers regarding forests are poorly linked with the spatial scales covered by conventional forest inventory methods. Remote sensing is seen as one of the key data sources to fill existing forest monitoring information gaps, particularly in many developing countries [3]. Information retrieved through remote sensing, especially through space- and airborne acquisition methods, can offer a synoptic view over large or inaccessible areas. The Forest Resources Assessment [4] supports global tree cover and forest land use monitoring. Monitoring programs implement a systematic framework to obtain information about changes in forest cover and forest land use changes on a global scale. On the global scale, the land use of approximately 12 million hectares of forest ecosystems changed in 2018 (World Resources Institute), making geographically extensive and spatially detailed forest monitoring an important task. Furthermore, deforestation and forest degradation account for about 12% of global anthropogenic carbon emissions, which is second only to fossil fuel combustion [5,6]. However, this estimate is quite uncertain due to inadequate estimates of forest carbon stocks and is expected to range from 6% to 17%. Carbon emissions are partially compensated by forest growth, forestation and the rebuilding of soil carbon pools following afforestation. However, the global distribution of terrestrial carbon sinks and sources is highly uncertain. Initiatives constraining the inaccuracy in forest carbon estimates are essential to the development of new techniques and methodologies for supporting information needs of effective forest management and future climate mitigation actions [7,8].

Advances in close-range and remote sensing technologies drive innovations in forest resource assessments and monitoring at varying scales. Data acquired with airborne and spaceborne platforms provide us with higher spatial resolution, more frequent coverage and increased spectral information. Recent developments in ground-based sensors have advanced 3D measurements, low-cost permanent systems and community-based monitoring of forests. REDD+ is a climate change mitigation programme that is supported by implementation initiatives such as the United Nations Collaborative Programme on Reducing Emissions from Deforestation and Forest Degradation (UN-REDD) [9]. The commitments and requirements for countries to participate in the REDD+ mechanism have encouraged the remote sensing community in advancing and developing forest geospatial products that can be used by countries for national forest monitoring and international reporting. However, there still is an urgent need to better understand the methodological options and limitations of remote and close-range sensing techniques in the field of forest degradation and change monitoring.

This Special Issue contains 12 studies that provided insight into new advances in the field of remote sensing for forest management and REDD+. This included developments into (1) algorithm development using satellite data [10–16]; (2) synthetic aperture radar (SAR) [11,17]; (3) airborne [18] and terrestrial [19,20] LiDAR; and (4) forest reference emissions level (FREL) frameworks [21]. Chen et al. [11] combine texture characteristics and backscatter coefficients of Sentinel-1 with multispectral information derived from Sentinel-2 and traditional field inventory data to develop above-ground biomass (AGB) prediction models using machine learning. Shen et al. [12] apply machine learning techniques, using Landsat-5 Thematic Mapper (TM) and Landsat-8 Operational Land Imager (OLI) images to monitor the five-year change in AGB over three regions with different topographic conditions in Zhejiang Province, China. Li et al. [13] test various statistical frameworks on Landsat-8 OLI data to improve AGB mapping over a subtropical forest in Western Hunan in Central China. Zhang et al. [17] explore the use of advanced land observing satellite-2 (ALOS-2) phased array-type L-band synthetic aperture radar (PALSAR-2) full polarimetric SAR data to estimate forest growing stock volume in a region with challenging terrain conditions. Blinn et al. [16] derive leaf area index (LAI) from Landsat-7 (ETM+) and Landsat-8 (OLI) vegetation indices. Zhao et al. [10] use machine learning on both structural and spectral indices from QuickBird multispectral and panchromatic images to map forest canopy cover. Spracklen and Spracklen [14] demonstrate the use of machine learning with Sentinel-2 images for identifying old-growth forests in Europe. Gigovic et al. [15] create a remote sensing (MODIS, Landsat-8 OLI and Worldview-2) derived forest inventory map to train a machine learning algorithm to predict forest fire susceptibility.

Random forest (RF) has been the most popular machine learning algorithm to link remote sensing data to forest structural attributes such as AGB [11,12], canopy cover [10] or forest fire susceptibility [15]. Spracklen and Spracklen [14] apply RF to identify old-growth forests using remote sensing data. Chen et al. [11] and Gigovic et al. [15] test different machine learning approaches for predicting AGB. Whereas Chen et al. [11] identify that RF was not always the most suitable method for predicting and mapping AGB, Gigovic et al. [15] obtain comparable results between RF and support vector machine approaches.

The LiDAR studies in this Special Issue present work on a regional scale using airborne LiDAR [18], as well as tree-level assessment of LAI [19] and AGB [20] from terrestrial LiDAR. Often, LiDAR is not available over large continuous areas but can be essential for the calibration and validation (cal/val) of many forest map products that have been derived using coarser resolution satellite observations. Duncanson et al. [22] argue that spatially continuous maps of forest attributes are essential for programs as REDD+, but their accuracy might be challenged if appropriate reference data, such as airborne or terrestrial LiDAR, is not available for calibration and validation. Advances in remote and close-range sensing techniques will be critical to implement scientific output in developing a forest reference emissions level (FREL) for countries in the REDD+ context to account for likely future developments. In this Special Issue, Pirker et al. [21] demonstrate this approach for Southern Cameroon. Their work identifies the priorities for creating and improving the necessary data, information and infrastructure for improving each element of the FREL with the ultimate objective of developing a FREL for a performance-based payment program.

Finally, we would like to thank the authors of the Special Issue for their contributions and additionally thank the reviewers and the Forests Editorial Office for conducting a high-quality review process for all of the published papers. We hope that this Special Issue will foster the remote sensing science and policies related to forestry and REDD+.

Author Contributions: All authors have read and agreed to the published version of the manuscript. Writing—original draft preparation, K.C.; writing—review and editing, all the authors.

Conflicts of Interest: The authors declare no conflict of interest.

References

1. United Nations. The Sustainable Development Goals Report. 2017. Available online: https://unstats.un.org/sdgs/files/report/2017/TheSustainableDevelopmentGoalsReport2017.pdf (accessed on 6 February 2020).
2. Waring, R.H.; Running, S.W. Forest Ecosystem Analysis at Multiple Time and Space Scales, 3rd ed. 2007, pp. 1–16. Available online: https://booksite.elsevier.com/samplechapters/9780123706058/Sample_Chapters/02~{}Chapter_1.pdf (accessed on 6 February 2020).
3. Romijn, E.; Herold, M.; Kooistra, L.; Murdiyarso, D.; Verchot, L. Assessing capacities of non-Annex I countries for national forest monitoring in the context of REDD+. *Environ. Sci. Policy* **2012**, *19–20*, 33–48. [CrossRef]
4. Food and Agriculture Organization of the United Nations. *Global Forest Resources Assessment 2015: How are the World's Forests Changing?* 2nd ed.; Food & Agriculture Organization of the United Nations: Rome, Italy, 2018; ISBN 9789251092835.
5. Van der Werf, G.R.; Morton, D.C.; DeFries, R.S.; Olivier, J.G.J.; Kasibhatla, P.S.; Jackson, R.B.; Collatz, G.J.; Randerson, J.T. CO_2 emissions from forest loss. *Nat. Geosci.* **2009**, *2*, 737–738. [CrossRef]
6. Le Quéré, C.; Raupach, M.R.; Canadell, J.G.; Marland, G.; Bopp, L.; Ciais, P.; Conway, T.J.; Doney, S.C.; Feely, R.A.; Foster, P.; et al. Trends in the sources and sinks of carbon dioxide. *Nat. Geosci.* **2009**, *2*, 831–836. [CrossRef]
7. Houghton, R.A.; Hall, F.; Goetz, S.J. Importance of biomass in the global carbon cycle. *J. Geophys. Res. Biogeosci.* **2009**, *114*, G00E03. [CrossRef]
8. Pan, Y.; Birdsey, R.A.; Fang, J.; Houghton, R.; Kauppi, P.E.; Kurz, W.A.; Phillips, O.L.; Shvidenko, A.; Lewis, S.L.; Canadell, J.G.; et al. A large and persistent carbon sink in the world's forests. *Science* **2011**, *333*, 988–993. [CrossRef] [PubMed]
9. United Nations. United Nations UN-REDD Programme. 2014. Available online: www.un-redd.org (accessed on 6 February 2020).
10. Zhao, Q.; Wang, F.; Zhao, J.; Zhou, J.; Yu, S.; Zhao, Z. Estimating Forest Canopy Cover in Black Locust (Robinia pseudoacacia L.) Plantations on the Loess Plateau Using Random Forest. *Forests* **2018**, *9*, 623. [CrossRef]
11. Chen, L.; Ren, C.; Zhang, B.; Wang, Z.; Xi, Y. Estimation of Forest Above-Ground Biomass by Geographically Weighted Regression and Machine Learning with Sentinel Imagery. *Forests* **2018**, *9*, 582. [CrossRef]
12. Shen, A.; Wu, C.; Jiang, B.; Deng, J.; Yuan, W.; Wang, K.; He, S.; Zhu, E.; Lin, Y.; Wu, C. Spatiotemporal Variations of Aboveground Biomass under Different Terrain Conditions. *Forests* **2018**, *9*, 778. [CrossRef]

13. Li, C.; Li, Y.; Li, M. Improving Forest Aboveground Biomass (AGB) Estimation by Incorporating Crown Density and Using Landsat 8 OLI Images of a Subtropical Forest in Western Hunan in Central China. *Forests* **2019**, *10*, 104. [CrossRef]

14. Spracklen, B.D.; Spracklen, D.V. Identifying European Old-Growth Forests using Remote Sensing: A Study in the Ukrainian Carpathians. *Forests* **2019**, *10*, 127. [CrossRef]

15. Gigović, L.; Pourghasemi, H.R.; Drobnjak, S.; Bai, S. Testing a New Ensemble Model Based on SVM and Random Forest in Forest Fire Susceptibility Assessment and Its Mapping in Serbia's Tara National Park. *Forests* **2019**, *10*, 408. [CrossRef]

16. Blinn, C.E.; House, M.N.; Wynne, R.H.; Thomas, V.A.; Fox, T.R.; Sumnall, M. Landsat 8 Based Leaf Area Index Estimation in Loblolly Pine Plantations. *Forests* **2019**, *10*, 222. [CrossRef]

17. Zhang, H.; Zhu, J.; Wang, C.; Lin, H.; Long, J.; Zhao, L.; Fu, H.; Liu, Z. Forest Growing Stock Volume Estimation in Subtropical Mountain Areas Using PALSAR-2 L-Band PolSAR Data. *Forests* **2019**, *10*, 276. [CrossRef]

18. Zörner, J.; Dymond, J.R.; Shepherd, J.D.; Wiser, S.K.; Jolly, B. LiDAR-Based Regional Inventory of Tall Trees—Wellington, New Zealand. *Forests* **2018**, *9*, 702. [CrossRef]

19. Deng, Y.; Yu, K.; Yao, X.; Xie, Q.; Hsieh, Y.; Liu, J. Estimation of Pinus massoniana Leaf Area Using Terrestrial Laser Scanning. *Forests* **2019**, *10*, 660. [CrossRef]

20. Lau, A.; Calders, K.; Bartholomeus, H.; Martius, C.; Raumonen, P.; Herold, M.; Vicari, M.; Sukhdeo, H.; Singh, J.; Goodman, R.C. Tree Biomass Equations from Terrestrial LiDAR: A Case Study in Guyana. *Forests* **2019**, *10*, 527. [CrossRef]

21. Pirker, J.; Mosnier, A.; Nana, T.; Dees, M.; Momo, A.; Muys, B.; Kraxner, F.; Siwe, R. Determining a Carbon Reference Level for a High-Forest-Low-Deforestation Country. *Forests* **2019**, *10*, 1095. [CrossRef]

22. Duncanson, L.; Armston, J.; Disney, M.; Avitabile, V.; Barbier, N.; Calders, K.; Carter, S.; Chave, J.; Herold, M.; Crowther, T.W.; et al. The Importance of Consistent Global Forest Aboveground Biomass Product Validation. *Surv. Geophys.* **2019**, *40*, 979–999. [CrossRef] [PubMed]

 © 2020 by the authors. Licensee MDPI, Basel, Switzerland. This article is an open access article distributed under the terms and conditions of the Creative Commons Attribution (CC BY) license (http://creativecommons.org/licenses/by/4.0/).

 *forests*

Article

Estimation of Forest Above-Ground Biomass by Geographically Weighted Regression and Machine Learning with Sentinel Imagery

Lin Chen [1,2], Chunying Ren [1], Bai Zhang [1,*], Zongming Wang [1] and Yanbiao Xi [1,2]

[1] Key Laboratory of Wetland Ecology and Environment, Northeast Institute of Geography and Agroecology, Chinese Academy of Sciences, Changchun 130102, China; chenlin@neigae.ac.cn (L.C.); renchy@neigae.ac.cn (C.R.); zongmingwang@neigae.ac.cn (Z.W.); xiyanbiao@neigae.ac.cn (Y.X.)

[2] University of Chinese Academy of Sciences, Beijing 100049, China

* Correspondence: zhangbai@neigae.ac.cn; Tel.: +86-431-8554-2222

Received: 31 August 2018; Accepted: 19 September 2018; Published: 20 September 2018

Abstract: Accurate forest above-ground biomass (AGB) is crucial for sustaining forest management and mitigating climate change to support REDD+ (reducing emissions from deforestation and forest degradation, plus the sustainable management of forests, and the conservation and enhancement of forest carbon stocks) processes. Recently launched Sentinel imagery offers a new opportunity for forest AGB mapping and monitoring. In this study, texture characteristics and backscatter coefficients of Sentinel-1, in addition to multispectral bands, vegetation indices, and biophysical variables of Sentinal-2, based on 56 measured AGB samples in the center of the Changbai Mountains, China, were used to develop biomass prediction models through geographically weighted regression (GWR) and machine learning (ML) algorithms, such as the artificial neural network (ANN), support vector machine for regression (SVR), and random forest (RF). The results showed that texture characteristics and vegetation biophysical variables were the most important predictors. SVR was the best method for predicting and mapping the patterns of AGB in the study site with limited samples, whose mean error, mean absolute error, root mean square error, and correlation coefficient were 4×10^{-3}, 0.07, 0.08 Mg·ha^{-1}, and 1, respectively. Predicted values of AGB from four models ranged from 11.80 to 324.12 Mg·ha^{-1}, and those for broadleaved deciduous forests were the most accurate, while those for AGB above 160 Mg·ha^{-1} were the least accurate. The study demonstrated encouraging results in forest AGB mapping of the normal vegetated area using the freely accessible and high-resolution Sentinel imagery, based on ML techniques.

Keywords: sentinel imagery; above-ground biomass; predictive mapping; machine learning; geographically weighted regression

---

1. Introduction

As the largest carbon sinks on land, forest ecosystems account for about 80% of terrestrial biosphere carbon storage, and play a pivotal role in mitigating climate change [1,2]. Above-ground biomass (AGB), accounting for between 70% and 90% of total forests biomass [3], is one of the important carbon pools in forest ecosystems, and it is a key indicator of forest vegetal health, as well as related seral stages [4,5]. The spatially explicit measurement of forests' AGB also supports REDD+ (reducing emissions from deforestation and forest degradation, plus the sustainable management of forests, and the conservation and enhancement of forest carbon stocks) processes [6]. Therefore, the rapid and accurate estimation and monitoring of AGB over various scales of space and time are crucial for greatly reducing the uncertainty in carbon stock assessments, and for informing strategic forest management plans [7–9].

Traditional field-based measurements provide the most accurate AGB values, but they are destructive and spatially limited [10,11]. Uncertainty and bias in field measurements obviously exist, particularly those with large trees and tropical issues [4,5]. Combining remote sensing and sample plot data has become a popular method to generate spatially explicit estimations of forest AGB [12,13]. Various types of remote-sensing data are used for forest biomass estimation such as optical sensor data, radio detection and ranging (radar) data, and light detection and ranging (LiDAR) data, with each one having certain advantages over the others [14,15]. Optical sensors were first applied to retrieve the horizontal forest structure and AGB assessments through field sampling, due to their aggregate spectral signatures (reflectance or vegetation indices) with global coverage, repetitiveness, and cost-effectiveness [16,17]. Optical remote sensing data from a number of platforms, such as IKONOS, Quickbird, Worldview, ZY-3, systeme probatoire d'observation de la terre (SPOT), Sentinel, Landsat, and moderate-resolution imaging spectroradiometer (MODIS), with spatial resolutions varying from less than one meter to hundreds of meters, have been used by numerous researchers for biomass estimation [18–20]. However, the widespread usage of optical data is limited by its poor penetration capacity through clouds and forest canopies, as well as data saturation problems [21]. Radar data, available internationally from airborne or space-borne systems with different frequency bands, polarizations, and variable imaging geometries, such as Terra-SAR (Terra-Synthetic Aperture Radar), advanced land observing satellite phased array L-band synthetic aperture radar (ALOS PALSAR), and Sentinel, have gained prominence for AGB estimation because of their better penetration ability and detailed vegetation structural information, but these still suffer from signal saturation problems [6,14,22]. LiDAR, an active remote-sensing technology, captures forests' vertical structures in great detail and provides 3D information, such as the geoscience laser altimeter system (GLAS), which has found favor in biomass estimation with an improved accuracy, but with complex data-processing, and the lack of space continuity problems [23,24]. Additionally, some research using 3-D terrestrial LiDAR has shown bias in biomass, especially from tropical species, mainly because of the underestimation of tree height [25,26]. In other words, the accuracy of forest AGB estimates could be improved by a combined use of multi-source remote sensing data. The above-mentioned Sentinel satellite constellation series, including the Sentinel-1 C-band Synthetic Aperture Radar (SAR) and the Sentinel-2 multispectral instrument by the European Space Agency (ESA), provide new capabilities for AGB mapping with wide coverage, a short return cycle, and a long lifespan as the same data format [27–29]. In other words, the Sentinel series may have a high synergetic potential for overcoming the limitation of single remote sensing techniques for forest AGB estimation. The Sentinel imagery have been applied in a number of previous vegetation studies, focusing on classification [30–32], vegetation parameters on agricultural fields [33–35], grassland [36–38], and forests [39,40], as well as the damage extent of disasters [41–43], while forest AGB mapping based on Sentinel imagery is still insufficient.

The techniques for estimating forests' AGB based on remote sensing data have allowed for 'scale-up' or extrapolation of the field data collected for larger scales [8,44]. It is a predictive mapping process for an estimation of the value at a location without direct observation. It depends on the values of points at nearby sites where observations were made, and/or values of other factors at the sites, through various methods. Those methods can be divided into two categories: parametric and non-parametric algorithms [45,46]. The former refers to statistical regression methods, such as the stepwise regression models (SWR) and geographically weighted regression (GWR), by which the expression relating to the dependent variable (AGB) and the independent variables is easy to calculate [2]. However, there is no simple global linear relationship like the SWR model, between remote sensing data and forest AGB, because it is affected by many factors. The GWR method explores spatial heterogeneity, as well as the non-stationarity of relationships, and it estimates the parameters for each sample location, which makes it a very attractive tool for remotely-sensed biomass modeling [47,48]. Non-parametric techniques, including machine learning (ML) methods such as the k-nearest neighbor (KNN), artificial neural network (ANN), support vector machine for regression

(SVR), and random forest (RF), have a better ability for identifying complex relationships between predictors and the forest AGB [2,49]. Despite a variety of forest AGB models, quite a few research studies merely focused on one parametric or non-parametric model are unpersuasive. Thus, a systematic comparison of GWR, ANN, SVR, and RF for mapping forest AGB based on Sentinel imagery is fairly urgent, but also rare in the literature.

In this study, the ability of Sentinel-1 and Sentinel-2 imagery for the retrieval and predictive mapping of forests' AGB estimation was evaluated. The specific objectives included the following: (1) to determine and model the relationship between field-measured forests AGB and Sentinel-based predictors, including Sentinel-1 SAR backscatter information and Sentinel-2 multispectral indices based on GWR and ML; (2) to evaluate and compare the accuracy of the biomass prediction models, including GWR, ANN, SVR, and RF models; and (3) to map forest AGB spatial distribution by four optimal models. The novelty of this paper is the use of Sentinel-1 (texture characteristics and backscatter coefficients) and Sentinel-2 (multispectral bands, vegetation indices, and biophysical variables) imagery in the mapping of forest AGB and AGB model development, as well as their comparison. This study attempted to contribute to the development of remote sensing-based predictive mapping techniques for forest AGB using freely accessible multi-source remote sensing data with a relatively high spatial resolution.

2. Materials and Methods

2.1. Study Site

This study was conducted in a sample area (42°17′–42°49′ N, 127°35′–128°20′ E), which spanned over 2500 km². It is located in the southeast region of Jilin Province, northeast China, in the center of the Changbai Mountains (Figure 1). The study area consists of five towns: Yanjiang and Lushuihe towns of the Fusong County of Baishan City; and Yongqing, Liangjiang, and Erdaobaihe towns of the Antu County of the Yanbian Korean Autonomous Prefecture. This area has a northern temperate continental monsoon climate, with an annual average temperature of 2.8 °C and an annual precipitation of 8000 mm [50,51]. Characterized by high forest cover, the spatial distribution of forest types in the study area obtains obvious vertical zonality, with *Pinus koraiensis* Sieb. et Zucc., *Larix gmelinii* var. japonica, *Betula platyphylla* Suk., *Fraxinus mandschurica* Rupr., and *Juglans mandshurica* Maxim. as typical tree species [51].

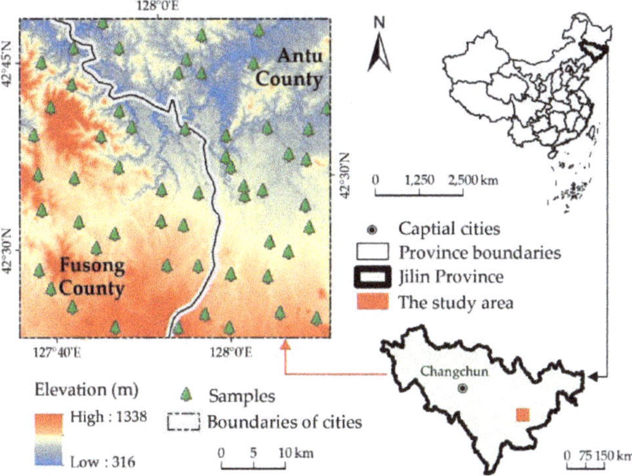

Figure 1. The location of the study area and surveyed forest quadrats.

2.2. Field Observations

The field campaign was conducted from July to August, 2017. Before that, the distribution of sampling plots was generated using ArcGIS (version 10.0, ESRI, RedLands, CA, USA), with the non-forest area being masked out. Non-forested areas were derived from 2015 land use and a land cover map [52] by visual interpretation and manual modification based on Sentinel-2 images (Table 1). In the field, nearby preset plots, a total of 56 sampling quadrats measuring 10 m × 10 m were laid out at each representative sampling plot (Figure 1), including two evergreen coniferous forests, 30 broadleaved deciduous forests, five deciduous coniferous forests, and 19 mixed broadleaf-conifer forests.

The sampling equipment were the diameter at breast height ruler for measuring the diameter at 1.3 m from the ground and the laser altimeter (TruPulse200, Laser Technology Inc., Norristown, PA, USA) for the tree height measurement. Based on allometric equations (Table 2), some AGB are the sum of tree trunks, branches, and leaf biomass, and others are directly calculated by diameters at breast height (cm) and tree height (m). The AGB of 56 samples ranged from 14.64~317.40 Mg·ha^{-1}, with mean and median values of 80.44 and 66.00 Mg·ha^{-1}, respectively.

Table 1. List of Sentinel imagery acquired for the study.

Mission	Product	Observation Date	Cell Size (m)	Uniform Resource Identifier (URI)
Sentinel-1B	Level-1 GRD-HR	22 September 2017	10	S1B_IW_GRDH_1SDV_20170922T213003_20170922T213028_007510_00D425_4962.SAFE
Sentinel-2A	Multispectral image Level-1C	3 May 2017	10	S2A_MSIL1C_20170503T021611_N0205_R003_T52TDN_20170503T022350.SAFE
Sentinel-2A	Multispectral image Level-1C	25 July 2017	10	S2A_MSIL1C_20170725T022551_N0205_R046_T52TCM_20170725T023524.SAFE
Sentinel-2A	Multispectral image Level-1C	23 September 2017	10	S2A_MSIL1C_20170923T022551_N0205_R046_T52TCN_20170923T023519.SAFE
Sentinel-2A	Multispectral image Level-1C	23 September 2017	10	S2A_MSIL1C_20170923T022551_N0205_R046_T52TDM_20170923T023519.SAFE

Table 2. Main allometric equations for the above-ground biomass calculation of each tree species [53,54].

Tree Species	Allometric Equations
Betula platyphylla Suk.	$\mathrm{AGB = T + B + L} = 0.04939 \times (D^2 \times H)^{0.9011} + 0.01417 \times (D^2 \times H)^{0.7686} + 0.0109 \times (D^2 \times H)^{0.6472}$
Acer mono Maxim.	$\mathrm{AGB = T + B + L} = 0.3274 \times (D^2 \times H)^{0.7218} + 0.01349 \times (D^2 \times H)^{0.7198} + 0.02347 \times (D^2 \times H)^{0.6929}$
Tilia amurensis Rupr.	$\mathrm{AGB = T + B + L} = 0.01275 \times (D^2 \times H)^{1.0094} + 0.00182 \times (D^2 \times H)^{0.9746} + 0.00024 \times (D^2 \times H)^{0.9907}$
Mongolian oak (Quercus spp.)	$\mathrm{AGB = T + B + L} = 0.03147 \times (D^2 \times H)^{0.7329} + 0.002127 \times D^{2.9504} + 0.00321 \times D^{2.47349}$
Ulmus japonica Sarg.	$\mathrm{AGB = T + B + L} = 0.031457 \times (D^2 \times H)^{1.032} + 0.007429 \times D^{2.6745} + 0.002754 \times D^{2.4965}$
Fraxinus mandschurica Rupr	$\mathrm{AGB = T + B + L} = 1.416 \times D^{1.71} + 1.154 \times D^{1.549} + 0.7655 \times D^{0.886}$
Populus cathayana Rehd.	$\mathrm{AGB = T + B + L} = 0.3642 \times D^{2.0043} + 0.0317 \times D^{2.6398} + 0.0149 \times D^{2.2541}$
Juglans mandshurica Maxim.	$\mathrm{AGB} = 0.099 \times (D^2 H)^{0.841}$
Prunus padus L.	$\mathrm{AGB} = 0.09 \times D^{2.696}$
Pinus koraiensis Sieb. et Zucc.	$\mathrm{AGB = T + B + L} = 0.0144 \times (D^2 \times H)^{1.0004} + 0.0332 \times (D^2 \times H)^{0.6941} + 0.0866 \times (D^2 \times H)^{0.4696}$
Larix gmelinii var. japonica	$\mathrm{AGB = T + B + L} = 0.025 \times (D^2 \times H)^{0.96} + 0.0021 \times (D^2 \times H)^{0.9638} + 0.00126 \times (D^2 \times H)^{0.9675}$

AGB, T, B, L, D, and H represent above-ground biomass, tree trunk biomass, branch biomass, leaf biomass, diameter at breast height, and tree height, respectively.

2.3. Satellite Data Pre-Processing and Predictors Derived

Sentinel-1 Synthetic Aperture Radar (SAR) and cloud-free Sentinel-2 multispectral imagery from the European Space Agency used in this study (Table 1) were downloaded from the agency's Copernicus Sentinel Scientific Data Hub (https://scihub.copernicus.eu/dhus/#/home). Sentinel-1 C-band images adopted in this study were collected in the interferometric wide swath mode of the VH (vertical transmit-Horizontal receive) and VV (vertical transmit-vertical receive) polarizations. With a pixel size of 10 m, the SAR images are at a high-resolution (HR) Level-1 ground range detected (GRD) processing level. The Sentinel-2 Level 1C data involved were top-of-atmosphere reflectance images, and they were processed for orthorectification and spatial registration. The imagery had 13 spectral bands in the visible, near infrared, and short-wave infrared regions, and had 10 m (band 2–4, 8), 20 m (band 5–7, 8a, 11–12), and 60 m (band1, 9–10) spatial resolutions. In addition, elevation data (30 m) from the Shuttle Radar Topography Mission (SRTM) product was acquired from the United States Geological Service's Earth Explorer (https://earthexplorer.usgs.gov/) for inclusion in the analysis of the Sentinel imagery [55].

The processing steps used in the study are summarized in Figure 2. Sentinel application platform (SNAP) software (version 6.0, European Space Agency) was used to pre-process Sentinel-1 and Sentinel-2 imagery. The steps for the SAR imagery based on the Sentinel-1 Toolbox consisted of image calibration, speckle reduction using the Refined Lee Filter, and terrain correction using the Range-Doppler to acquire an accurate radar intensity backscatter coefficient with a map projection [56,57]. Multispectral imagery was atmospherically corrected and processed by the radiative transfer model-based Sen2Cor atmospheric correction processor (version 2.5.5, European Space Agency) to a Level-2A product, a bottom-of-atmosphere-corrected reflectance image. The pre-processed Sentinel images, as well as the elevation data, were brought into a common map projection, universal transverse mercator (UTM) Zone 52 WGS84, and resampled to 10 m pixel sizes. Subsetting and mosaicking were done to cover the study area.

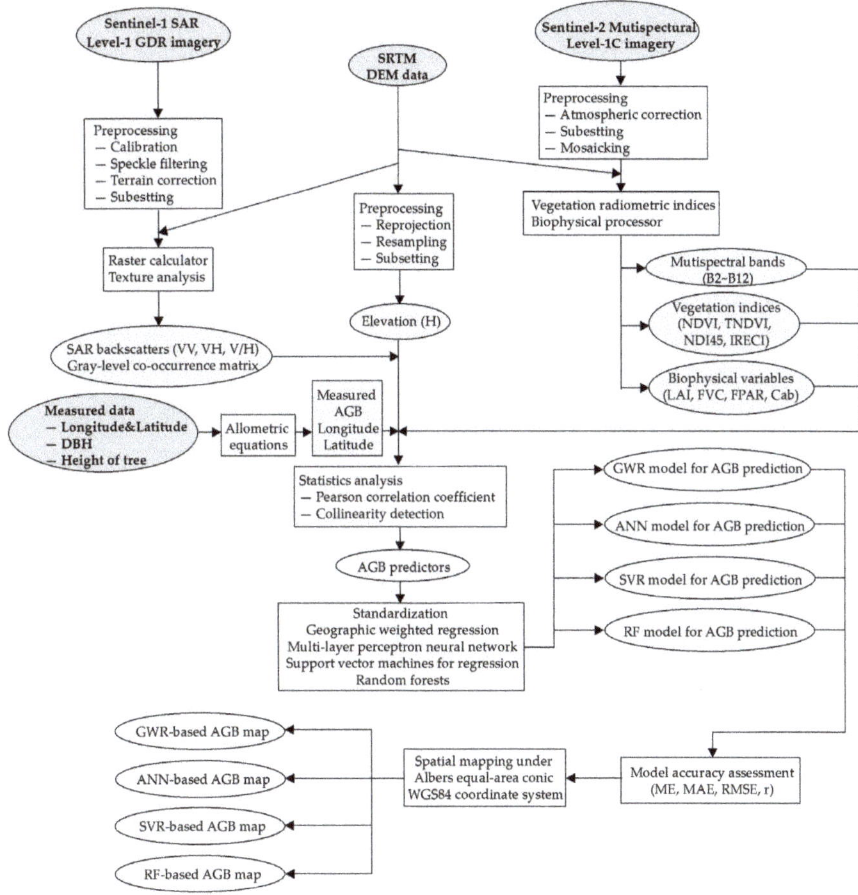

Figure 2. Flowchart of steps used for forest above-ground biomass mapping using Sentinel SAR and multispectral imagery. DBH is the abbreviation of diameter at breast height.

In this study, 44 predictors were selected and extracted according to previous research [55,58]. Shown in Table 3, 23 predictors were from Sentinel-1 and 18 variables were from Sentinel-2, as well as elevation (H) from SRTM. Additionally, the other two variables were longitude and latitude. The first and second part derived from the Sentinel-1 imagery consisted of relating the field AGB with Sentinel SAR polarization channels, and their calculation and texture characteristics. The third to fifth parts based on the Sentinel-2 images proceeded with relating the field AGB to the multispectral bands, vegetation indices, and biophysical variables. The biophysical variables were also calculated in SNAP from its biophysical processor, which uses a neural network algorithm based on the PROSPECT+SAIL (PROSAIL) radiative transfer model [59]. Except for serving as a predictor, the elevation data from SRTM digital elevation model (DEM) were supplemented with Sentinel imagery processing to improve the accuracy (Figure 2).

Table 3. Sentinel-based imagery data predictors of above-ground biomass.

Source Image	Relevant Predictors		Description
Sentinel-1	Polarization	VV	vertical transmit-vertical channel
		VH	vertical transmit-Horizontal channel
		V/H [1]	VV/VH
	Texture [2]	VH_CON, VV_CON	Contrast
		VH_DIS, VV_DIS	Dissimilarity
		VH_HOM, VV_HOM	Homogeneity
		VH_ASM, VV_ASM	Angular Second Moment
		VH_ENE, VV_ENE	Energy
		VH_MAX, VV_MAX	Maximum Probability
		VH_ENT, VV_ENT	Entropy
		VH_MEA, VV_MEA	GLCM Mean
		VH_VAR, VV_VAR	GLCM Variance
		VH_COR, VV_COR	GLCM Correlation
Sentinel-2	Multispectral bands	B2	Blue, 490 nm
		B3	Green, 560 nm
		B4	Red, 665 nm
		B5	Red edge, 705 nm
		B6	Red edge, 749 nm
		B7	Red edge, 783 nm
		B8	Near Infrared, 842 nm
		B8a	Near Infrared, 865 nm
		B11	Short Wave IR, 1610 nm
		B12	Short Wave IR, 2190 nm
	Vegetation indices	NDVI [3]	(Band 8 − Band 4)/(Band 8 + Band 4)
		NDI45 [4]	(Band 5 − Band 4)/(Band 5 + Band 4)
		IRECI [5]	(Band 7 − Band 4)/(Band 5/Band 6)
		TNDVI [6]	$[(\text{Band } 8 - \text{Band } 4)/(\text{Band } 8 + \text{Band } 4) + 0.5]^{1/2}$
	Vegetation biophysical variables	LAI	Leaf Area Index
		FVC	Fraction of Vegetation Cover
		FPAR	Fraction of Absorbed Photosynthetically Active Radiation
		Cab	Chlorophyll content in the leaf
SRTM DEM	Elevation	H	Elevation, 30 m resolution

[1] Pan and Sun (2018) [58]; [2] GLCM = Gray-level Co-occurrence Matrix with a nine by nine-pixel window, Haralick et al. (1973) [60]; [3] NDVI = Normalized Difference Vegetation Index, Rouse et al. (1973) [61]; [4] NDI45 = Normalized Difference Vegetation Index with band 4 and 5, Delegido et al. (2011) [62]; [5] IRECI = Inverted Red-Edge Chlorophyll Index, Clevers et al. (2002) [63]; [6] TNDVI = Transformed Normalised Difference Vegetation Index, Deering et al. (1975) [64].

3. Modeling the Relationship between Field AGB and Satellite Data

Firstly, the pairwise Pearson's product-moment correlation analysis was conducted to determine the correlation of observed above-ground biomass and Sentinel-based predictors, as well as the collinearity between predictors. Predictors that were highly correlated to each other ($r \geq 0.8$), and had high variance inflation factors (VIFs ≥ 10) in regression analysis, were excluded from modeling [65,66]. These analysis steps were performed using SPSS (version 21.0, IBM, Armonk, NY, USA). Then, all of the explanatory variables were transformed to a Z-score to eliminate the effect of index dimension and quantity of data. The formula was defined as $x^* = (x - \mu)/\sigma$, where μ is the mean value of a specific explanatory variable and σ is its standard deviation [67]. After that, GWR, ANN, SVR, and RF were used in this study to model ABG based on Sentinel-derived predictors (Figure 2).

3.1. Geographically Weighted Regression

Originally proposed by Brunsdon et al. (1998), GWR is a powerful approach for modeling spatially heterogeneous processes at a local scale [68–70]. It estimates the individual parameters for each estimation location, and the parameter estimation at any location obeys the distance decay. In other words, the closer to the location of an observation, the greater the weight that is allotted for the observation. The GWR form is regularly expressed as [47]:

$$\hat{y}_i = \beta_0(u_i, v_i) + \sum_{k=1}^{p} \beta_k(u_i, v_i)x^*_{ik} + \varepsilon_i \tag{1}$$

where $\hat{y}_i$ is the dependent variable value of observation i considered in the parameter estimation at the location (u_i, v_i); $\beta_0(u_i, v_i)$ is the intercept; $\beta_k(u_i, v_i)$ is the coefficient of k explanatory variables, indicating a parameter estimate that explains the relationship around location (u_i, v_i), which varies with the location; x^*_{ik} represents the independent variables of observation i; p is the total number of explanatory variables; and ε_i is the error term that is generally assumed to be explanatory and normally distributed with zero mean and constant variance. The parameter estimator $\beta_k(u_i, v_i)$ is identical to the weighted least squares regression, where the weights are computed based on the distance between the observations [48]. The parameters are estimated as [71]:

$$\hat{\beta}(u_i, v_i) = (X^T W(u_i, v_i) X)^{-1} X^T W(u_i, v_i) Y \tag{2}$$

where X is the matrix formed by x^*_{ik}; Y is the vector formed by values of the dependent variables; $W(u_i, v_i)$ is a weight matrix to ensure that those observations near the point i have more influence on the results than those farther away; and the weights are calculated based on a kernel weighting scheme such as fixed Gaussian, fixed bi-square, adaptive bi-square, and adaptive Gaussian [72]. In this study, we used the fixed Gaussian kernel as [73]:

$$W(u_i, v_i) = e^{-0.5(d_i(u_i, v_i)/h)^2} \tag{3}$$

where $d_i(u_i, v_i)$ is the distance between the observation i and the location (u_i, v_i); and h is a quantity called bandwidth, which controls the effect of the distance on the weight value.

GWR was conducted using GWR (version 4.0, Ritsumeikan University, Kyoto, Japan), by which the weight function (a geographic kernel type) and the minimum value of the corrected Akaike information criterion (AICc, small sample bias corrected AIC) are determined to find the optimal bandwidth by a golden section search [74].

3.2. Machine Learning Methods

The machine learning methods adopted in this study were modeled in WEKA software (version 3.8, The University of Waikato, Hamilton, New Zealand). The models defined the best parameters with the highest correlation coefficient (r) and the lowest root mean square error (RMSE) for the prediction of AGB, and then AGB mapping was implemented in ArcGIS.

3.2.1. Multi-Layer Perception Neural Network

As a nonparametric mathematical model, ANN is inspired by biological neural networks and it has strong abilities for linear and nonlinear fitting [75,76]. The ANN considered in this study was the multi-layer perception neural network (MLPNN). The architecture of the MLPNN consists of an input layer containing predictors, one or more hidden layers, and an output layer containing the response variable, along with interconnection weights characterizing the connection strength between these layers (Figure 3). The algorithm chosen was the back-propagation (BP) learning rule, an iterative gradient descent algorithm that was designed to minimize the mean square error between the desired target and the actual output vectors [77,78]. The initial weights were assigned randomly, and when developing the network, the interconnection weights were adjusted to minimize the prediction error [79,80].

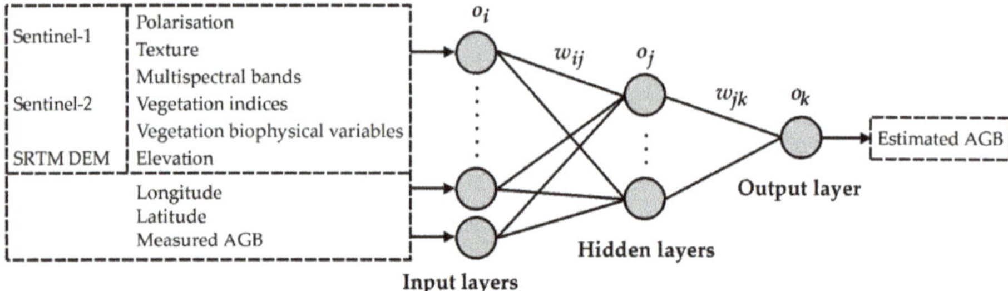

Figure 3. Schematic representation of an example multi-layer perception neural network (MLPNN) model structure to predict forest above-ground biomass. Shown are the inputs with their neurons o_i, and interlinked connections from each input to all hidden layer neurons o_j, along with the selected weightings. The weighted outputs were then merged and fed into the output neuron o_k to form the output values.

3.2.2. Support Vector Machines for Regression

SVR is a regression version of support vector machines that project the training dataset from a lower dimensional space into a higher dimensional feature space using kernel functions to separate groups of input data in a linearized manner, based on the Vapnik-Chervonenkis (VC) dimension theory and structural risk minimization [81–83]. An SVR function for AGB estimation is defined as [84]:

$$AGB = \sum_{k=1}^{P} (\alpha_k - \alpha_k^*) \cdot K(x_k, x_j) + b \tag{4}$$

where x is a vector of the input predictors; $K(x_k, x_j)$ is a kernel function; b is a constant threshold; and α_k and α_k^* are the weights (Lagrange multipliers), with the constraints given in Equation (5):

$$\begin{cases} \sum_{k=1}^{p} (\alpha_k + \alpha_k^*) = 0 \\ 0 \le \alpha_k, \alpha_k^* \le C \end{cases} \tag{5}$$

where C is the regularization parameter for balancing between the training error and model complexity. The sequential minimal optimization (SMO) algorithm was used to solve the quadratic programming optimization problem step-by-step and to update Equation (4) to reflect the new values until the Lagrange multipliers converged [66,85].

Among the various kernel functions, the radial basis function (RBF) shows a superior performance and robust results [86,87], and was used in this study [88]:

$$K(x_k, x_j) = \exp \left[\frac{(x_k - x_j)^2}{\sigma^2} \right] \tag{6}$$

where σ is a scale parameter chosen based on the training data, and a unit vector could be concatenated with kernels as the intercept. In a word, the training of the SVR model required finding the best values for the two meta-parameters: the regularization parameter (C) and the kernel width (σ).

3.2.3. Random Forests

As a classification and regression tree (CART) technique proposed by Breiman (2001), RF combines bagging [89,90] with random variable selections at each node [91] to iteratively generate a large group of CARTs. The classification output represents a majority vote (classification), or an average (regression) from the whole ensemble, and hence achieves a more robust model than a single classification tree that is produced by a single model run [89,92,93]. A number of decision trees in RF choose their best splitting attributes from a random subset of predictors at each internal node without pruning. Based on the bootstrap sampling procedure, RF ensures at the same time the smallest obtainable bias and very low data variance [94]. There are two main important parameters: numFeatures, which means the number of features for splitting the nodes, whose default value is int[ln(numbers of predictors) + 1] in WEKA; and numIterations, which means the number of trees to be optimized in the modeling process, depending on specific application objectives [95].

3.3. Evaluation of ABG Models

Based on measured AGB samples, the mean error (ME), mean absolute error (MAE), and RMSE as defined by Isaaks and Srivastava (1989), with r between the measured and estimated ABG, were used to evaluate the performances of different interpolation methods [55,96].

4. Results

4.1. Statistics Analysis

The poor correlation of the observed AGB and the predictors was acquired from the low value of r (-0.288–0.263). Among predictors, VV_MAX ($r = -0.288$), VV_ENE ($r = -0.284$), VV_HOM ($r = -0.277$), VV_ASM ($r = -0.276$), and VV_ENT ($r = 0.263$) were significantly correlated (p-values < 0.05) with AGB. Predictors from the texture characteristics of Sentinel-1 VV polarization were significantly associated with AGB, as similarly found by Pan and Sun (2018) [58]. Those r values represent an average global correlation; thus, it also indicated that the global linear regression was inappropriate for AGB modeling in this study.

Among the 44 explanatory variables, VV_ASM ($r_{vv\_ASM,MAX} = 0.99$, VIFs = 41.3), VV_ENE ($r_{vv\_ENE,MAX} = 0.995$, VIFs = 106.8), VV_ENT ($r_{vv\_ENT,MAX} = -0.97$, VIFs = 18.2), and VV_HOM ($r_{vv\_HOM,MAX} = 0.98$, VIFs = 20.1) were excluded from model building because their VIFs exceeded the above-mentioned threshold (VIFs $\geq$ 10), and this reduced the number of explanatory variables from 44 to 40.

4.2. Models of GWR and ML

The optimal bandwidth for GWR in this study was 0.023, with the minimum value of AICc being 694.91, and the adjusted R^2 of this GWR model being 0.79. For a given environmental variable, its coefficient from GWR varied across the study area. The top five predictors for the mean magnitude of the coefficients were VV_MEA (16.5, negative), VV_VAR (16.1, positive), LAI (1.6, positive), Cab (0.9, positive), and FVC (Fraction of Vegetation Cover) (0.8, positive). Predictors from the texture characteristics of Sentinel-1 VV polarization and the vegetation biophysical variables of Sentinel-2 showed a relatively strong association with AGB at a local regression.

As for the MLPNN model, the accuracy for various numbers of neurons in the hidden layer is shown in Figure 4. The results revealed that the optimized MLPNN architecture was 40-10-1, indicating that there were 40 input nodes in the input layer, 10 nodes in the hidden layer with the unipolar sigmoid as the transfer function, and one node in the output layer. Using the Levenberg-Marquardt learning algorithm, the best learning rate, momentum, and training time (iterations) obtained were determined to be 0.2, 0.3, and 500, respectively. In the SVR model, the best parameters C and σ obtained were 1 and 2, respectively. With iteration (trees) numbers of 50 and feature numbers of 5, the top five predictors for attribute importance in the selected RF model were VV_CON, VH_HOM, VH_DIS, VH_ASM, and VH_ENT. The texture characteristics of Sentinel-1 VV and VH polarizations were relatively vital for modeling AGB by RF in this study.

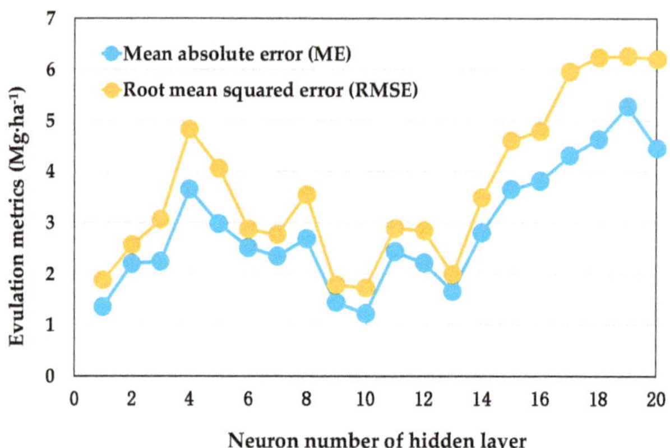

Figure 4. Training errors associated with a given number of neurons in the hidden layer.

4.3. Models Evaluation and Mapping of AGB

4.3.1. Models Assessment by Evaluation Indices

Table 4 presents the accuracies of four models for estimating the AGB of 56 forest quadrats. All four models overestimated the AGB. The ANN model resulted in an ME of 0.84 Mg·ha^{-1}, and had the highest tendency for overestimation; the SVR model with an ME of 0.004 Mg·ha^{-1} showed the lowest tendency for overestimation. The MAE, indicating the extent to which the process leads to error, was lower with SVR (0.07 Mg·ha^{-1}), and higher with the other methods, ranging from 1.21 (ANN) to 4.01 Mg·ha^{-1} (GWR). The values of the RMSE suggested that ML methods, whose RMSE ranged from 0.08 (SVR) to 4.43 (RF) Mg·ha^{-1}, produced less error than GWR. A better consistency between the measured AGB and the estimated one was discovered by the *r* values from the ML models (0.999 of RF to 1 of SVR and ANN) compared to the GWR model (0.995). The SVR model gave rise to the lowest RMSE and the closest-to-zero ME and MAE values, as well as the highest *r* value. Hence, in this study, the SVR model was the most accurate model for estimating AGB. Besides, the accuracy ranking of the four methods from high to low was SVR, ANN, RF, and GWR. To further analyze the modeling accuracy, the estimated values of AGB were plotted against the measured values (Figure 5). An estimation from the SVR model showed the best agreement along the 1:1 line, followed by those from ANN, GWR, and RF.

Table 4. Performance evaluation of the GWR, ANN, SVR, and RF models.

Evaluation Index	GWR	ANN	SVR	RF
ME	0.04	0.84	4×10^{-3}	0.55
MAE	4.01	1.21	0.07	3.48
RMSE	5.26	1.73	0.08	4.43
r	0.995	1	1	0.999

ME, MAE, RMSE, and r are the abbreviations of mean error, mean absolute error, root mean squared error, and correlation coefficient, respectively. The *p*-values of r were all below 0.01.

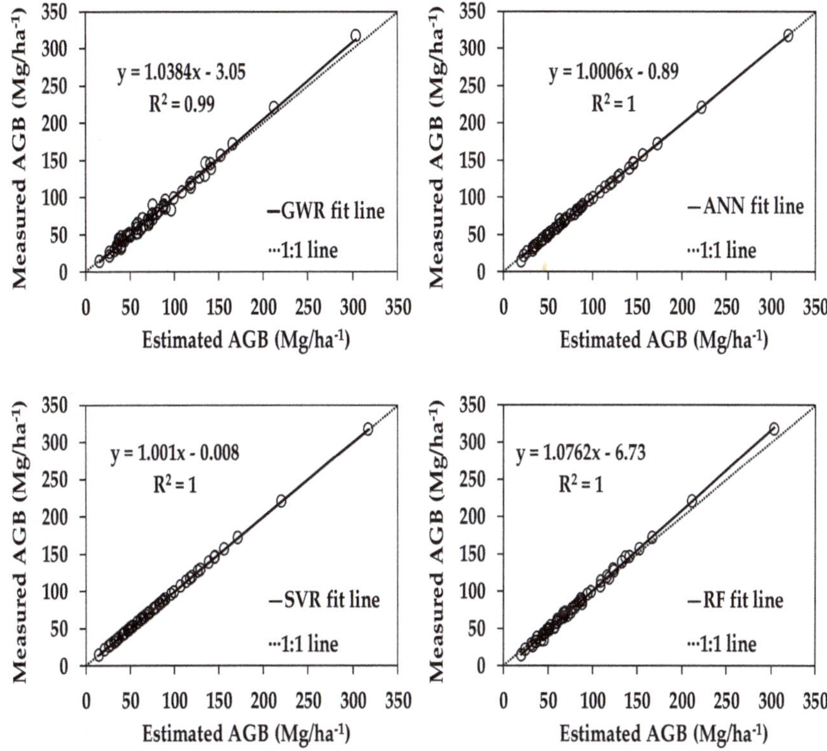

Figure 5. Scatterplots of the measured and estimated above-ground biomass (AGB, Mg·ha^{-1}).

4.3.2. Mapping of Four AGB Models

The predicted values of AGB from the four models ranged from 11.80 to 324.12 Mg·ha^{-1}. For a better comparison, the values were divided into five levels by equal intervals of 62.46 Mg·ha^{-1} (Figure 6). Maps show the various spatial distributions of AGB. All of the six maps show that the western part of Lushuihe town was a high AGB region, with values ranging from 199.19 to 324.12 Mg·ha^{-1}, while low AGB regions were located near the highway connecting Lushuihe and Yangjiang towns, with values ranging from 11.80 to 136.72 Mg·ha^{-1}. The map resulting from ANN was characterized by more explicit spatial variation than the others. Comparing the estimated and measured AGB (14.64~317.40 Mg·ha^{-1}, mainly ranging from 11.80 to 136.72 Mg·ha^{-1} with 87.5%) resulted in the following performance ranking for the four algorithms from strong to weak: SVR, ANN, RF, and GWR, meaning that ML performed better than GWR. These maps can guide resource allocation for carbon sequestration and forest management. The evaluation of the forest AGB mapping results by the four models was insufficient for this study as we were limited by the sample size. Future verification work should be conducted; this is conventionally done by independent sample sets or by acknowledged high-accuracy results such as airborne data, especially unmanned aerial vehicle LiDAR data [17,97].

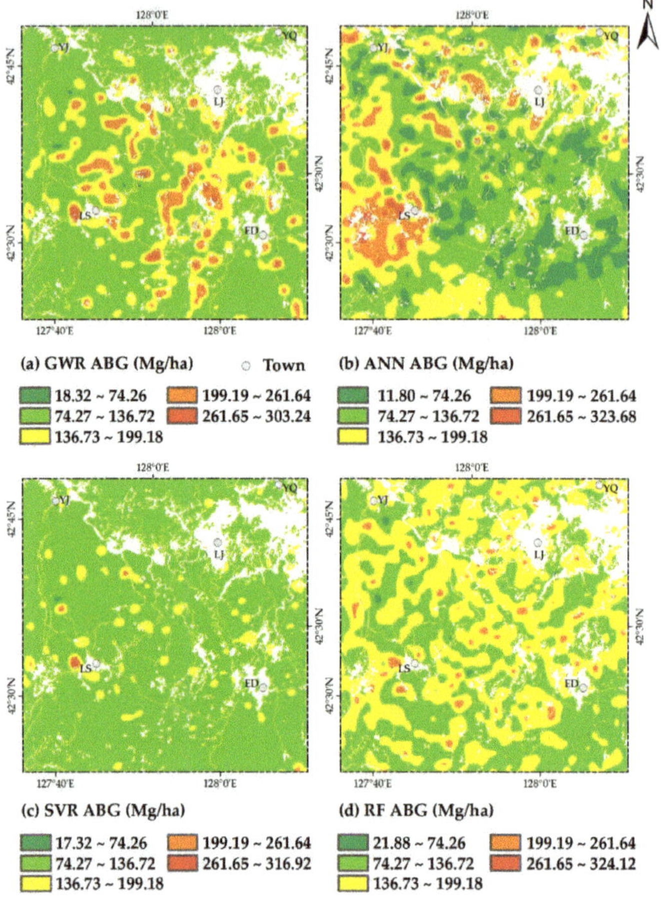

Figure 6. Predicted maps of above-ground biomass in the study site derived from (**a**) geographically weighted regression (GWR), (**b**) multi-layer perception neural network (ANN), (**c**) support vector machines for regression (SVR), and (**d**) random forests (RF). YJ, YQ, LJ, LS, and ED are the abbreviations of Yanjiang, Yongqing, Liangjiang, Lushuihe, and Erdaobaihe towns, respectively.

5. Discussion

5.1. Sentinel-Derived Predictors

By comparing the results of the correlation analysis, the coefficients from GWR, and the attribute importance from RF, it was indicated that texture characteristics of Sentinel-1 had great potential for estimating AGB, which was also shown in previous studies [98,99]. Additionally, it was a pioneering finding that the vegetation biophysical variables of Sentinel-2 were very helpful for AGB estimation using a local regression, which was found previously by non-parametric prediction [55]. The backscatter coefficient of Sentinel-1 and the vegetation indices of Sentinel-2 were useful and common predictors, as confirmed by other researchers [55,99–101], but their roles were assisted and not apparent for forest AGB mapping in this study. This may have resulted from a mixture of forest types in the study area, while previous studies mainly aimed at a certain type of forest, or modeling by forests types. Besides, the texture characteristics and backscatter coefficients of Sentinel-1, and the multispectral bands, vegetation indices, and biophysical variables of Sentinel-2, were first applied simultaneously in AGB modeling, so that the texture characteristics and biophysical variables were outstanding compared to the other kinds of Sentinel-based predictors in this study. In a word, this study dug out vital and new information from the Sentinel series about forest AGB estimation.

5.2. The Comparison of Models

Similarly, previous researchers have used these four methods to estimate forest AGB and achieve good accuracies, while results of the models' comparison vary compared with this study. Cao et al. (2018), integrating airborne LiDAR and optical data, compared the accuracies of forest AGB models in the upper Heihe River Basin in northwest China, and found that RF was the best (R^2 = 0.9, RMSE = 13.4 Mg·ha^{-1}), following by ANN and SVR [17]. Based on Landsat satellite imagery, Wu et al. (2016) implemented the optimal spatial forest AGB estimation in northwestern Zhejiang Province, China, and RF (R^2 = 0.6, RMSE = 26.4 Mg·ha^{-1}) also performed better than SVR [11]. Liu et al. (2017) developed forest AGB models using GLAS and Landsat data, and also found that RF (R^2 = 0.95, RMSE = 17.73 Mg·ha^{-1}) had a better estimation than SVR [2]. Gao et al. (2018) concluded that ANN performed (RMSE = 27.6 Mg·ha^{-1}) better than SVR and RF, by conducting a comparative analysis of algorithms for forest AGB estimation with ALOS PALSAR and Landsat data [49]. In a word, ANN and SVR models all showed a close performance for forest AGB modeling in this study, and in previous research. The RF models generally obtained the highest accuracies among the three ML methods, while SVR showed the best performance in this study. This may be due to the smaller sample sizes in this study, and the uniform random distribution of samples in the study area. It also highlighted the powerful capacity of SVR for limited samples. Additionally, due to direct evaluation and the accuracy of the training data, rather than the independent validation set or by cross-validation from limited samples, the models of this study were obviously much more accurate than in previous research. Because the GWR and ML models have not been compared in any previous research, this finding can provide a reference for mapping forest AGB in the future.

5.3. Model Evaluation by Forest Types and Measured AGB

The mean errors of AGB prediction using the four models for different forest types and the measured values of AGB were also calculated to analyze the prediction accuracy of each forest type, and the data saturation in the Sentinel data was discussed (Figure 7). Generally, the estimated AGB values of the four forest types of 56 quadrats by the four models were all higher than the measured values (Figure 7a). Among that, the AGB estimation of the deciduous coniferous forests obtained the maximum error of 0.7 Mg·ha^{-1}, and all models, excluding the GWR with ME values of −0.6 Mg·ha^{-1}, performed the worst for deciduous coniferous forests compared to the other three forest types, with ME values ranging from 0.04 (SVR) to 1.8 (RF) Mg·ha^{-1}. The AGB estimation of mixed broadleaf-conifer forests had the second-large error, with 0.6 Mg·ha^{-1}, followed by that of broadleaved deciduous forests (0.2 Mg·ha^{-1}) and evergreen coniferous forests (0.02 Mg·ha^{-1}). As for the four models, the GWR performed best in the AGB estimation of mixed broadleaf-conifer forests, and the worst for evergreen coniferous forests. The ANN, SVR, and RF all gave the most accurate assessments for broadleaved deciduous forests. The ANN showed the least precise assessments for evergreen coniferous forests, but SVR and RF gave the worst for deciduous coniferous forests. In a word, the AGB estimation of broadleaved deciduous forests was the most accurate and stable in the study area using the four models based on Sentinel imagery. Among the five levels of AGB values, the last level with AGB above 160 Mg·ha^{-1} (from 160 to 320 Mg·ha^{-1}) had the least accuracy and the most fluctuated errors, based on Sentinel imagery, while AGB from 40 to 120 Mg·ha^{-1} obtained relatively higher accuracies (Figure 7b). The saturation level shown in this study was much higher than other studies (at around 60–70 Mg·ha^{-1}), using SAR C band data [102]. This could be attributed to the integration of abundant predictors from Sentinel-1 and Sentinel-2 in the study area with normal forest coverage, which was human-dominated zones with nearby towns and villages.

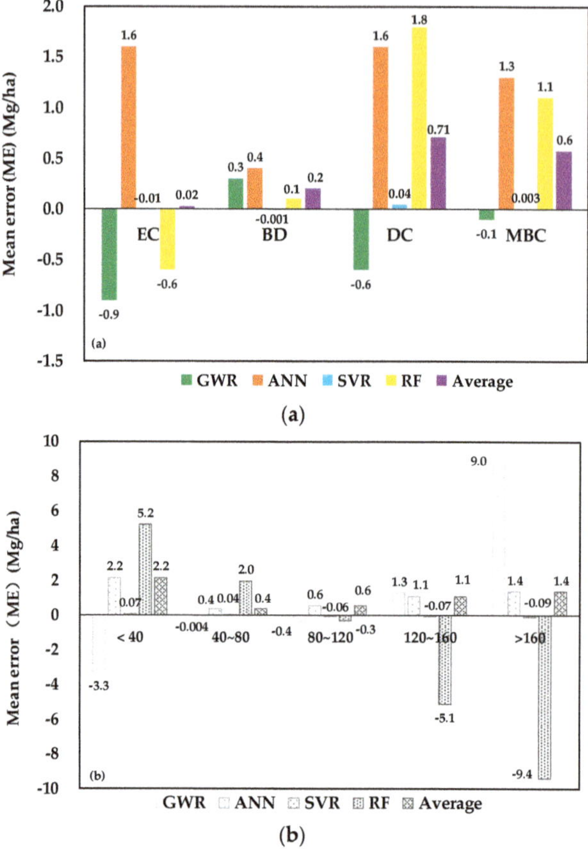

Figure 7. The mean errors of the above-ground biomass predictions for (**a**) different forest types, (**b**) different AGB using geographically weighted regression (GWR), the multi-layer perception neural network (ANN), support vector machines for regression (SVR), and random forests (RF). EC, BD, DC, and MBC represent evergreen coniferous, broadleaved deciduous, deciduous coniferous, and mixed broadleaf-conifer forests, respectively.

6. Conclusions

To map the distribution of forest AGB at a regional scale, Sentinel SAR and multispectral imagery were selected for a group of field quadrats with a resolution of 10 m. Four AGB models, one GWR model, and three ML models were built using these field measurements and remote-sensing datasets. The results demonstrated that SVR with SMO algorithms are the best for spatially predicting and mapping the patterns of AGB in the study site. The results also showed that the texture characteristics of Sentinel-1 and the vegetation biophysical variables of Sentinel-2 were the most relative and important predictors for explaining the observed variability of AGB in the area, and that the contributions of the other Sentinel-derived factors were only marginal. The AGB estimation of broadleaved deciduous forests was the most accurate, while the AGB above 160 Mg·ha^{-1} had the least accuracy, indicating data saturation of Sentinel imagery. Overall, the performance of the models in this study will inform the selection of predictive mapping techniques for forest AGB modeling, while the map that is generated will be instrumental for formulating spatially-targeted climate change mitigation and sustainable land management strategies. In the future, the model performance will be improved by incorporating other important environmental data (e.g., distance to the city center and roads, as well as human disturbance) and other up-to-date remote sensing techniques (e.g., Tandem-X and LiDAR), as well as the stochastic component of AGB.

Although SAR C band and optical multispectral techniques have few advantages for detecting the sensibility of forest AGB compared to SAR P band or LiDAR, the available free Sentinel series at a relatively high spatial resolution with full coverage is indeed useful information for applications in global forest AGB estimation. The study demonstrated encouraging results in forest AGB mapping of the normal vegetated area using Sentinel imagery; thus, it is helpful and valuable for vital information mining from the Sentinel series when it is applied to global forest AGB estimation.

Author Contributions: L.C., C.R., and B.Z. designed this research. L.C. and Y.X. conducted field sampling and performed the experiments. L.C. conducted the analysis and wrote the paper. C.R., B.Z., and Z.W. drafted the paper and revised it critically. All authors reviewed the manuscript.

Funding: This study was supported by the National Key Research and Development Project of China (No. 2016YFC0500300).

Acknowledgments: The authors are grateful to the ESA (https://scihub.copernicus.eu/) and USGS (http://glovis.usgs.gov/) for providing the Sentinel imagery and SRTM DEM.

Conflicts of Interest: The authors declare no conflict of interest.

References

1. Olson, J.S.; Watts, J.; Allison, L.J. *Carbon in Live Vegetation of Major World Ecosystems*; Oak Ridge National Laboratory: Oak Ridge, TN, USA, 1983.
2. Liu, K.; Wang, J.D.; Zeng, W.S.; Song, J.L. Comparison and evaluation of three methods for estimating forest above ground biomass using TM and GLAS data. *Remote Sens.* **2017**, *9*, 341. [CrossRef]
3. Cairns, M.A.; Brown, S.; Helmer, E.H.; Baumgardner, G.A. Root biomass allocation in the world's upland forests. *Oecologia* **1997**, *111*, 1–11. [CrossRef] [PubMed]
4. Brown, S.; Schroeder, P.; Birdsey, R. Aboveground biomass distribution of US eastern hardwood forests and the use of large trees as an indicator of forest development. *For. Ecol. Manag.* **1997**, *96*, 37–47. [CrossRef]
5. Deb, D.; Singh, J.P.; Deb, S.; Datta, D.; Ghosh, A.; Chaurasia, R.S. An alternative approach for estimating above ground biomass using Resourcesat-2 satellite data and artificial neural network in Bundelkhand region of India. *Environ. Monit. Assess.* **2017**, *189*, 576. [CrossRef] [PubMed]
6. Kaasalainen, S.; Holopainen, M.; Karjalainen, M.; Vastaranta, M.; Kankare, V.; Karila, K.; Osmanoglu, B. Combining Lidar and Synthetic Aperture Radar data to estimate forest biomass: Status and prospects. *Forests* **2015**, *6*, 252–270. [CrossRef]
7. Pan, Y.; Birdsey, R.A.; Fang, J.; Houghton, R.; Kauppi, P.E.; Kurz, W.A.; Phillips, O.L.; Shvidenko, A.; Lewis, S.L.; Canadell, J.G.; et al. A large and persistent carbon sink in the world's forests. *Science* **2011**, *333*, 988–993. [CrossRef] [PubMed]
8. Saatchi, S.S.; Harris, N.L.; Brown, S.; Lefsky, M.; Mitchard, E.T.A.; Salas, W.; Zutta, B.R.; Buermann, W.; Lewis, S.L.; Hagen, S.; et al. Benchmark map of forest carbon stocks in tropical regions across three continents. *Proc. Natl. Acad. Sci. USA* **2011**, *108*, 9899–9904. [CrossRef] [PubMed]
9. Deo, R.K.; Russell, M.B.; Domke, G.M.; Andersen, H.E.; Cohen, W.B.; Woodall, C.W. Evaluating site-specific and generic spatial models of aboveground forest biomass based on Landsat time-series and LiDAR strip samples in the Eastern USA. *Remote Sens.* **2017**, *9*, 598. [CrossRef]
10. Ene, L.T.; Naesset, E.; Gobakken, T.; Gregoire, T.G.; Stahl, G.; Holm, S. A simulation approach for accuracy assessment of two-phase post-stratified estimation in large-area LiDAR biomass surveys. *Remote Sens. Environ.* **2013**, *133*, 210–224. [CrossRef]
11. Wu, C.F.; Shen, H.H.; Shen, A.H.; Deng, J.S.; Gan, M.Y.; Zhu, J.X.; Xu, H.W.; Wang, K. Comparison of machine-learning methods for above-ground biomass estimation based on Landsat imagery. *J. Appl. Remote Sens.* **2016**, *10*, 035010. [CrossRef]
12. McRoberts, R.E.; Næsset, E.; Gobakken, T. Inference for lidar-assisted estimation of forest growing stock volume. *Remote Sens. Environ.* **2013**, *128*, 268–275. [CrossRef]

13. Zhao, P.P.; Lu, D.S.; Wang, G.X.; Liu, L.J.; Li, D.Q.; Zhu, J.R.; Yu, S.Q. Forest aboveground biomass estimation in Zhejiang Province using the integration of Landsat TM and ALOS PALSAR data. *Int. J. Appl. Earth Obs.* **2016**, *53*, 1–15. [CrossRef]

14. Kumar, L.; Sinha, P.; Taylor, S.; Alqurashi, A.F. Review of the use of remote sensing for biomass estimation to support renewable energy generation. *J. Appl. Remote Sens.* **2015**, *9*, 097696. [CrossRef]

15. Lin, Y.; West, G. Reflecting conifer phenology using mobile terrestrial LiDAR: A case study of Pinus sylvestris growing under the Mediterranean climate in Perth, Australia. *Ecol. Indic.* **2016**, *70*, 1–9. [CrossRef]

16. Blackard, J.A.; Finco, M.V.; Helmer, E.H.; Holden, G.R.; Hoppus, M.L.; Jacobs, D.M.; Lister, A.J.; Moisen, G.G.; Nelson, M.D.; Riemann, R.; et al. Mapping us forest biomass using nationwide forest inventory data and moderate resolution information. *Remote Sens. Environ.* **2008**, *112*, 1658–1677. [CrossRef]

17. Cao, L.D.; Pan, J.J.; Li, R.J.; Li, J.L.; Li, Z.F. Integrating airborne LiDAR and optical data to estimate forest aboveground biomass in arid and semi-arid regions of China. *Remote Sens.* **2018**, *10*, 532. [CrossRef]

18. Thenkabail, P.S.; Stucky, N.; Griscom, B.W.; Ashton, M.S.; Diels, J.; van der Meer, B.; Enclona, E. Biomass estimations and carbon stock calculations in the oil palm plantations of African derived savannas using IKONOS data. *Int. J. Remote Sens.* **2004**, *25*, 5447–5472. [CrossRef]

19. Sun, G.; Ni, W.; Zhang, Z.; Xiong, C. Forest Aboveground Biomass Mapping using Spaceborne Stereo Imagery Acquired by Chinese ZY-3. In Proceedings of the AGU Fall Meeting, San Francisco, CA, USA, 14–18 December 2015; Volume 12, p. 2089.

20. Kumar, L.; Mutanga, O. Remote sensing of above-ground biomass. *Remote Sens.* **2017**, *9*, 935. [CrossRef]

21. Avitabile, V.; Baccini, A.; Friedl, M.A.; Schmullius, C. Capabilities and limitations of Landsat and land cover data for aboveground woody biomass estimation of Uganda. *Remote Sens. Environ.* **2012**, *117*, 366–380. [CrossRef]

22. Santi, E.; Paloscia, S.; Pettinato, S.; Fontanelli, G.; Mura, M.; Zolli, C.; Maselli, F.; Chiesi, M.; Bottai, L.; Chirici, G. The potential of multifrequency SAR images for estimating forest biomass in Mediterranean areas. *Remote Sens. Environ.* **2017**, *200*, 63–73. [CrossRef]

23. Laurin, G.V.; Chen, Q.; Lindsell, J.A.; Coomes, D.A.; Del Frate, F.; Guerriero, L.; Pirotti, F.; Valentini, R. Above ground biomass estimation in an African tropical forest with LiDAR and hyperspectral data. *ISPRS J. Photogramm. Remote Sens.* **2014**, *89*, 49–58. [CrossRef]

24. Chi, H.; Sun, G.Q.; Huang, J.L.; Guo, Z.F.; Ni, W.J.; Fu, A.M. National forest aboveground biomass mapping from ICESat/GLAS data and MODIS imagery in China. *Remote Sens.* **2015**, *7*, 5534–5564. [CrossRef]

25. Guo, Q.H.; Liu, J.; Tao, S.L.; Xue, B.L.; Li, L.; Xu, G.C.; Li, W.K.; Wu, F.F.; Li, Y.M.; Chen, L.H.; et al. Perspectives and prospects of LiDAR in forest ecosystem monitoring and modeling. *Chin. Sci. Bull.* **2014**, *59*, 459–478. [CrossRef]

26. Wilkes, P.; Disney, M.; Vicari, M.B.; Calders, K.; Burt, A. Estimating urban above ground biomass with multi-scale LiDAR. *Carbon Balance Manag.* **2018**, *13*, 10. [CrossRef] [PubMed]

27. Malenovsky, Z.; Rott, H.; Cihlar, J.; Schaepman, M.E.; Garcia-Santos, G.; Fernandes, R.; Berger, M. Sentinels for science: Potential of Sentinel-1, -2, and -3 missions for scientific observations of ocean, cryosphere, and land. *Remote Sens. Environ.* **2012**, *120*, 91–101. [CrossRef]

28. Torres, R.; Snoeij, P.; Geudtner, D.; Bibby, D.; Davidson, M.; Attema, E.; Potin, P.; Rommen, B.; Floury, N.; Brown, M.; et al. GMES Sentinel-1 mission. *Remote Sens. Environ.* **2012**, *120*, 9–24. [CrossRef]

29. Chang, J.S.; Shoshany, M. Mediterranean Shrublands Biomass Estimation using Sentinel-1 and Sentinel-2. In Proceedings of the 36th IEEE International Geoscience and Remote Sensing Symposium (IGARSS), Beijing, China, 10–15 July 2016; pp. 5300–5303.

30. Laurin, G.V.; Puletti, N.; Hawthorne, W.; Liesenberg, V.; Corona, P.; Papale, D.; Chen, Q.; Valentini, R. Discrimination of tropical forest types, dominant species, and mapping of functional guilds by hyperspectral and simulated multispectral Sentinel-2 data. *Remote Sens. Environ.* **2016**, *176*, 163–176. [CrossRef]

31. Liu, Y.A.; Gong, W.S.; Hu, X.Y.; Gong, J.Y. Forest type identification with random forest using Sentinel-1A, Sentinel-2A, multi-temporal Landsat-8 and DEM data. *Remote Sens.* **2018**, *10*, 946. [CrossRef]

32. Tesfamichael, S.G.; Newete, S.W.; Adam, E.; Dubula, B. Field spectroradiometer and simulated multispectral bands for discriminating invasive species from morphologically similar cohabitant plants. *GISci. Remote Sens.* **2018**, *55*, 417–436. [CrossRef]

33. Battude, M.; Al Bitar, A.; Morin, D.; Cros, J.; Huc, M.; Sicre, C.M.; Le Dantec, V.; Demarez, V. Estimating maize biomass and yield over large areas using high spatial and temporal resolution Sentinel-2 like remote sensing data. *Remote Sens.* **2016**, *184*, 668–681. [CrossRef]

34. Su, W.; Hou, N.; Li, Q.; Zhang, M.Z.; Zhao, X.F.; Jiang, K.P. Retrieving leaf area index of corn canopy based on Sentinel-2 remote sensing image. *Trans. Chin. Soc. Agric. Mach.* **2018**, *49*, 151–156.

35. Sanches, I.D.; Feitosa, R.Q.; Diaz, P.M.A.; Soares, M.D.; Luiz, A.J.B.; Schultz, B.; Maurano, L.E.P. Campo verde database: Seeking to improve agricultural remote sensing of tropical areas. *IEEE Geosci. Remote Sens. Lett.* **2018**, *15*, 369–373. [CrossRef]

36. Sibanda, M.; Mutanga, O.; Rouget, M. Examining the potential of Sentinel-2 MSI spectral resolution in quantifying above ground biomass across different fertilizer treatments. *ISPRS J. Photogramm.* **2015**, *110*, 55–65. [CrossRef]

37. Sakowska, K.; Juszczak, R.; Gianelle, D. Remote sensing of grassland biophysical parameters in the context of the Sentinel-2 satellite mission. *J. Sens.* **2016**, *2016*, 4612809. [CrossRef]

38. Ali, I.; Cawkwell, F.; Dwyer, E.; Green, S. Modeling managed grassland biomass estimation by using multitemporal remote sensing data-A machine learning approach. *IEEE J. Sel. Top. Appl. Earth Obs. Remote Sens.* **2017**, *10*, 3254–3264. [CrossRef]

39. Hawrylo, P.; Wezyk, P. Predicting growing stock volume of scots pine stands using Sentinel-2 satellite imagery and airborne image-derived point clouds. *Forests* **2018**, *9*, 274. [CrossRef]

40. Mura, M.; Bottalico, F.; Giannetti, F.; Bertani, R.; Giannini, R.; Mancini, M.; Orlandini, S.; Travaglini, D.; Chirici, G. Exploiting the capabilities of the Sentinel-2 multi spectral instrument for predicting growing stock volume in forest ecosystems. *Int. J. Appl. Earth Obs.* **2018**, *66*, 126–134. [CrossRef]

41. Plank, S. Rapid damage assessment by means of multi-temporal SAR-A comprehensive review and outlook to Sentinel-1. *Remote Sens.* **2014**, *6*, 4870–4906. [CrossRef]

42. Chemura, A.; Mutanga, O.; Dube, T. Separability of coffee leaf rust infection levels with machine learning methods at Sentinel-2 MSI spectral resolutions. *Precis. Agric.* **2017**, *18*, 859–881. [CrossRef]

43. Mallinis, G.; Mitsopoulos, I.; Chrysafi, I. Evaluating and comparing Sentinel 2A and Landsat-8 Operational Land Imager (OLI) spectral indices for estimating fire severity in a Mediterranean pine ecosystem of Greece. *GISci. Remote Sens.* **2018**, *55*, 1–18. [CrossRef]

44. Baccini, A.; Goetz, S.J.; Walker, W.S.; Laporte, N.T.; Sun, M.; Sulla-Menashe, D.; Hackler, J.; Beck, P.S.A.; Dubayah, R.; Friedl, M.A.; et al. Estimated carbon dioxide emissions from tropical deforestation improved by carbon-density maps. *Nat. Clim. Chang.* **2012**, *2*, 182–185. [CrossRef]

45. Lu, D. The potential and challenge of remote sensing–based biomass estimation. *Int. J. Remote Sens.* **2006**, *27*, 1297–1328. [CrossRef]

46. Fassnacht, F.E.; Hartig, F.; Latifi, H.; Berger, C.; Hernandez, J.; Corvalan, P.; Koch, B. Importance of sample size, data type and prediction method for remote sensing-based estimations of aboveground forest biomass. *Remote Sens. Environ.* **2014**, *154*, 102–114. [CrossRef]

47. Fotheringham, A.S.; Brunsdon, C.; Charlton, M.E. *Geographically Weighted Regression: The Analysis of Spatially Varying Relationships*; Wiley: Chichester, UK, 2002.

48. Propastin, P. Modifying geographically weighted regression for estimating aboveground biomass in tropical rainforests by multispectral remote sensing data. *Int. J. Appl. Earth Obs.* **2012**, *18*, 82–90. [CrossRef]

49. Gao, Y.K.; Lu, D.S.; Li, G.Y.; Wang, G.X.; Chen, Q.; Liu, L.J.; Li, D.Q. Comparative analysis of modeling algorithms for forest aboveground biomass estimation in a subtropical region. *Remote Sens.* **2018**, *10*, 627. [CrossRef]

50. Gao, Y.; Bian, J.M.; Song, C. Study on the dynamic relation between spring discharge and precipitation in Fusong County, Changbai Mountain, Jilin Province of China. *Water Sci. Technol. Water Supply* **2016**, *16*, 428–437. [CrossRef]

51. Yang, F.; Yao, Z.F.; Sun, J.L.; Zhu, Y.Q.; Wang, Z.M. The landscape pattern changes analysis of Changbai Mountain forest based on RS and GIS—A case study in Fusong and Antu Counties. *Syst. Sci. Compr. Stud. Agric.* **2010**, *26*, 431–437. (In Chinese)

52. Chen, L.; Ren, C.Y.; Zhang, B.; Wang, Z.M.; Liu, M.Y. Quantifying urban land sprawl and its driving forces in Northeast China from 1990 to 2015. *Sustainability* **2018**, *10*, 188. [CrossRef]

53. Li, X.N.; Guo, Q.X.; Wang, X.C.; Zheng, H.F. Allometry of understory tree species in a natural secondary forest in Northeast China. *Sci. Silvae Sin.* **2010**, *46*, 22–32. (In Chinese)

54. Tang, X.G. Estimation of Forest Aboveground Biomass by Integrating ICESat/GLAS Waveform and TM Data. Ph.D. Thesis, University of Chinese Academy of Sciences, Beijing, China, 2013. (In Chinese)

55. Castillo, J.A.A.; Apan, A.A.; Maraseni, T.N.; Salmo, S.G. Estimation and mapping of above-ground biomass of mangrove forests and their replacement land uses in the Philippines using Sentinel imagery. *ISPRS J. Photogramm. Remote Sens.* **2017**, *134*, 70–85. [CrossRef]

56. Veci, L. *Sentinel-1 Toolbox: SAR Basics Tutorial*; ARRAY Systems Computing, Inc. and European Space Agency: Paris, France, 2015.

57. Liu, C. Analysis of Sentinel-1 SAR Data for Mapping Standing Water in the Twente Region. Master's Thesis, University of Twente, Enschede, The Netherlands, 2016.

58. Pan, L.; Sun, Y.J. Estimation of Cunninghamia lanceolata forest biomass based on Sentinel-1 image texture information. *J. Northeast For. Univ.* **2018**, *46*, 58–62. (In Chinese)

59. Jacquemoud, S.; Verhoef, W.; Baret, F.; Bacour, C.; Zarco-Tejada, P.J.; Asner, G.P.; François, C.; Ustin, S.L. PROSPECT + SAIL models: A review of use for vegetation characterization. *Remote Sens. Environ.* **2009**, *113*, S56–S66. [CrossRef]

60. Haralick, R.M.; Shanmugam, K.; Denstien, I. Textural features for image classification. *IEEE Trans. Syst. Man Cybern.* **1973**, *3*, 610–621. [CrossRef]

61. Rouse, J.W.; Haas, R.H.; Schell, J.A.; Deering, D.W. Monitoring Vegetation Systems in the Great Plains with ERTS. In Proceedings of the Third Earth Resources Technology Satellite-1 Symposium, Washington, DC, USA, 10–14 December 1973; Volume 1, pp. 309–317.

62. Delegido, J.; Verrelst, J.; Alonso, L.; Moreno, J. Evaluation of Sentinel-2 red-edge bands for empirical estimation of green LAI and chlorophyll content. *Sensors* **2011**, *11*, 7063–7081. [CrossRef] [PubMed]

63. Clevers, J.G.P.W.; De Jong, S.M.; Epema, G.F.; Van der Meer, F.D.; Bakker, W.H.; Skidmore, A.K.; Scholte, K.H. Derivation of the red edge index using the MERIS standard band setting. *Int. J. Remote Sens.* **2002**, *23*, 3169–3184. [CrossRef]

64. Deering, D.W.; Rouse, J.W.; Haas, R.H.; Schell, J.A. Measuring Forage Production of Grazing Units from Landsat MSS data. In Proceedings of the Tenth International Symposium on Remote Sensing of Environment, Ann Arbor, MI, USA, 6 October 1975; Volume 2, pp. 1169–1178.

65. O'brien, R.M. A caution regarding rules of thumb for variance inflation factors. *Qual. Quant.* **2007**, *41*, 673–690. [CrossRef]

66. Were, K.; Bui, D.T.; Dick, Ø.B.; Singh, B.R. A comparative assessment of support vector regression, artificial neural networks, and random forests for predicting and mapping soil organic carbon stocks across an Afromontane landscape. *Ecol. Indic.* **2015**, *52*, 394–403. [CrossRef]

67. Cheadle, C.; Vawter, M.P.; Freed, W.J.; Becker, K.G. Analysis of microarray data using Z score transformation. *J. Mol. Diagn.* **2003**, *5*, 73–81. [CrossRef]

68. Brunsdon, C.; Fotheringham, S.; Charlton, M. Geographically Weighted Regression–Modelling Spatial Non-stationarity. In *Workshop on Local Indicators of Spatial Association*; University of Leicester: Leicester, UK, 1998; Volume 47, pp. 431–443.

69. Kumar, S.; Lal, R.; Liu, D. A geographically weighted regression kriging approach for mapping soil organic carbon stock. *Geoderma* **2012**, *189*, 627–634. [CrossRef]

70. Shin, J.; Temesgen, H.; Strunk, J.L.; Hilker, T. Comparing modeling methods for predicting forest attributes using LiDAR metrics and ground measurements. *Can. J. Remote Sens.* **2016**, *42*, 739–765. [CrossRef]

71. Zhang, C.S.; Tang, Y.; Xu, X.L.; Kiely, G. Towards spatial geochemical modelling: Use of geographically weighted regression for mapping soil organic carbon contents in Ireland. *Appl. Geochem.* **2011**, *26*, 1239–1248. [CrossRef]

72. Peter, M. Efficient statistical classification of satellite measurements. *Int. J. Remote Sens.* **2011**, *32*, 6109–6132.

73. Ahmed, M.A.A.; Abd-Elrahman, A.; Escobedo, F.J.; Cropper, W.P.; Martin, T.A.; Timilsina, N. Spatially-explicit modeling of multi-scale drivers of aboveground forest biomass and water yield in watersheds of the Southeastern United States. *J. Environ. Manag.* **2017**, *199*, 158–171. [CrossRef] [PubMed]

74. Nakaya, T.; Charlton, M.; Lewis, P.; Brunsdon, C.; Yao, J.; Fotheringham, S. *GWR4 User Manual, Windows Application for Geographically Weighted Regression Modelling*; Ritsumeikan University: Kyoto, Japan, 2014.

75. Haykin, S.S. *Neural Networks: A Comprehensive Foundation*; Tsinghua University Press: Beijing, China, 2001.

76. Zhu, Y.H.; Liu, K.; Liu, L.; Wang, S.G.; Liu, H.X. Retrieval of mangrove aboveground biomass at the individual species level with WorldView-2 images. *Remote Sens.* **2015**, *7*, 12192–12214. [CrossRef]

77. Hornik, K. Multilayer feed forward network are universal approximators. *Neural Netw.* **1989**, *2*, 359–366. [CrossRef]

78. Santi, E.; Paloscia, S.; Pettinato, S.; Chirici, G.; Mura, M.; Maselli, F. Application of Neural Networks for the retrieval of forest woody volume from SAR multifrequency data at L and C bands. *Eur. J. Remote Sens.* **2015**, *48*, 673–687. [CrossRef]

79. Lee, S.; Evangelista, D.G. Earthquake-induced landslide susceptibility mapping using an artificial neural network. *Nat. Hazards Earth Syst. Sci.* **2006**, *6*, 687–695. [CrossRef]

80. Ottoy, S.; De Vos, B.; Sindayihebura, A.; Hermy, M.; Van Orshoven, J. Assessing soil organic carbon stocks under current and potential forest cover using digital soil mapping and spatial generalisation. *Ecol. Indic.* **2017**, *77*, 139–150. [CrossRef]

81. Vapnik, V.N. *The Nature of Statistical Learning Theory. Statistics for Engineering and Information Science*; Springer: New York, NY, USA, 2000.

82. Li, M.; Im, J.; Quackenbush, L.J.; Liu, T. Forest biomass and carbon stock quantification using airborne LiDAR data: A case study over Huntington wildlife forest in the Adirondack Park. *IEEE J. Sel. Top. Appl. Earth Obs. Remote Sens.* **2014**, *7*, 3143–3156. [CrossRef]

83. Meng, S.L.; Pang, Y.; Zhang, Z.J.; Jia, W.; Li, Z.Y. Mapping aboveground biomass using texture indices from aerial photos in a temperate forest of Northeastern China. *Remote Sens.* **2015**, *7*, 12192–12214. [CrossRef]

84. Sharifi, A.; Amini, J.; Tateishi, R. Estimation of forest biomass using multivariate relevance vector regression. *Photogramm. Eng. Remote Sens.* **2016**, *82*, 41–49. [CrossRef]

85. Platt, J. Fast Training of Support Vector Machines Using Sequential Minimal Optimization. In *Advances in Kernel Methods–Support Vector Learning*; MIT Press: Cambridge, MA, USA, 1999.

86. Zeng, Z.; Hsieh, W.W.; Shabbar, A.; Burrows, W.R. Seasonal prediction of winter extreme precipitation over Canada by support vector regression. *Hydrol. Earth Syst. Sci.* **2011**, *15*, 65–74. [CrossRef]

87. Singh, M.; Evans, D.; Friess, D.A.; Tan, B.S.; Nin, C.S. Mapping above-ground biomass in a tropical forest in Cambodia using canopy textures derived from Google Earth. *Remote Sens.* **2015**, *7*, 5057–5076. [CrossRef]

88. Rabe, A.; van der Linden, S.; Hostert, P. Simplifying Support Vector Machines for Regression Analysis of Hyperspectral Imagery. In Proceedings of the 1st Workshop on Hyperspectral Image and Signal Processing–Evolution in Remote Sensing, Grenoble, France, 26–28 August 2009.

89. Breiman, L. Random Forests. *Mach. Learn.* **2001**, *45*, 5–32. [CrossRef]

90. Breiman, L. Bagging predictors. *Mach. Learn.* **1996**, *24*, 123–140. [CrossRef]

91. Amit, Y.; Geman, D. Shape quantization and recognition with randomized trees. *Neural Comput.* **1997**, *9*, 1545–1588. [CrossRef]

92. Dhanda, P.; Nandy, S.; Kushwaha, S.P.S.; Ghosh, S.; Murthy, Y.V.N.K.; Dadhwal, V.K. Optimizing spaceborne LiDAR and very high resolution optical sensor parameters for biomass estimation at ICESat/GLAS footprint level using regression algorithms. *Prog. Phys. Geogr.* **2017**, *41*, 247–267. [CrossRef]

93. Genuer, R.; Poggi, J.-M.; Tuleau-Malot, C.; Villa-Vialaneix, N. Random forests for big data. *Big Data Res.* **2017**, *9*, 28–46. [CrossRef]

94. Jovic, A.; Bogunovic, N. Electrocardiogram analysis using a combination of statistical, geometric, and nonlinear heart rate variability features. *Artif. Intell. Med.* **2011**, *51*, 175–186. [CrossRef] [PubMed]

95. Mutanga, O.; Adam, E.; Cho, M.A. High density biomass estimation for wetland vegetation using WorldView-2 imagery and random forest regression algorithm. *Int. J. Appl. Earth. Obs.* **2012**, *18*, 399–406. [CrossRef]

96. Isaaks, E.H.; Srivastava, R.M. *An Introduction to Applied Geostatistics*; Oxford University Press: Oxford, UK, 1989.

97. Wang, D.L.; Xin, X.P.; Shao, Q.Q.; Brolly, M.; Zhu, Z.L.; Chen, J. Modeling aboveground biomass in Hulunber grassland ecosystem by using unmanned aerial vehicle discrete Lidar. *Sensors* **2017**, *17*, 180. [CrossRef] [PubMed]

98. Bourgoin, C.; Blanc, L.; Bailly, J.S.; Cornu, G.; Berenguer, E.; Oszwald, J.; Tritsch, I.; Laurent, F.; Hasan, A.F.; Sist, P.; et al. The potential of multisource remote sensing for mapping the biomass of a degraded Amazonian forest. *Forests* **2018**, *9*, 303. [CrossRef]

99. Ghosh, S.M.; Behera, M.D. Aboveground biomass estimation using multi-sensor data synergy and machine learning algorithms in a dense tropical forest. *Appl. Geogr.* **2018**, *96*, 29–40. [CrossRef]

100. Berninger, A.; Lohberger, S.; Stangel, M.; Siegert, F. SAR-based estimation of above-ground biomass and its changes in tropical forests of Kalimantan using L- and C-band. *Remote Sens.* **2018**, *10*, 831. [CrossRef]
101. Vafaei, S.; Soosani, J.; Adeli, K.; Fadaei, H.; Naghavi, H.; Pham, T.D.; Bui, D.T. Improving accuracy estimation of forest aboveground biomass based on incorporation of ALOS-2 PALSAR-2 and Sentinel-2A imagery and machine learning: A case study of the Hyrcanian forest area (Iran). *Remote Sens.* **2018**, *10*, 172. [CrossRef]
102. Sinha, S.; Jeganathan, C.; Sharma, L.K. A review of radar remote sensing for biomass estimation. *Int. J. Environ. Sci. Technol.* **2015**, *12*, 1779–1792. [CrossRef]

© 2018 by the authors. Licensee MDPI, Basel, Switzerland. This article is an open access article distributed under the terms and conditions of the Creative Commons Attribution (CC BY) license (http://creativecommons.org/licenses/by/4.0/).

Article

Estimating Forest Canopy Cover in Black Locust (*Robinia pseudoacacia* L.) Plantations on the Loess Plateau Using Random Forest

Qingxia Zhao [1,2], Fei Wang [3], Jun Zhao [4], Jingjing Zhou [5], Shichuan Yu [1,2] and Zhong Zhao [1,2,*]

[1] College of Forestry, Northwest A&F University, Yangling 712100, Shaanxi, China; zhaoqingxia@nwafu.edu.cn (Q.Z.); yushichuan@nwafu.edu.cn (S.Y.)

[2] Key Comprehensive Laboratory of Forestry of Shaanxi Province, Northwest A&F University, Yangling 712100, Shaanxi, China

[3] Northwest Institute of Forest Inventory, Planning and Design, State Forestry Administration, Xi'an 710048, Shaanxi, China; wangfei2008@nwsuaf.edu.cn

[4] Research Center for Eco-Environmental Science, Chinese Academy of Sciences, Beijing 100085, China; junzhao@rcees.ac.cn

[5] College of Horticulture & Forestry Sciences/Hubei Engineering Technology Research Center for Forestry Information, Huazhong Agricultural University, Wuhan 430070, Hubei, China; hupodingxiangyu@mail.hzau.edu.cn

* Correspondence: zhaozh@nwafu.edu.cn; Tel./Fax: +86-029-8708-2801

Received: 10 September 2018; Accepted: 7 October 2018; Published: 10 October 2018

Abstract: The forest canopy is the medium for energy and mass exchange between forest ecosystems and the atmosphere. Remote sensing techniques are more efficient and appropriate for estimating forest canopy cover (CC) than traditional methods, especially at large scales. In this study, we evaluated the CC of black locust plantations on the Loess Plateau using random forest (RF) regression models. The models were established using the relationships between digital hemispherical photograph (DHP) field data and variables that were calculated from satellite images. Three types of variables were calculated from the satellite data: spectral variables calculated from a multispectral image, textural variables calculated from a panchromatic image (T_{pan}) with a 15 × 15 window size, and textural variables calculated from spectral variables (T_{B+VIs}) with a 9 × 9 window size. We compared different *mtry* and *ntree* values to find the most suitable parameters for the RF models. The results indicated that the RF model of spectral variables explained 57% (root mean square error (RMSE) = 0.06) of the variability in the field CC data. The soil-adjusted vegetation index (SAVI) and enhanced vegetation index (EVI) were more important than other spectral variables. The RF model of T_{pan} obtained higher accuracy (R^2 = 0.69, RMSE = 0.05) than the spectral variables, and the grey level co-occurrence matrix-based texture measure—Correlation (COR) was the most important variable for T_{pan}. The most accurate model was obtained from the T_{B+VIs} (R^2 = 0.79, RMSE = 0.05), which combined spectral and textural information, thus providing a significant improvement in estimating CC. This model provided an effective approach for detecting the CC of black locust plantations on the Loess Plateau.

Keywords: canopy cover (CC); spectral; texture; digital hemispherical photograph (DHP); random forest (RF); gray level co-occurrence matrix (GLCM)

---

1. Introduction

Forest canopy cover (CC), which is defined as the proportion of the forest floor that is covered by the vertical projection of the tree crown [1], plays a major role in understanding the structure and health condition of forest ecosystems [2]. It is also a useful measure for evaluating the leaf area index (LAI), carbon stocks, tree density, and wildlife microhabitat [3–6]. The growth and diversity of understory vegetation is also related to CC [7]. In addition, the forest CC and understory vegetation play an important role in minimize the rate of soil loss by reducing the erosive effects of rainfall with interception [8]. The Loess Plateau of China has experienced severe soil erosion, vegetation degradation, and desertification [9]. Black locust (*Robinia pseudoacacia* L.) represents the most abundant type of plantation on the Loess Plateau, and these plantations provide a wide range of ecological and socio-economic functions [10,11]. Accurate and regular measurements of the CC of black locust plantations are important and can be used to monitor forest degradation and desertification on the Loess Plateau [12].

There are two common approaches for CC estimation: field measurements and remote sensing. Field-based methods are the most accurate estimation approaches. Optical instruments, such as digital hemispherical photographs (DHPs), LAI-2200 plant canopy analyzers, and AccuPAR Ceptometers are widely adopted for CC acquisition due to their non-destructive nature [13]. However, these methods suffer from many shortcomings; for example, they are time-consuming, labor-intensive, and difficult to apply for large areas [5,14]. Thus, a method is needed that can be used to easily extract forest CC over a large area. Remote sensing techniques can represent such a method, because they offer new strategies for measuring forest CC from local to global scales [3,12,15–18]. Different type of sensors such as aerial photo, satellite images and active sensors (e.g., LiDAR, SAR, and RADAR) have been applied to estimate forest CC [18–20]. The unmanned aerial vehicle (UAV) with high acquisition flexibility and resolution appears to be very promising for the assessment of CC, but only if the forest is widely open and a precise digital terrain model is available [21]. Ma et al. [20] compared LiDAR, aerial imagery and satellite imagery in CC estimation, and found that LiDAR-derived CC were marginally influenced by the estimation algorithms and generate comparable results. The major limitations in aerial photo were distortion and the presence of shadows. While the distortion and shadow have less impact on satellite imagery since these images were collected at high altitude [20]. The active sensors also have some limitations, such as challenging processing requirements and complex interactions with forest structure. Wallis et al. [22] noted that optical remote sensing data can be substituted for LiDAR data in habitat diversity studies. Optical sensors are still a popular choice for forest parameter estimation. In this research, we select a satellite image to estimate CC.

A range of variables is calculated from satellite images for CC estimation. Spectral vegetation indices (VIs), such as the normalized difference vegetation index (NDVI), the soil-adjusted vegetation index (SAVI), and the enhanced vegetation index (EVI), have been used to estimate CC in monospecific and mixed forests [18,19,23,24]. For example, Korhonen et al. [18] used NDVI, atmospherically resistant vegetation index (ARVI), and simple ration (SR) to estimate CC in tropical forests based on ALOS AVNIR-2 image. The NDVI, SR, and SAVI derived from SPOT 5 satellite data were used to estimate CC by Chasmer et al. [25], their results indicated that the NDVI and SAVI are more comparable to CC. In addition, González-Roglich and Swenson [5] studied the tree cover and carbon of Argentine savannas using seven VIs that are based on the China Brazil Earth Resources Satellite series and Landsat images. Karlson et al. [24] estimated the tree CC and aboveground biomass using six VIs based on the Landsat 8 OLI in a woodland landscape. However, spectral variables suffer from saturation and multiple layering problems in high canopy regions [11,26].

With recent high-resolution imagery, such as QuickBird and IKONOS, forest CC can be recognized at a crown level. When spatial resolution increases, the objects on the ground tend to be represented by few pixels; therefore, the spatial information becomes increasingly important when compared with spectral information. The texture calculated from high spatial resolution images can enhance the discrimination of spatial information and improve detection levels by increasing saturation levels [11,27,28]. Li et al. [19], Sarker and Nichol [27], and Wood et al. [29] all found that textural variables provided a larger contribution than spectral variables for forest parameter estimations. Tuanmu and Jetz [30] demonstrated that the texture measures of the EVI were superior to the conventional topography- and land-cover-based metrics in terms of their ability to capture fine-grain habitat heterogeneity and predict key biodiversity patterns across a large extent. Combining spatial detail and unique spectral information leverages complementary information, which could improve the accuracy of estimating canopy properties [11,31]. Gu et al. [16] reported that a combination of VIs and texture can improve the accuracy of estimating vegetation fractional coverage of planted and natural forests. Halperin et al. [12] used spectral and textural variables to estimate the CC in the Miombo woodlands of Zambia, and they found that the texture was more prominent in the imagery and that a combination of spectral and textural variables provided the best estimation. Pu et al. [28] demonstrated similar results that the combination of spectral and textural variables could generate higher accuracy than their either one separately in mapping forest LAI.

Many parametric and non-parametric methods have been used for predicting forest structure parameters. Parametric methods, such as traditional linear regression models, were the most widely used models in the last decades [3,11,26,28], and they are simple and easy to explain [32]. The traditional linear models have an explicit model structure and they can be specified by parameters [33]. The relationship between predictors and response variables can be explained from an ecological perspective [34]. However, these models make strong assumptions about the data, and multicollinearity problems may occur [35]. The models usually obtain moderate accuracy, and their performance mainly depends on the goodness of these assumptions [34]. In contrast, non-parametric methods have fewer assumptions, higher methodological accuracy, and high non-linear adaptation [36]. The major drawback of these models is uneasy interpreting, because it seems more of a "black box" [32,34]. The structure developed for many remote sensing regression applications can be very complex and it is impossible to derive meaningful interpretation. Even so, the non-parametric approaches are becoming more popular, especially since there a diverse array of spatially-explicit explanatory variables that are available to researchers. One of the most widely adopted approaches used for CC estimation is random forest (RF) regression models. The RF algorithm can be used to reduce input data dimension and the variable importance measurement could identify the most relevant remote sensing variables [37,38]. Pullanagari et al. [39] evaluated pasture quality using RF combined with recursive feature elimination, and obtained stable result with improved accuracy. Karlson et al. [24] used RF and Landsat 8 OLI to map the CC of trees and the aboveground biomass in the Sudano-Sahelian woodlands. In addition, the RF has been successfully applied in land use and land cover classification, this method provide higher accuracy when compared to maximum likelihood classifier, CART and SVM [38,40,41]. Shataee et al. [42] and Cracknell and Reading [43] indicated that RF was superior to other non-parametric approaches in predicting forest parameters and supervised classification for lithology. This method can easily train and stabilize a range of model parameter values [44]. The potential of RF to predict the CC of black locust plantations on the Loess Plateau has not received much attention.

The objectives of this study are to demonstrate a potential approach for modeling and mapping black locust plantations CC on the Loess Plateau using QuickBird imagery. Specifically, the goals are to (1) identify the optimal window size and suitable parameters of RF models; (2) compare the performance of three types of variables in modeling the CC of black locust plantations, i.e., spectral variables (Bands + VIs), textural variables calculated from panchromatic image (T_{pan}) and textural variables calculated from spectral variables (T_{B+VIs}); and, (3) map the CC in the black locust plantations using the most accurate RF regression model. The results obtained in this study are conducive to efficiently estimating the CC. A CC map is an effective tool for detecting the state of forest areas and the associated forest health conditions, both of which can be used in developing forest management plans on the Loess Plateau of China.

2. Materials and Methods

2.1. Study Area

The study area is located in Yongshou County of Shaanxi Province on the southern Loess Plateau of China (34°44′–34°56′ N and 107°56′–108°09′ E) (Figure 1). The region has a semi-humid, temperate continental climate with a mean temperature of 7–13.3 °C. The average annual precipitation is 601.6 mm, of which more than 53% falls between July and September. The soil type at the study site is cinnamon soil (based on the Chinese Soil Taxonomy). The study area is located in the loess hilly-gully region and it has an elevation ranging from 1060 to 1508 m above sea level. The forest is distributed across two sites, the Huaiping forest farm and the Maliantan forest farm, and it is dominated by black locust plantations (*Robinia pseudoacacia* L.), which account for about 90% of the forested area.

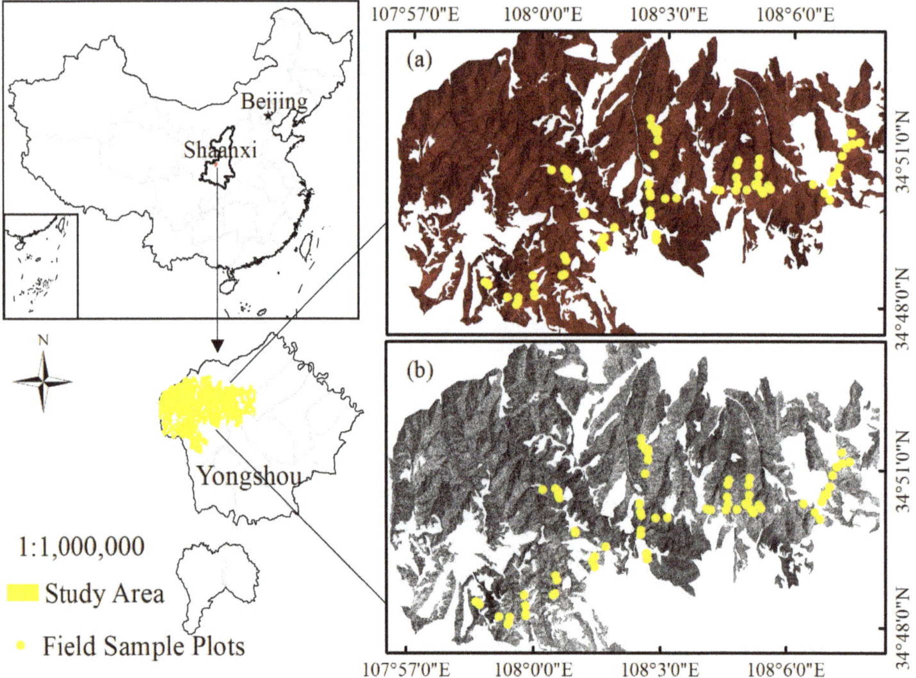

Figure 1. Location of the study area and the field sample plots identified from the multispectral (**a**) and panchromatic (**b**) data of QuickBird imagery.

2.2. Field Data

A field survey of the sample plots was performed from 16 June to 15 July 2012. Overall, 74 plots were randomly distributed within pure black locust plantation forest areas, where the forest area was delimited by the satellite image based on the National Forest Inventory (NFI) data. Each plot was 20 × 20 m in size. The CC, diameter at breast height (DBH), tree height, crown diameter, and stand density were measured in each plot. Trees with a DBH less than 5 cm were not included. A differential global positioning system was used to determine the center and the four corners of each plot, thus allowing for the plots to be geo-referenced with satellite data. The field data characteristics are summarized in Table 1.

Table 1. Summary of the characteristics of the forest field plots.

Variable (Unit)	Minimum	Maximum	Mean	Standard Deviation
CC	0.28	0.88	0.67	0.10
DBH (cm)	5.38	26.41	12.58	4.80
Crown Diameter (m)	2.02	5.81	3.51	0.88
Density (N/ha)	250	2775	1228	676
Height (m)	5.38	19.98	11.98	3.03

DHPs were obtained to calculate the forest CC. The DHPs were collected from five randomly selected locations in each plot using a Nikon Coolpix 4500 (Nikon Corp., Tokyo, Japan) digital camera in combination with an FC-E8 Circular Fisheye lens. The camera and lens provided a focal length equivalent of 6 mm, a combined f-stop of f/2.8, and a full 180-degree field-of-view. The camera was mounted on a tripod, leveled using a two-axis camera-mounted bubble level, oriented to magnetic north, and positioned 1.3 m above ground level. Overhead branches were avoided, and the camera exposure time was set to automatic. To avoid the effects of sunlight, we chose a time close to sunrise (or sunset) under uniform sky conditions and not during common working hours to avoid the interference of direct sunlight. All of the images were shot at "fine" quality and the maximum resolution (2048 pixels × 1536 pixels) of the camera. Images were saved in JPEG format.

Each DHP was analyzed with Gap Light Analyzer 2.0 (Simon Fraser University, Burnaby, BC, Canada) [45]. In this study, we used the blue channel and a threshold of 128 to calculate the CC of each photograph. The blue band is preferred because this portion of the spectrum is superior for separating pixels into forest canopy and sky classes [45–47]. Several authors have noted that subjective adjustment of the threshold can be a source of error because it is somewhat arbitrary [47,48]. To overcome these subjectivity issues, we use a constant threshold of 128 (half of a 256-bit image) to separate the sky and canopy of each photograph [49,50]. Then, the CC was calculated, as follows.

$$CC = \frac{\text{Total pixels} - \text{Sky pixels}}{\text{Total pixels}} \tag{1}$$

The CC was calculated for each DHP, and the average of five photographs was calculated in each plot.

2.3. Remote Sensing Data

The QuickBird multispectral and panchromatic images used in this study were obtained on 22 June 2012, and the spatial resolution was 0.61 m for the panchromatic image and 2.44 m for the multispectral image. The multispectral image had four bands, including blue (450–520 nm), green (520–600 nm), red (630–690 nm), and near-infrared (760–900 nm) bands. The panchromatic image was ortho-rectified using the ENVI 5.1 software package (Exelis Visual Information Solutions, Boulder, CO, USA). Fifty high-quality and well-distributed ground control points (GCPs) that were obtained via the field survey using differential global positioning system equipment were used for ortho-rectification. A digital elevation model (DEM) (1:10,000) derived from stereo aerial photographs with a resolution of 5 m was also used to ortho-rectify the QuickBird panchromatic image. The overall error was 0.68 pixels. Then, the corrected panchromatic image was used to rectify the multispectral image, resulting in an overall error of 0.34 pixels. The raw digital numeric values of the multispectral image were converted to spectral radiance and subsequently to top of atmosphere (TOA) reflectance. Atmospheric correction was performed using the Fast Line-of-sight Atmospheric Analysis of Spectral Hypercubes (FLAASH) approach to remove the scattering and absorption effects from the atmosphere and to obtain the surface reflectance character. The non-black locust area was manually masked from the image based on the NFI data.

2.4. Predictor Variables

2.4.1. Spectral Variables

The spectral variables that were extracted from the QuickBird multispectral image included four reflectance bands (B1-blue, B2-green, B3-red, and B4-NIR) and eight VIs (Table 2). The VIs combine information from two or more spectral bands to enhance the vegetation signal while minimizing soil, atmospheric, and solar irradiance effects [51]. This study analyzed the correlations between the different spectral variables and the forest CC obtained from the DHP data.

Table 2. Selected vegetation indices (VIs) used for canopy cover (CC) estimation [11].

Spectral Vegetation Indices
1. Simple Ratio (SR) = $\frac{NIR}{R}$
2. Soil Adjusted Vegetation Index (SAVI) = $(1 + L)\frac{NIR-R}{NIR+R+L}$
3. Enhanced Vegetation index (EVI) = $G\frac{NIR-R}{NIR+C_1R-C_2B+L}$
4. Atmospherically Resistant Vegetation Index (ARVI) = $\frac{NIR-RB}{NIR+RB}$, $RB = R - \gamma(B - R)$
5. Modified Soil Adjusted Vegetation Index (MSAVI) = $\left[(2NIR + 1) - \sqrt{(2NIR + 1)^2 - 8(NIR - R)}\right]/2$
6. Non-linear Vegetation index (NLI) = $\frac{NIR^2-R}{NIR^2+R}$
7. Difference Vegetation index (DVI) = $NIR - R$
8. Normalized Difference Vegetation Index (NDVI) = $\frac{NIR-R}{NIR+R}$

B, R, and NIR represent the QuickBird reflectance for the blue, red and near-infrared wavelengths, respectively. Parameters L and γ represent the SAVI term (set equal to 0.5) and the ARVI term (set equal to 1), respectively. The coefficients adopted in the EVI algorithm are L = 1.0, C_1 = 6.0, C_2 = 7.5, and G (gain factor) = 2.5 [51].

2.4.2. Textural Variables Calculated from Panchromatic Image (T_{pan})

Image texture, which is defined by Haralick et al. [52] as "the pattern of spatial distributions of grey-tone", describes the relationships among surface cover elements. The texture of an image contains valuable information about the spatial and structural arrangement of objects [26,52]. In our research, eight widely used Grey Level Co-occurrence Matrix (GLCM) measures that were proposed by Haralick et al. [52] were selected (Table 3). To find the optimal window size for T_{pan}, we compared the model accuracy of the eight GLCM measures with different moving window sizes ranging from 3×3 to 15×15 pixels (discussed below). As a result, eight T_{pan} (each GLCM with optimal window size) variables were selected to analyze their relationships with CC.

Table 3. Formulas for the texture measurements used in this study [52].

Grey Level Co-occurrence Matrix (GLCM) Based Texture Parameter Estimation
1. Mean (MEAN) = $\sum_{i,j=0}^{N-1} p(i,j)$
2. Homogeneity (HOM) = $\sum_i \sum_j \frac{p(i,j)}{1+(i-j)^2}$
3. Contrast (CON) = $\sum_{n=0}^{N-1} n^2 \{\sum_{i=1}^{N} \sum_{j=1}^{N} p(i,j)\}$
4. Dissimilarity (DIS) = $\sum_{n=0}^{N-1} n\{\sum_{i=1}^{N} \sum_{j=1}^{N} p(i,j)\}$
5. Entropy (ENT) = $-\sum_i \sum_j p(i,j) \log(p(i,j))$
6. Variance (VAR) = $p_{i,j}(i - u_i)^2$
7. Angular Second Moment (ASM) = $\sum_i \sum_j \{p(i,j)\}^2$

Table 3. *Cont.*

Grey Level Co-occurrence Matrix (GLCM) Based Texture Parameter Estimation
8. Correlation (COR) = $\dfrac{\sum_i \sum_j (ij)\, p(i,j) - \mu_x \mu_y}{\sigma_x \sigma_y}$ $\mu_x = \sum_{i=0}^{N-1} i \sum_{j=0}^{N-1} P_{i,j}$ $\mu_y = \sum_{i=0}^{N-1} j \sum_{j=0}^{N-1} P_{i,j}$ ${\sigma_x}^2 = \sum_{i=0}^{N-1} (i - \mu_x)^2 \sum_{j=0}^{N-1} P_{i,j}$ ${\sigma_y}^2 = \sum_{j=0}^{N-1} (j - \mu_y)^2 \sum_{i=0}^{N-1} P_{i,j}$ Here, $P(i,j)$ is the normalized co-occurrence matrix.

2.4.3. Textural Variables Calculated from Spectral Variables (T_{B+VIs})

The texture calculated from the spectral variables (T_{B+VIs}), which included texture calculated from reflectance bands (T_B) and texture calculated from VIs (T_{VIs}), combines spectral and textural information. To find the optimal window size for T_{B+VIs}, we compared the model accuracy of the eight GLCM measures with different moving window sizes ranging from 3×3 to 15×15 pixels based on B4 (T_{B4}) (discussed below). Then, the optimal window size was applied to other spectral variables. The reason for choosing B4 as the deputation of spectral variables was that B4 was the most important and effective band to correlate with forest canopy [44]. Finally, 96 T_{B+VIs} variables (12 spectral variables $\times$ 8 GLCM measures) were selected to analyze their relationships with CC.

2.5. Random Forest (RF) Prediction of CC

The RF method, which was originally proposed by Breiman [53], is an ensemble of many classification or regression trees that can reduce the overfitting of models. RF does not make assumptions about the nature of the data distribution, and this function is simply learned from training samples [44]. The algorithm trains each tree on an independently randomized subset of the predictors and determines the number of variables to be used at each node by the evaluation of a random vector. By growing each tree to its maximum size without pruning and selecting only the best split among a random subset at each node, RF tries to maintain some prediction strength while inducing diversity among the trees [53]. The result is an ensemble of low bias and high variance regression trees, where the final predictions are derived by averaging the predictions of the individual trees [53–55].

The number of predictor variables has an effect on the model accuracy. Removing irrelevant variables could result in a more parsimonious model and to obtain higher accuracy. The Boruta method can be used to choose the optimal number of predictor variables based on the RF model. The Boruta method proposed by Kursa and Rudnichi [56] is an all-relevant feature selection wrapper algorithm. The method compares the importance of the original attributes with the randomly achievable importance, uses attribute replacement copies for estimation, and gradually eliminates the irrelevant features to stabilize the test, thereby performing a top-down search for relevant features.

RF only requires users to make decisions about two tuning parameters: the number of trees to grow (*ntree*) and number of variables randomly sampled as candidates at each split (*mtry*). The *ntree* values were tested from 100 to 5000 trees with intervals of 100. A suggested starting value of *mtry* included one-third of the predictive variables, followed by checking half this number and twice this number [57]. The mean squared error (MSE) was plotted for *ntree* and *mtry*.

To test the accuracy of different kind of variables, three RF models were developed:

Model 1—spectral variables
Model 2—textural variables calculated from the panchromatic image (T_{pan})
Model 3—textural variables calculated from the spectral variables (T_{B+VIs})

The field sample plots were randomly split into two unequal subsets: 70% for model construction and 30% for model validation. The coefficient of determination (R^2) and the root mean square error

(RMSE) were used to identify the best prediction model. The formulas of these statistical parameters are as follows:

$$R^2 = \frac{\sum_{i=1}^{n}(\hat{y}_i - \overline{y}_i)^2}{\sum_{i=1}^{n}(y_i - \overline{y})^2} \tag{2}$$

$$RMSE = \sqrt{\frac{\sum_{i=1}^{n}(\hat{y}_i - y_i)^2}{n}} \tag{3}$$

where y_i is the observed value, $\hat{y}_i$ is the predicted value, $\overline{y}$ is the mean of the observed values, and n is the number of observations for prediction model.

All the statistical analyses were completed using R software (version 3.4.3) [58].

3. Results

3.1. Determining the Optimal Window Size

Figure 2 shows the optimal window size with T_{pan} and T_{B4} in the selected seven window sizes. In T_{pan}, the model accuracy increases as the window sizes increases and the optimal window size is 15 × 15. In T_{B4}, the model accuracy increases from 3 × 3 to 9 × 9 and then a slight decrease is observed as the window size further increases. Therefore, we choose a window size of 15 × 15 as the optimal window size to calculate the texture from the panchromatic image and a window size of 9 × 9 as the optimal window size to calculate the texture from the spectral variables.

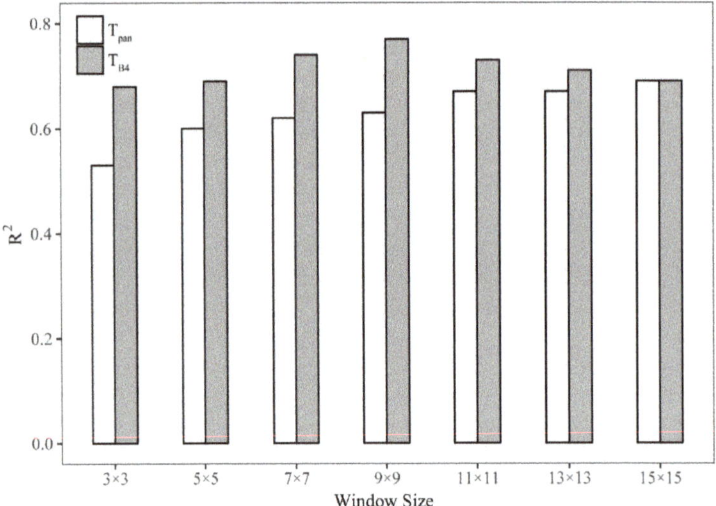

Figure 2. Illustration of the window size effect on the prediction of the forest CC (based on texture calculated from panchromatic image and Band 4).

3.2. Variable Selection and Parameter Tuning for the Final Three RF Models

Based on the Boruta algorithm, the explanatory variables that are relevant to the response variables were selected. For Model 1, the optimal number of explanatory variables was eight. In the selected spectral variables, SAVI had the highest importance value. In addition, the EVI, B4 and DVI also had relatively higher values than other spectral variables (Figure 3a). Model 2 was performed based on T_{pan} with a 15 × 15 window size. All the eight texture measures were selected as relevant variables, and the COR and MEAN hold higher importance values than other texture measures (Figure 3b). Model 3 was performed based on a 9 × 9 window size, and the optimal number of variables was 16. The top five variables in the variable importance were calculated based on the MEAN texture measure

(Figure 3c). The MEAN measure calculated from SAVI (MEAN$_{SAVI}$) and MSAVI (MEAN$_{MSAVI}$) had higher importance values than that of the other variables.

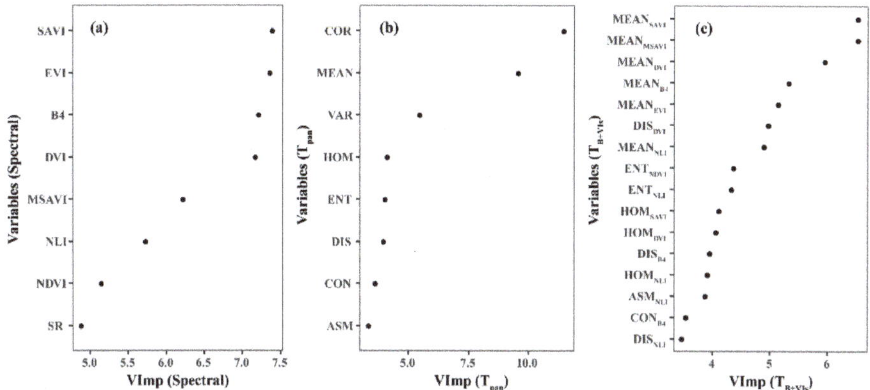

Figure 3. Number of explanatory variables selected by the Boruta algorithm and their importance values: (**a**) spectral variables; (**b**) T$_{pan}$; (**c**) T$_{B+VIs}$. (VImp represent the variable's importance values).

The results indicated that 8, 8, and 16 relevant variables were included in Model 1, Model 2, and Model 3, respectively. Considering the selection rules of *mtry* [57] and the selected number of explanatory variables in the three RF models, we considered three values for *mtry*: 2, 4, and 8. After tuning the RF models, the optimal parameters for Model 1 were *mtry* equal to 2 and *ntree* equal to 400 (Figure 4a), for Model 2 were *mtry* equal to 4 and *ntree* equal to 600 (Figure 4b), and for Model 3 were *mtry* equal to 2 and *ntree* equal to 2500 (Figure 4c).

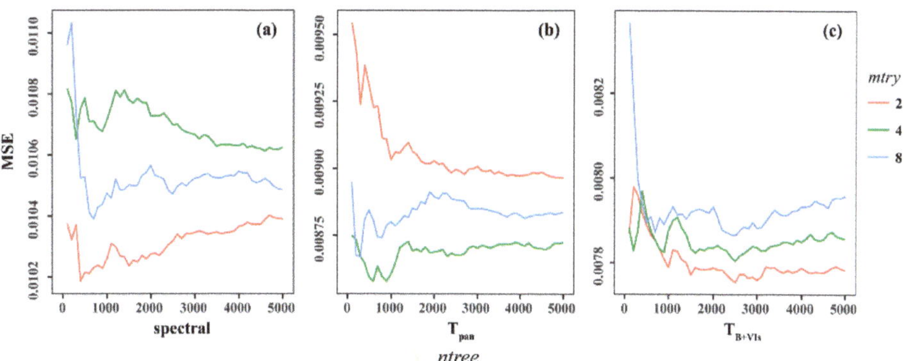

Figure 4. Effect of *mtry* and *ntree* on the random forest (RF) models of (**a**) spectral variables; (**b**) T$_{pan}$; (**c**) T$_{B+VIs}$.

3.3. Model Comparison and CC Mapping

Satisfactory agreement was observed when the remaining 30% of the validation field data were compared with the satellite image predicted data. Figure 5 shows the estimated accuracy of the final three RF models. Model 1 with spectral variables presented the lowest accuracy ($R^2 = 0.57$, RMSE = 0.06, Figure 5a), and a higher accuracy was obtained when using T$_{pan}$, (R^2 of 0.69 and RMSE of 0.05, Figure 5b). Model 3 with T$_{B+VIs}$ had the highest accuracy ($R^2 = 0.79$, RMSE = 0.05, Figure 5c). Therefore, Model 3 was used for the final estimation and mapping of the CC of black locust plantations.

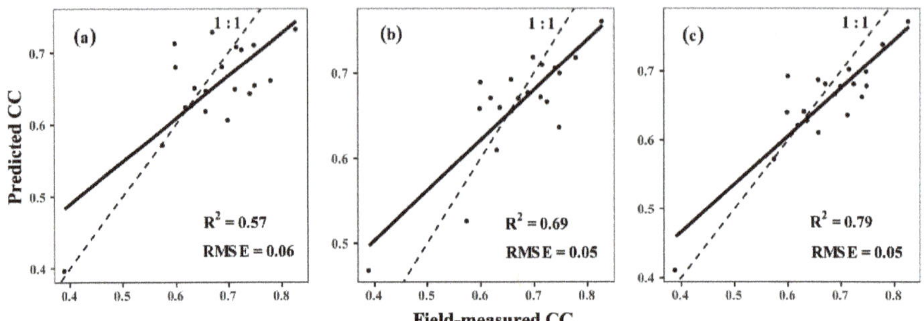

Figure 5. Plots of the observed and predicted CC using RF with (**a**) Model 1; (**b**) Model 2; and, (**c**) Model 3.

The developed RF model (Model 3) was used for calculating pixel-based CC values from the corresponding raster layers of the predictor variables. Figure 6 shows the CC distribution map of the black locust plantations predicted based on Model 3. The eastern region had higher CC values while the western part had lower CC values. In the study area, the average value of CC was approximately 0.6. Most forest CC values were between 0.4 and 0.8, which accounted for more than 99% of the study area (Table 4). Among these, more than half of the forest CC ranged from 0.6 to 0.8 and 40% of the forest CC were between 0.4 and 0.6.

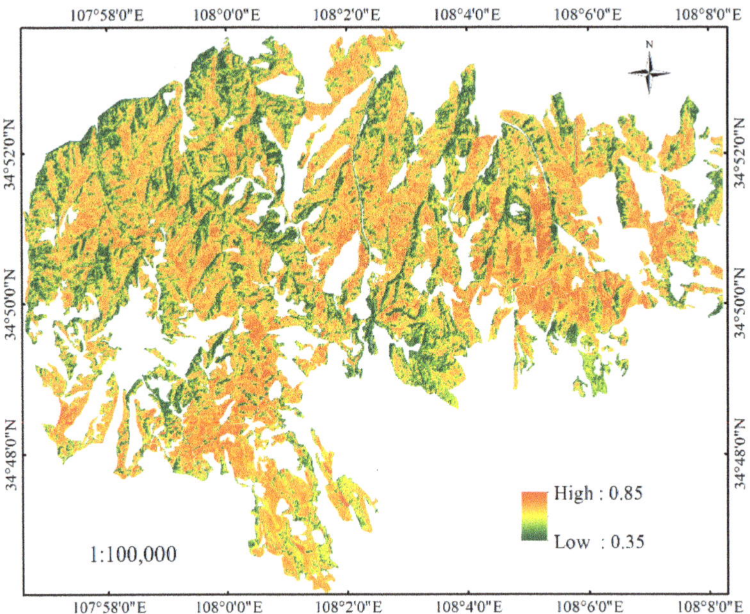

Figure 6. Predicted CC map of the black locust plantations based on Model 3 using QuickBird imagery.

Table 4. Summary the proportion of CC in different grades.

CC	Percent (%)
<0.4	0.75
0.4–0.6	40.38
0.6–0.8	58.82
0.8–1.0	0.05

4. Discussion

The importance of window size has been stressed in evaluations of texture measures [59]. Generally, for the eight GLCM texture measures in the selected seven window sizes, the 15 × 15 window size is optimal for panchromatic image and the 9 × 9 window size is optimal for the spectral variables. Furthermore, the T_{B+VIs} with 9 × 9 windows obtained higher accuracy than T_{pan} with 15 × 15 windows. Kamal et al. [60] observed that a pixel window size corresponding to the field plot size or slightly larger could generate high accuracies in LAI estimation. Chen et al. [61] concluded that images at a finer spatial resolution needed a larger window size than at a coarse resolution. In our study, the 15 × 15 window size of T_{pan} (equivalent to 9 m × 9 m) was still smaller than the sample plot size (20 m × 20 m). For T_{B+VIs}, the window size of 9 × 9 (equivalent to 21.6 m × 21.6 m), which corresponded to the extent of the field plots, produced higher accuracy than T_{pan}. This result was consistent with that of Wood et al. [29] and Gomez et al. [59], who suggested that the window size should match the sample plot size to achieve high accuracy.

After filtering the explanatory variables by selecting the optimal window sizes, there are still variables that have a weak relationship with response variable retained. The Boruta algorithm based on RF can select variables that are relevant to the response variables. Wu [62] compared three variable selection methods, i.e., stepwise regression analysis, Pearson correlation analysis, and Boruta algorithm, and found that the Boruta method selected variables capable of obtaining the highest accuracy. Among the spectral variables, the SAVI and EVI were well-correlated with field response variables. The SAVI and EVI could minimize the influences of the soil background, sun angle, and atmosphere [51,63]. Campos et al. [64] compared the performance of SAVI and NDVI in evaluating fraction of ground cover, and found that the SAVI could improve the accuracy, since it was less sensitive to the sun azimuth and row directions. In the aboveground biomass estimation, SAVI showed higher relationship with biomass by adjustment the effect of soil background [65]. Eckert [66] also demonstrated that the EVI is particularly suitable for mapping and monitoring tropical rainforest biomass. Among the four bands, the NIR band was more conducive to CC estimation, and the other three bands were not selected as relevant variables. This result was expected because the NIR band is sensitive to vegetation stress and the chlorophyll content of vegetation [44]. Additionally, NIR reflectance is strong due to the scattering of radiation in the mesophyll cell of leaves and its minimal absorption [23]. However, our study produced only moderately accurate results while using spectral variables, which may have suffered from saturation and multiple layering problems because the CC in our study area was in its peak growth period and had high vegetation cover [26,27].

Compared with spectral variables, the RF model with T_{pan} presented a significant improvement. All of the texture measures were selected as relevant to the response variables. The eight texture measures explained the variation of CC from different aspects, and only one type of texture measure contained insufficient information to explain the CC variance. Kim [67] demonstrated that adding individual texture measure to spectral bands did not improve forest classification accuracy. However, when incorporated multiple texture measures, the forest classification accuracy increased to 83% in overall accuracy. St.-Louis et al. [6] also found that multiple texture measures explained a higher proportion of the variability in bird species richness than single measures. The higher accuracy of Model 2 was consistent with numerous prior studies, which indicated that T_{pan} was particularly useful in measuring complex structures, such as tropical forests [29,59]. The usefulness of T_{pan} may be due to the high resolution of the panchromatic image used for the texture analysis, which increased the scope for distinguishing specific forest structure parameters, especially crown attributes (e.g., CC, crown diameter, etc.) [5,26,27,66].

T_{B+VIs} yielded the highest accuracy in estimating forest CC compared with the spectral variables and T_{pan}. The improved performance may be related to their combination of spatial and spectral information, which is consistent with the findings of many previous studies [6,12,27,44,52]. Gu et al. [16], Pfeifer et al. [68], and Pu and Cheng [28] demonstrated that by including texture information into spectral data models, the models' predictive capacity could be improved, especially for the canopy structure at the stand level, which is mainly because the information associated with spectral and textural signatures is complementary in the estimation of forest parameters [59]. In addition, texture was credible in detecting varying forest canopy structural characteristics and is efficient in addressing saturation problems that are associated with vegetation indices when mapping CC, especially in dense canopies [26].

Our results suggest that QuickBird imagery effectively captured the CC of black locust plantations. The generated map displayed the continuous distribution of CC over a large spatial area (Figure 6), which highlights the convenience of using satellite data for mapping large areas. Such maps can be used to improve the planning and management of tasks, such as land-cover mapping and land-use classification, among others. High CC forests, especially those with a young forest stage, need thinning and pruning to decrease the space and resource competition among individual trees [69]. Young forests with low CC values can lead to enclosures and replanting. Moreover, quantitative maps of forest resources can be used for decision-making by managers and for monitoring a variety of forest inventory parameters, such as forest area changes, biomass accumulation, and health conditions [5,44,68].

Multiple sources of error can lead to uncertainties in forest CC estimation. First, the site CC data were collected and analyzed based on DHPs. To avoid subjective errors when adjusting the threshold, a constant threshold of 128 was used to separate the sky and canopy values. However, the threshold of 128 may not be suitable for all photographs, which may lead to errors in certain photographs. Second, satellite image observations capture an aerial view and obtain the CC by the vertical projection of tree crowns. However, the DHP-obtained CC represented an under the crown measurement and the proportion of sky hemisphere obscured by vegetation when viewed from a single point [70]. This mismatch can be a source of error in the model. Furthermore, shrub and grass in the forest understory will affect the reflectance of the overstory layer, especially in forests with lower CC values [71]. Third, although the satellite images were corrected, errors may remain, and precise co-registration might not be obtained between the images and field plots. Fourth, errors are observed in the RF model itself because the model tends to be overestimated at lower values and underestimated at higher values. However, these errors cannot be avoided. Avitabile and Camia [72] suggest that overestimation may occur in open or young forests while underestimation may be due to the optical saturation under the high biomass of dense forests.

Considering the uncertainties in CC estimation with satellite images, perhaps active sensor and UAV can be used to reduce the effects of these uncertainties in the future. Ma et al. [20] indicated that satellite images were limited by penetration capability in forest area. The active sensors, such as LiDAR and SAR, can penetrate forest canopy and generate vertical structure of vegetation [14,20]. The UAV, which offer high acquisition flexibility and resolution at relatively low costs, have been used to estimate forest cover and basal area successfully [21]. An attractive next step is to using UAV to evaluate forest CC on the Loess Plateau. In addition, the combination of satellite images and UAV or LiDAR is warranted in future research.

5. Conclusions

This study explored the potential use of QuickBird imagery for CC estimations of black locust plantations on the Loess Plateau. We compared the spectral variables, T_{pan} and T_{B+VIs} to estimate black locust plantation CC based on RF regression models. The optimal window size for T_{pan} and T_{B+VIs} were 15×15 and 9×9, respectively. The experimental results demonstrated that both T_{pan}

and T_{B+VIs} performed better than spectral variables. The RF model of T_{B+VIs}, which reflected the complementary relationship between spectral and textural information, provided the most useful approach to investigating and characterizing black locust plantations CC. This model can be applied for mapping black locust plantations CC on the Loess Plateau of China.

Author Contributions: Z.Z. developed and supervised the work. Q.Z. pre-processed QuickBird imagery, analyzed the data and wrote the paper. F.W. and J.Z. (Jun Zhao) were contributed to data analysis and prepare figures. J.Z. (Jingjing Zhou) and S.Y. helped the field data collection. All authors read and approved the final manuscript.

Funding: This research and the APC were funded by the Key Techniques and Demonstration of Plantation Landscape Management in the Gullied-hilly Area, grant number 2017YFC0504605.

Acknowledgments: Thanks to the anonymous reviewers for their constructive and valuable comments, and the editors for their assistance in refining this article.

Conflicts of Interest: The authors declare no conflict of interest.

References

1. Jennings, S.B.; Brown, N.D.; Sheil, D. Assessing forest canopies and understorey illumination: Canopy closure, canopy cover and other measures. *Forestry* **1999**, *72*, 59–73. [CrossRef]
2. Chopping, M.; North, M.; Chen, J.Q.; Schaaf, C.B.; Blair, J.B.; Martonchik, J.V.; Bull, M.A. Forest canopy cover and height from MISR in topographically complex southwestern US landscapes assessed with high quality reference data. *IEEE J. Sel. Top. Appl. Earth Obs. Remote Sens.* **2012**, *5*, 44–58. [CrossRef]
3. Crowther, T.W.; Glick, H.B.; Covey, K.R.; Bettigole, C.; Maynard, D.S.; Thomas, S.M.; Smith, J.R.; Hintler, G.; Duguid, M.C.; Amatulli, G.; et al. Mapping tree density at a global scale. *Nature* **2015**, *525*, 201–205. [CrossRef] [PubMed]
4. Gonsamo, A. Leaf area index retrieval using gap fractions obtained from high resolution satellite data: Comparisons of approaches, scales and atmospheric effects. *Int. J. Appl. Earth Obs. Geoinf.* **2010**, *12*, 233–248. [CrossRef]
5. González-Roglich, M.; Swenson, J.J. Tree cover and carbon mapping of Argentine savannas: Scaling from field to region. *Remote Sens. Environ.* **2016**, *172*, 139–147. [CrossRef]
6. St-Louis, V.; Pidgeon, A.M.; Radeloff, V.C.; Hawbaker, T.J.; Clayton, M.K. High-resolution image texture as a predictor of bird species richness. *Remote Sens. Environ.* **2006**, *105*, 299–312. [CrossRef]
7. Peterson, D.W.; Reich, P.B. Fire frequency and tree canopy structure influence plant species diversity in a forest-grassland ecotone. *Plant Ecol.* **2008**, *194*, 5–16. [CrossRef]
8. Vatandaşlar, C.; Yavuz, M. Modeling cover management factor of RUSLE using very high-resolution satellite imagery in a semiarid watershed. *Environ. Earth Sci.* **2017**, *76*, 65. [CrossRef]
9. Xiao, J.F. Satellite evidence for significant biophysical consequences of the "Grain for Green" Program on the Loess Plateau in China. *J. Geophys. Res. Biogeosci.* **2014**, *119*, 2261–2275. [CrossRef]
10. Burner, D.M.; Pote, D.; Ares, A. Management effects on biomass and foliar nutritive value of *Robinia pseudoacacia* and *Gleditsia triacanthos* f. inermis in Arkansas, USA. *Agrofor. Syst.* **2005**, *65*, 207–214. [CrossRef]
11. Zhou, J.J.; Zhao, Z.; Zhao, J.; Zhao, Q.X.; Wang, F.; Wang, H.Z. A comparison of three methods for estimating the LAI of black locust (*Robinia pseudoacacia* L.) plantations on the Loess Plateau, China. *Int. J. Remote Sens.* **2014**, *35*, 171–188. [CrossRef]
12. Halperin, J.; LeMay, V.; Coops, N.; Verchot, L.; Marshall, P.; Lochhead, K. Canopy cover estimation in miombo woodlands of Zambia: Comparison of Landsat 8 OLI versus RapidEye imagery using parametric, nonparametric, and semiparametric methods. *Remote Sens. Environ.* **2016**, *179*, 170–182. [CrossRef]
13. Fang, H.; Li, W.; Wei, S.; Jiang, C. Seasonal variation of leaf area index (LAI) over paddy rice fields in NE China: Intercomparison of destructive sampling, LAI-2200, digital hemispherical photography (DHP), and AccuPAR methods. *Agric. For. Meteorol.* **2014**, *198–199*, 126–141. [CrossRef]
14. Stojanova, D.; Panov, P.; Gjorgjioski, V.; Kohler, A.; Dzeroski, S. Estimating vegetation height and canopy cover from remotely sensed data with machine learning. *Ecol. Inf.* **2010**, *5*, 256–266. [CrossRef]
15. Castillo-Santiago, M.A.; Ricker, M.; de Jong, B.H.J. Estimation of tropical forest structure from SPOT-5 satellite images. *Int. J. Remote Sens.* **2010**, *31*, 2767–2782. [CrossRef]

16. Gu, Z.J.; Ju, W.M.; Li, L.; Li, D.Q.; Liu, Y.B.; Fan, W.L. Using vegetation indices and texture measures to estimate vegetation fractional coverage (VFC) of planted and natural forests in Nanjing City, China. *Adv. Space Res.* **2013**, *51*, 1186–1194. [CrossRef]

17. Hansen, M.C.; Potapov, P.V.; Moore, R.; Hancher, M.; Turubanova, S.A.; Tyukavina, A.; Thau, D.; Stehman, S.V.; Goetz, S.J.; Loveland, T.R.; et al. High-resolution global maps of 21st-century forest cover change. *Science* **2013**, *342*, 850–853. [CrossRef] [PubMed]

18. Korhonen, L.; Ali-Sisto, D.; Tokola, T. Tropical forest canopy cover estimation using satellite imagery and airborne lidar reference data. *Silva Fenn.* **2015**, *49*. [CrossRef]

19. Li, W.; Niu, Z.; Liang, X.; Li, Z.; Huang, N.; Gao, S.; Wang, C.; Muhammad, S. Geostatistical modeling using LiDAR-derived prior knowledge with SPOT-6 data to estimate temperate forest canopy cover and above-ground biomass via stratified random sampling. *Int. J. Appl. Earth Obs. Geoinf.* **2015**, *41*, 88–98. [CrossRef]

20. Ma, Q.; Su, Y.; Guo, Q. Comparison of canopy cover estimations from airborne LiDAR, aerial imagery, and satellite imagery. *IEEE J. Sel. Top. Appl. Earth Obs. Remote Sens.* **2017**, *10*, 4225–4236. [CrossRef]

21. Rossi, F.; Fritz, A.; Becker, G. Combining satellite and UAV imagery to delineate forest cover and basal area after mixed-severity fires. *Sustainability* **2018**, *10*, 2227. [CrossRef]

22. Wallis, C.I.B.; Paulsch, D.; Zeilinger, J.; Silva, B.; Fernandez, G.F.C.; Brandl, R.; Farwig, N.; Bendix, J. Contrasting performance of Lidar and optical texture models in predicting avian diversity in a tropical mountain forest. *Remote Sens. Environ.* **2016**, *174*, 223–232. [CrossRef]

23. Calvao, T.; Palmeirim, J.M. Mapping Mediterranean scrub with satellite imagery: Biomass estimation and spectral behaviour. *Int. J. Remote Sens.* **2004**, *25*, 3113–3126. [CrossRef]

24. Karlson, M.; Ostwald, M.; Reese, H.; Sanou, J.; Tankoano, B.; Mattsson, E. Mapping tree canopy cover and aboveground biomass in Sudano-Sahelian woodlands using Landsat 8 and Random Forest. *Remote Sens.* **2015**, *7*, 10017–10041. [CrossRef]

25. Chasmer, L.; Baker, T.; Carey, S.K.; Straker, J.; Strilesky, S.; Petrone, R. Monitoring ecosystem reclamation recovery using optical remote sensing: Comparison with field measurements and eddy covariance. *Sci. Total Environ.* **2018**, *642*, 436–446. [CrossRef] [PubMed]

26. Dube, T.; Mutanga, O. Investigating the robustness of the new Landsat-8 Operational Land Imager derived texture metrics in estimating plantation forest aboveground biomass in resource constrained areas. *ISPRS J. Photogramm. Remote Sens.* **2015**, *108*, 12–32. [CrossRef]

27. Sarker, L.R.; Nichol, J.E. Improved forest biomass estimates using ALOS AVNIR-2 texture indices. *Remote Sens. Environ.* **2011**, *115*, 968–977. [CrossRef]

28. Pu, R.L.; Cheng, J. Mapping forest leaf area index using reflectance and textural information derived from WorldView-2 imagery in a mixed natural forest area in Florida, US. *Int. J. Appl. Earth Obs. Geoinf.* **2015**, *42*, 11–23. [CrossRef]

29. Wood, E.M.; Pidgeon, A.M.; Radeloff, V.C.; Keuler, N.S. Image texture as a remotely sensed measure of vegetation structure. *Remote Sens. Environ.* **2012**, *121*, 516–526. [CrossRef]

30. Tuanmu, M.N.; Walter, J. A global, remote sensing-based characterization of terrestrial habitat heterogeneity for biodiversity and ecosystem modelling. *Glob. Ecol. Biogeogr.* **2015**, *24*, 1329–1339. [CrossRef]

31. Levesque, J.; King, D.J. Spatial analysis of radiometric fractions from high-resolution multispectral imagery for modelling individual tree crown and forest canopy structure and health. *Remote Sens. Environ.* **2003**, *84*, 589–602. [CrossRef]

32. Song, L.; Langfelder, P.; Horvath, S. Random generalized linear model: A highly accurate and interpretable ensemble predictor. *BMC Bioinform.* **2013**, *14*, 5. [CrossRef] [PubMed]

33. Zhang, C.; Denka, S.; Cooper, H.; Mishra, D.R. Quantification of sawgrass marsh aboveground biomass in the coastal everglades using object-based ensemble analysis and Landsat data. *Remote Sens. Environ.* **2018**, *204*, 366–379. [CrossRef]

34. Ali, I.; Greifeneder, F.; Stamenkovic, J.; Neumann, M.; Notarnicola, C. Review of machine learning approaches for biomass and soil moisture retrievals from remote sensing data. *Remote Sens.* **2015**, *7*, 16398–16421. [CrossRef]

35. Were, K.; Bui, D.T.; Dick, Ø.B.; Singh, B.R. A comparative assessment of support vector regression, artificial neural networks, and random forests for predicting and mapping soil organic carbon stocks across an Afromontane landscape. *Ecol. Indic.* **2015**, *52*, 394–403. [CrossRef]

36. Pham, L.T.H.; Brabyn, L. Monitoring mangrove biomass change in Vietnam using SPOT images and an object-based approach combined with machine learning algorithms. *ISPRS J. Photogramm. Remote Sens.* **2017**, *128*, 86–97. [CrossRef]

37. Belgiu, M.; Drăguţ, L. Random forest in remote sensing: A review of applications and future directions. *ISPRS J. Photogramm. Remote Sens.* **2016**, *114*, 24–31. [CrossRef]

38. Rodriguez-Galiano, V.F.; Chica-Olmo, M.; Abarca-Hernandez, F.; Atkinson, P.M.; Jeganathan, C. Random forest classification of mediterranean land cover using multi-seasonal imagery and multi-seasonal texture. *Remote Sens. Environ.* **2012**, *121*, 93–107. [CrossRef]

39. Pullanagari, R.R.; Kereszturi, G.; Yule, I. Integrating airborne hyperspectral, topographic, and soil data for estimating pasture quality using recursive feature elimination with random forest regression. *Remote Sens.* **2018**, *10*, 1117. [CrossRef]

40. Duro, D.C.; Franklin, S.E.; Dube, M.G. A comparison of pixel-based and object-based image analysis with selected machine learning algorithms for the classification of agricultural landscapes using SPOT-5 HRG imagery. *Remote Sens. Environ.* **2012**, *118*, 259–272. [CrossRef]

41. Son, N.-T.; Chen, C.-F.; Chen, C.-R.; Minh, V.-Q. Assessment of sentinel-1a data for rice crop classification using random forests and support vector machines. *Geocarto Int.* **2018**, *33*, 587–601. [CrossRef]

42. Shataee, S.; Kalbi, S.; Fallah, A.; Pelz, D. Forest attribute imputation using machine-learning methods and ASTER data: Comparison of *k*-NN, SVR and random forest regression algorithms. *Int. J. Remote Sens.* **2012**, *33*, 6254–6280. [CrossRef]

43. Cracknell, M.J.; Reading, A.M. Geological mapping using remote sensing data: A comparison of five machine learning algorithms, their response to variations in the spatial distribution of training data and the use of explicit spatial information. *Comput. Geosci.* **2014**, *63*, 22–33. [CrossRef]

44. Wang, H.; Zhao, Y.; Pu, R.L.; Zhang, Z.Z. Mapping *Robinia pseudoacacia* forest health conditions by using combined spectral, spatial, and textural information extracted from IKONOS imagery and random forest classifier. *Remote Sens.* **2015**, *7*, 9020–9044. [CrossRef]

45. Frazer, G.W.; Canham, C.; Lertzman, K. *Gap Light Analyzer (GLA), Version 2.0: Imaging Software to Extract Canopy Structure and Gap Light Transmission Indices from True-Colour Fisheye Photographs, Users Manual and Program Documentation*; Simon Fraser University: Burnaby, BC, Canada; Institute of Ecosystem Studies: Millbrook, NY, USA, 1999; Volume 36.

46. Brusa, A.; Bunker, D.E. Increasing the precision of canopy closure estimates from hemispherical photography: Blue channel analysis and under-exposure. *Agric. For. Meteorol.* **2014**, *195*, 102–107. [CrossRef]

47. Pueschel, P.; Buddenbaum, H.; Hill, J. An efficient approach to standardizing the processing of hemispherical images for the estimation of forest structural attributes. *Agric. For. Meteorol.* **2012**, *160*, 1–13. [CrossRef]

48. Cescatti, A. Indirect estimates of canopy gap fraction based on the linear conversion of hemispherical photographs-Methodology and comparison with standard thresholding techniques. *Agric. For. Meteorol.* **2007**, *143*, 1–12. [CrossRef]

49. Nobis, M.; Hunziker, U. Automatic thresholding for hemispherical canopy-photographs based on edge detection. *Agric. For. Meteorol.* **2005**, *128*, 243–250. [CrossRef]

50. Seidel, D.; Fleck, S.; Leuschner, C. Analyzing forest canopies with ground-based laser scanning: A comparison with hemispherical photography. *Agric. For. Meteorol.* **2012**, *154*, 1–8. [CrossRef]

51. Huete, A.; Didan, K.; Miura, T.; Rodriguez, E.P.; Gao, X.; Ferreira, L.G. Overview of the radiometric and biophysical performance of the MODIS vegetation indices. *Remote Sens. Environ.* **2002**, *83*, 195–213. [CrossRef]

52. Haralick, R.M.; Shanmugam, K.; Dinstein, I.H. Textural features for image classification. *IEEE Trans. Syst. Man Cybern. Syst.* **1973**, *SMC-3*, 610–621. [CrossRef]

53. Breiman, L. Random forests. *Mach. Learn.* **2001**, *45*, 5–32. [CrossRef]

54. Geiß, C.; Aravena Pelizari, P.; Marconcini, M.; Sengara, W.; Edwards, M.; Lakes, T.; Taubenböck, H. Estimation of seismic building structural types using multi-sensor remote sensing and machine learning techniques. *ISPRS J. Photogramm. Remote Sens.* **2015**, *104*, 175–188. [CrossRef]

55. Cooner, A.J.; Shao, Y.; Campbell, J.B. Detection of urban damage using remote sensing and machine learning algorithms: Revisiting the 2010 Haiti earthquake. *Remote Sens.* **2016**, *8*, 868. [CrossRef]

56. Kursa, M.B.; Rudnicki, W.R. Feature selection with the Boruta package. *J. Stat. Softw.* **2010**, *36*, 1–13. [CrossRef]

57. Freeman, E.A.; Moisen, G.G.; Coulston, J.W.; Wilson, B.T. Random forests and stochastic gradient boosting for predicting tree canopy cover: Comparing tuning processes and model performance. *Can. J. For. Res.* **2016**, *46*, 323–339. [CrossRef]

58. R Core Team. *R: A Language and Environment for Statistical Computing*; R Foundation for Statistical Computing: Vienna, Austria, 2017.

59. Gomez, C.; Wulder, M.A.; Montes, F.; Delgado, J.A. Forest structural diversity characterization in Mediterranean pines of central Spain with QuickBird-2 imagery and canonical correlation analysis. *Can. J. Remote Sens.* **2011**, *37*, 628–642. [CrossRef]

60. Kamal, M.; Phinn, S.; Johansen, K. Assessment of multi-resolution image data for mangrove leaf area index mapping. *Remote Sens. Environ.* **2016**, *176*, 242–254. [CrossRef]

61. Chen, D.; Stow, D.A.; Gong, P. Examining the effect of spatial resolution and texture window size on classification accuracy: An urban environment case. *Int. J. Remote Sens.* **2004**, *25*, 2177–2192. [CrossRef]

62. Wu, C.F. Regional Biomass Estimation and Application Based on Remote Sensing. Ph.D. Thesis, Zhejiang University, Hangzhou, China, 2016. (In Chinese)

63. Xue, J.; Su, B. Significant remote sensing vegetation indices: A review of developments and applications. *J. Sens.* **2017**, *2017*, 1353691. [CrossRef]

64. Campos, I.; Neale, C.M.U.; Lopez, M.-L.; Balbontin, C.; Calera, A. Analyzing the effect of shadow on the relationship between ground cover and vegetation indices by using spectral mixture and radiative transfer models. *J. Appl. Remote Sens.* **2014**, *8*, 083562. [CrossRef]

65. Yan, F.; Wu, B.; Wang, Y. Estimating aboveground biomass in mu us sandy land using landsat spectral derived vegetation indices over the past 30 years. *J. Arid Land* **2013**, *5*, 521–530. [CrossRef]

66. Eckert, S. Improved forest biomass and carbon estimations using texture measures from WorldView-2 satellite data. *Remote Sens.* **2012**, *4*, 810–829. [CrossRef]

67. Kim, M. Object-Based Spatial Classification of Forest Vegetation with IKONOS Imagery. Ph.D. Thesis, University of Georgia, Athens, GA, USA, 2009.

68. Pfeifer, M.; Kor, L.; Nilus, R.; Turner, E.; Cusack, J.; Lysenko, I.; Khoo, M.; Chey, V.K.; Chung, A.C.; Ewers, R.M. Mapping the structure of Borneo's tropical forests across a degradation gradient. *Remote Sens. Environ.* **2016**, *176*, 84–97. [CrossRef]

69. Liu, J.L.; Yu, Z.Q.; Zhang, S.X.; Wang, D.H.; Zhao, Z. Establishment of forest health assessment system for black locust plantation in Weibei Loess Plateau. *J. Northwest A&F Univ. (Nat. Sci. Ed.)* **2014**, *42*, 93–99. (In Chinese)

70. Paletto, A.; Tosi, V. Forest canopy cover and canopy closure: Comparison of assessment techniques. *Eur. J. For. Res.* **2009**, *128*, 265–272. [CrossRef]

71. Hallik, L.; Kull, O.; Nilson, T.; Peñuelas, J. Spectral reflectance of multispecies herbaceous and moss canopies in the boreal forest understory and open field. *Can. J. Remote Sens.* **2009**, *35*, 474–485. [CrossRef]

72. Avitabile, V.; Camia, A. An assessment of forest biomass maps in Europe using harmonized national statistics and inventory plots. *For. Ecol. Manag.* **2018**, *409*, 489–498. [CrossRef] [PubMed]

© 2018 by the authors. Licensee MDPI, Basel, Switzerland. This article is an open access article distributed under the terms and conditions of the Creative Commons Attribution (CC BY) license (http://creativecommons.org/licenses/by/4.0/).

 *forests*

Article

LiDAR-Based Regional Inventory of Tall Trees—Wellington, New Zealand

Jan Zörner [1,*], John R. Dymond [2], James D. Shepherd [2], Susan K. Wiser [1] and Ben Jolly [2]

[1]   Landcare Research, Lincoln 7608, New Zealand; wisers@landcareresearch.co.nz
[2]   Landcare Research, Palmerston North 4410, New Zealand; dymondj@landcareresearch.co.nz (J.R.D.);
      shepherdj@landcareresearch.co.nz (J.D.S.); jollyb@landcareresearch.co.nz (B.J.)
*   Correspondence: zoernerj@landcareresearch.co.nz; Tel.: +64-3-321-9720

Received: 8 October 2018; Accepted: 10 November 2018; Published: 13 November 2018

Abstract: Indigenous forests cover 23.9% of New Zealand's land area and provide highly valued ecosystem services, including climate regulation, habitat for native biota, regulation of soil erosion and recreation. Despite their importance, information on the number of tall trees and the tree height distribution across different forest classes is scarce. We present the first region-wide spatial inventory of tall trees (>30 m) based on airborne LiDAR (Light Detection and Ranging) measurements in New Zealand—covering the Greater Wellington region. This region has 159,000 ha of indigenous forest, primarily on steep mountainous land. We implement a high-performance tree mapping algorithm that uses local maxima in a canopy height model (CHM) as initial tree locations and accurately identifies the tree top positions by combining a raster-based tree crown delineation approach with information from the digital surface and terrain models. Our algorithm includes a check and correction for over-estimated heights of trees on very steep terrain such as on cliff edges. The number of tall trees (>30 m) occurring in indigenous forest in the Wellington Region is estimated to be 286,041 (±1%) and the number of giant trees (>40 m tall) is estimated to be 7340 (±1%). Stereo-analysis of aerial photographs was used to determine the accuracy of the automated tree mapping. The giant trees are mainly in the beech-broadleaved-podocarp and broadleaved-podocarp forests, with density being 0.04 and 0.12 (trees per hectare) respectively. The inventory of tall trees in the Wellington Region established here improves the characterization of indigenous forests for management and provides a useful baseline for long-term monitoring of forest conditions. Our tree top detection scheme provides a simple and fast method to accurately map overstory trees in flat as well as mountainous areas and can be directly applied to improve existing and build new tree inventories in regions where LiDAR data is available.

Keywords: forest inventory; LiDAR; tall trees; overstory trees; tree mapping; crown delineation

---

1. Introduction

There are 6.3 million ha of indigenous forest in New Zealand, covering 23.9% of the land surface [1]. They provide highly valued cultural ecosystem services, including recreation, sense of belonging, tourism and important provisioning and regulating services, such as wild foods, fresh water, habitat for native biota, climate regulation and soil erosion regulation [2]. The evergreen temperate forests have three major physiognomic elements: species of the *Nothofagaceae* Kuprian (Southern beech), broadleaved angiosperm trees and conifers (predominantly of the *Podocarpaceae* Endl.) [3]. Where *Agathis australis* (D.Don) Loudon is absent, podocarp trees form the highest tier [3]. Since European settlement, which began in the early 19th century, much of the indigenous forest has been felled for pastoral agriculture and many introduced plants and animals have become pests, if not directly to the trees (possums, pigs and deer), then to the indigenous fauna in the forests (cats, rats, stoats and weasels). Recently, two tree diseases have raised concern in New Zealand, kauri dieback

(*Phytophthora agathidicida* B.S. Weir, Beever, Pennycook & Bellgard) and the South American myrtle rust (*Austropuccinia psidii* (G. Winter) Beenken) and these endanger some of the tall conifer and broadleaved trees (*Agathis australis, Metrosideros robusta* A.Cunn., *Metrosideros umbellata* Cav.).

Tall trees play critical roles in forests. They support a higher abundance and richness of epiphytes [4–6], are associated with increased bird richness [7,8] and can disperse their seeds further than short trees [9,10]. In New Zealand, 44 species of indigenous trees attain heights >15 m [11]. Because carbon density is in proportion to tree height to the power of 1.5 [12], tall trees make a higher contribution in terms of both timber recovery and carbon storage. The definitions of both *tall* and *giant* trees depend on the research context and may be different for different forest types, climatic zones and cultural definitions. In the context of New Zealand's natural forests we define *tall trees* as those with top heights >30 m and *giant trees* as those with top heights >40 m. Mapping individual trees can help inform models of forest dynamics [13], support forest inventory [14] and inform decision-making in public and non-governmental sectors [15]. Indeed, many tree planting initiatives are expressed as numbers of trees, such as the numerous million, billion and trillion trees projects worldwide [14,16]. Crowther et al. [17] integrated ground-based forest inventory data with geospatial covariates derived from satellite-based remote sensing to estimate the global number of trees (>10 cm diameter) to be 3.0 trillion.

Estimating individual tree heights and crown shapes from field measurements is both costly and labor-intensive and thus usually restricted to small study areas. Likewise, the analysis of stereo-models of aerial photographs to identify single trees is inefficient and time-consuming for region-wide or country-wide analysis. With the advent of airborne LiDAR (Light Detection And Ranging), also referred to as Airborne Laser Scanning (ALS), it is now possible to map the crowns of individual trees and infer tree heights over large areas accurately and systematically, measuring the 3-dimensional structure of the forest canopy. In ALS, high-frequency laser pulses are emitted downwards to the ground and reflected by canopy features or the ground surface back to the sensor [18]. The travel distance of the returned laser pulse can be directly inferred from the return time. Due to the divergence of the laser beam, it is often possible to penetrate into the canopy hitting multiple targets, such as leaves and branches and thereby capturing vertical information on canopy structure.

Numerous tree segmentation and crown delineation algorithms have been developed over the last decade ([19–28] and references therein), ranging from those that use a three dimensional point cloud to those that make use of a canopy height model (CHM) to detect individual trees. A CHM is defined as the interpolated surface, on a regular grid (i.e., a raster), of the high points of a LiDAR point cloud, corrected for ground elevation; alternatively as the difference of the digital surface model (DSM) and digital elevation model (DEM), which are interpolated surfaces of the first (top of canopy) and last (ground) returns of the pulse, respectively. This is often referred to as a normalized DSM (nDSM). Structure from Motion (SfM) photogrammetry using digital cameras mounted on Unmanned Aerial Vehicles (UAVs) is becoming a low-cost alternative to LiDAR measurements to derive top-of-canopy structure, that is, surface models [29–34]. Hybrid tree segmentation algorithms work both on the LiDAR point cloud and rasterized quantities. This has been recently demonstrated, for instance, in the 'Layer Stacking' technique [21], which aims at interpreting height resolved point clusters to identify and segment individual trees. Due to the high computational effort of point cloud-based methods, however, their application is mostly limited to smaller study areas with extent ~1000 ha [13,14,35], rather than to large areas such as whole regions, or countries, with extent ~100,000 ha. Fast raster-based algorithms are thus usually preferred for whole regions or country-wide analyses. During elevation normalization of the LiDAR point cloud or DSM a distortion of the canopy height can occur in steep terrain leading to an over-estimation of the actual tree height. To account for this effect, several previous studies [36–40] established theoretical and practical models to accurately determine the true tree top positions. It is reported that this effect can lead to systematic errors (over-estimation of the number of trees and incorrect geo-location of the tree top) and strongly depends on crown shape [38].

In 2013 and 2014, the Wellington Regional Government GIS Group conducted an extensive LiDAR survey with a minimum point density of 1.3 points per square meter (ppsm) and a vertical accuracy of ±0.15 m over the Wellington Region (812,000 ha) with the main intention of building new large-scale topographic maps of the region [41]. Many other sectors also profited from the data for other uses, such as vegetation mapping, infrastructure planning, hydrologic modelling and hazard mapping. The raw LiDAR point cloud is publicly available from the OpenTopography project at https://www.opentopography.org/. The derived DEM and DSM, at 1 × 1 m raster resolution and the LiDAR tile index can be downloaded from the Land Information New Zealand (LINZ) data service at https://data.linz.govt.nz/. The region-wide LiDAR point cloud provides an unprecedented opportunity to study forest composition and individual tree parameters on a large scale.

In this paper, we apply an adapted version of the CHM-based approach proposed by Dalponte and Coomes [28] and Hyyppä et al. [19] to the whole of the Wellington Region, to establish the first ever regional inventory of tall trees in New Zealand. The original tree mapping algorithm [28] has been previously used to map sections of forests in the Wellington Region [12], Southeast Asia [42] and in the Italian Alps [43] and has been validated against other methods [27]. We re-implemented the algorithm in Python code and further optimized it by: (i) growing the tree crown around the initial tree top in a circular fashion rather than quadrilateral as in the original implementation, which gives the resulting tree crowns a smoother shape and (ii) correcting the location of misplaced tree tops in steep terrain using the DSM and DTM (digital terrain model). The check and correction of misplaced tree tops in steep terrain is necessary for an accurate estimation of tree heights in mountainous areas with steep slopes. We map and count all *tall trees* greater than 30 m in height and *giant trees* greater than 40 m. The tall tree density is assessed for groups of forest alliances [44] and within altitudinal ranges to better characterize these forest types. The accuracy of the method is assessed by comparing results obtained from stereo analysis of aerial photographs with 30 cm pixels.

2. Materials and Methods

A LiDAR survey was conducted over the entire Wellington Region [45] primarily in early 2013 but with some additional aircraft flights later in 2013 and 2014, depending on weather. The LiDAR scanner was an Optech Airborne Laser Terrain Mapper ALTM 3100EA flown at a nominal height of 1000 m above the ground. Target point density was 1.7 ppsm with 50% swath overlap to ensure the minimum specification of 1.3 ppsm and a vertical accuracy of ±0.15 m. While these were the minimum requirements, the actual point density of the dataset ended up higher, ranging between 4–6 ppsm on average and >6 ppsm in regions with overlapping flight paths. Coincident with the LiDAR survey was an aerial photographic survey with blue, green, red and near-infrared bands, which was ortho-rectified to 30 cm pixels. The 1261 LiDAR flight lines of point cloud data were merged, tiled and further processed using the open-source LiDAR processing library, Sorted Pulse Data Software Library (SPDLib) [46].

The processing steps using SPDLib included: (i) removal of vertical noise, that is, undesired backscatter from atmospheric particles such as aerosols; (ii) 2-step classification of ground returns by applying the progressive morphology algorithm [47] first and the multiscale curvature algorithm [48] in the second step; (iii) definition of the height field within pulses and points by natural neighbor interpolation of the ground returns retrieving the absolute ground elevation; (iv) interpolation of ground and surface returns to a digital terrain model (DTM), digital surface model (DSM) and canopy height model (CHM) at 1 × 1 m pixel resolution using natural neighbor interpolation; and finally (v) removal of outliers in the CHM by applying a 5 × 5 m Gaussian median filter and screening out pixels below 0.5 m (undesired low vegetation) and above 60 m (erroneous high trees).

Individual trees were then identified from the CHM by finding the highest point in a 5 m circular moving window. If multiple pixels share the same CHM value, the center of mass of the pixel group is taken as the high point. These local maxima represent the tree top positions and the associated CHM value is extracted to define the tree top heights. During initial inspection using stereo mapping of the identified tree tops, it was found that some true tree top positions were different from the local maxima in the CHM if the tree was located in or close to steep terrain. This is the case, for instance, if the tree crown hangs over a cliff edge. The CHM-value assigned to the overhanging part of the crown is too high compared with the rest of the crown and the tree top height is thus overestimated. Figure 1 depicts an example scenario for a tree under such steep slope conditions. To account for this effect, we conducted a multi-step approach to identify a more realistic tree top position.

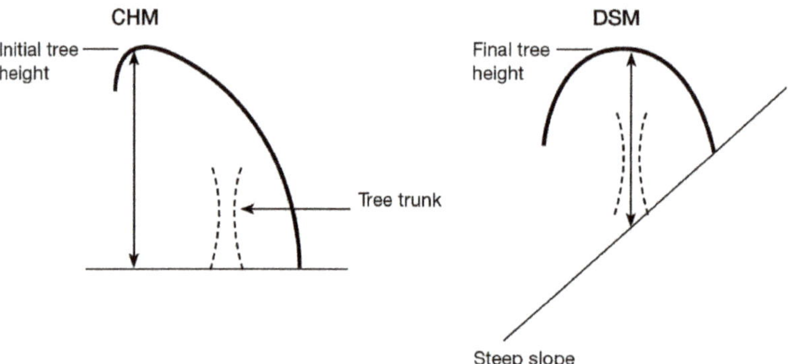

Figure 1. Sketch of initial tree top position as derived from the CHM (canopy height model) using local maxima filtering (**left panel**) and new tree top position (**right panel**) defined as the highest point in the DSM (digital surface model) within the tree crown boundary minus the DTM (digital terrain model).

First, tree crowns were delineated based on an adapted version of the raster-based tree crown-growing algorithm by Dalponte and Coomes [28], using the local maxima as initial seeds. The original implementation from the R-package itcSegment [49] was ported Python code (optimized with Numba/Cython) and made publicly available as the PyCrown package [50]. It performs one order of magnitude faster than the C++ implementation from the R package for Airborne LiDAR Data Manipulation and Visualization for Forestry Applications (lidR) [51]. The idea behind the algorithm is that the crown grows around the initial seed considering a set of threshold values which are calculated based on four manually set parameters and the current tree information in each step: (i) the neighboring pixel is higher than seed height * 0.7; (ii) the neighboring pixel is higher than the mean height of current crown * 0.55; (iii) the neighboring pixel is below seed height * 1.05 and (iv) a maximum distance to the seed of 10 m (crown radius). We used standard settings from itcSegment and lidR except for parameter (i). The latter was raised from the standard 0.45 and set to 0.7 (which was also used by Dalponte and Coomes [28]) as inspection with aerial imagery revealed that many delineated tree crowns exceeded their actual radius. We did not choose parameters specific to individual tree species in order to keep a generic crown growing ruleset that is appropriate given that canopies comprise a mixture of species in these forests. We forced the algorithm to walk around the initial seed in a circular fashion with increasing distance in each iteration. This was necessary because the quadrilateral growing pattern in the original implementation lead to block-shaped tree crowns at the edges when the maximum crown radius is reached. After all tree crowns were delineated, a check was performed on whether the tree top location was too far (>1 standard deviation) down-slope compared with the mean ground elevation of its tree crown (i.e., mean DTM value). In such a case, the location of the highest point from the DSM was taken as the new tree top. In cases where the new high point of the DSM was too close to the border of the tree crown, the center of mass (COM) of the tree crown was selected.

The entire processing chain was run for a total of 9480 LiDAR tiles (each 1 × 1 km with 100 m overlap) of the Wellington Region on the NeSI high performance computer [52] and a spatial database of trees higher than 20 m was established, storing information on the tree top location, tree top height and tree crown shape; 3D tree shapefiles were exported. Figure 2 shows the study area overlaid with the individual LiDAR tiles.

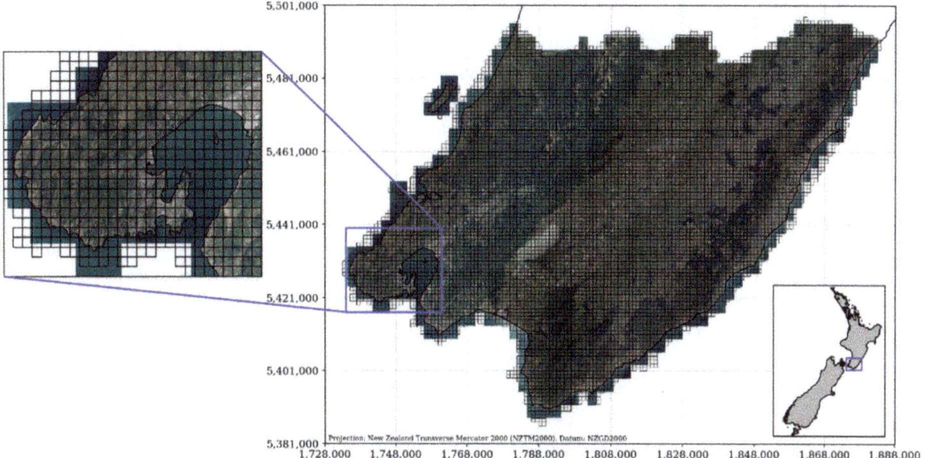

Figure 2. The Wellington Region of New Zealand and the mosaic of aerial photographs taken during the LiDAR survey in 2013 and 2014. The LiDAR data is subdivided into an equally-spaced grid of 9480 tiles with a 1 × 1 km spacing (outlined in black). Map projection is New Zealand Transverse Mercator (NZTM).

Wiser et al. [53,54] used data from an extensive network of vegetation plots in New Zealand to objectively define a classification of forest alliances. Allen et al. [44] classified these into seven broad groups designated as beech forest, beech-broadleaved forest, beech-broadleaved-podocarp forest, broadleaved-podocarp forest, podocarp forest and other forests. An existing digital map of indigenous forest types, EcoSat Forests [55], was recoded according to Table 1 to produce a national map of forest alliance groups ("broadleaved forest" was added to the forest alliance groups to match the EcoSat Forest class "broadleaved forest"). This map was intersected with the shape file of tree top positions and heights obtained from our tree mapping algorithm to determine the number of tall trees within each forest alliance group. Intersection of tree positions and heights with a 15 m pixel DEM gave the density of tall trees within elevation ranges.

Table 1. Table for recoding EcoSat forest class to forest alliance group.

Forest Alliance Group	EcoSat Forests Classes
Beech forest	Beech forest
Beech-broadleaved forest	Beech-broadleaved forest
Beech-broadleaved podocarp forest	Podocarp-broadleaved/Beech forest Beech/Podocarp-broadleaved forest
Broadleaved-podocarp forest	Podocarp-broadleaved forest Kauri forest
Podocarp forest	Podocarp forest
Broadleaved forest	Broadleaved forest Coastal forest
Unspecified indigenous forest	Unspecified indigenous forest

Assessing the accuracy of tall tree counting required three assumptions: (i) the CHM has a vertical random error of ±0.15 m (acquisition accuracy), no systematic error in the vertical and a small, random error component arising from spatial interpolation of the LiDAR points during the CHM generation (same order of magnitude as acquisition accuracy); (ii) the accuracy of identifying the tree top position by visually inspecting a stereo model of true color imagery (i.e., making use of the three dimensional view and color and textural differences between trees) is superior to the automated tree top identification using a local maximum filter; and (iii) manual analysis of stereo-imagery is more accurate at identifying tree top positions than ground-based techniques.

The accuracy assessment strategy does not try to validate the CHM itself but focuses on uncertainty in the geolocation of the automated tree mapping approach compared to a human pin-pointing the top of a tree using stereo vision. For validation of the CHM another independent measurement strategy is needed. This is typically conducted using ground-based measurements, which was beyond the scope of this study, as identifying overstory tree tops and estimating their heights from the ground in thick forest canopies is challenging and inherently uncertain. Therefore, we define the number of tall trees identified using stereo-imagery as the actual number of tall trees in absence of ground-based measurements.

We created stereo models of 25 permanent forest plots, using the raw digital aerial photographs, with stereo-analyst in Erdas Imagine [56] and compared the tree top position as identified by our algorithm to the visually perceived position in the stereo model. The tree height in both cases was taken from the CHM and compared. The forest plots are part of New Zealand's Land Use and Carbon Analysis System (LUCAS), which is a national network of 20 × 20 m plots set up on an 8 km grid intersected on a map of indigenous vegetation [12]. The accuracy assessment was performed for intermediate-scale landforms (area ~10 ha), that is, land components of relatively uniform slope and aspect such as ridge crests, head slopes or foot slopes [57], surrounding the LUCAS plots. For each such land component, the number of tall trees (>30 m) delineated by our algorithm was compared to the number of tall trees determined visually from the stereo imagery.

3. Results

Figure 3 shows the spatial distribution of forest alliance groups in the Wellington Region and Table 2 gives areas of the forest alliance groups for the Wellington Region, for North and South Islands and for all New Zealand. In total there is 6.3 million ha of indigenous forest in New Zealand, some 23.9% of the land mass. The most common groups of forest alliances in New Zealand are beech forest, beech-broadleaved-podocarp forest and broadleaved-podocarp forest, in that order. In the Wellington Region these three groups are also the most common but with beech-broadleaved podocarp forest being the most common. Most of the indigenous forest in the Wellington Region is in the Tararua mountains, which form a south-west to north axis in the region. Broadleaved-podocarp forest is common at low altitudes on the western side of the Tararuas and typically progresses through beech-broadleaved-podocarp forest and then beech forest up altitudinal gradients. On the eastern side of the Tararuas the progression is simpler, starting at beech-broadleaved-podocarp forest with broadleaved-podocarp forest largely absent. In the Aorangi mountains in the south-east of the region, beech-broadleaved forest is the most common group.

Table 2. Areas in hectares of forest alliance groups in the Wellington Region, North Island, South Island and New Zealand.

Forest Alliance Group	Wellington	North Island	South Island	New Zealand
Beech forest	37,402	360,104	1,826,940	2,187,044
Beech-broadleaved forest	7946	28,703	69,606	98,309
Beech-broadleaved podocarp forest	81,155	685,952	1,148,242	1,834,194
Broadleaved-podocarp forest	17,341	891,306	447,823	1,339,129
Podocarp forest	71	7939	57,362	65,301
Broadleaved forest	5172	241,297	112,423	353,720
Unspecified indigenous forest	9956	317,247	184,225	501,472
Total indigenous forest	159,043	2,532,548	3,846,621	6,379,169
Total land area	811,727	11,442,900	15,286,900	26,729,800

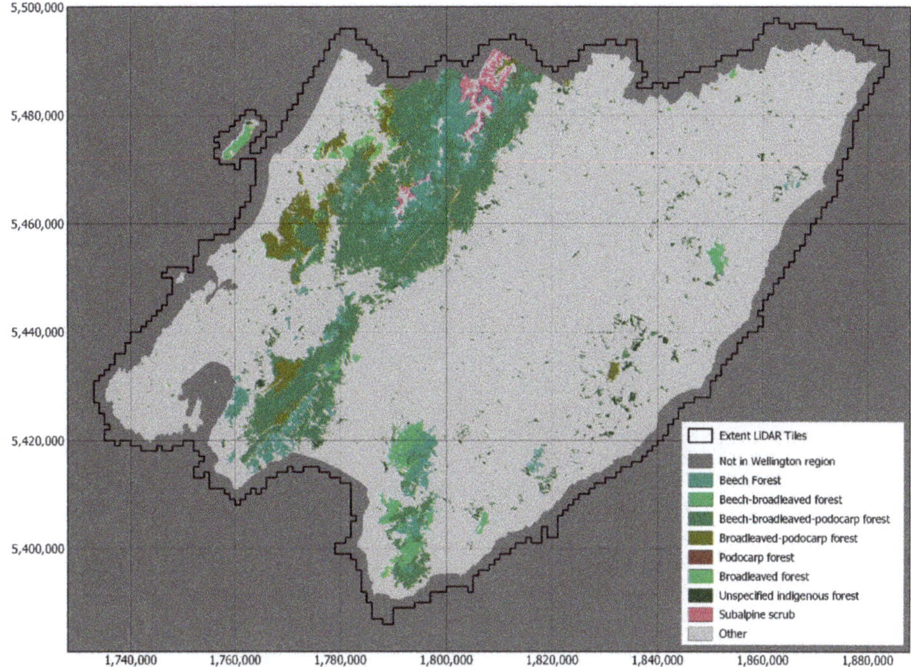

Figure 3. Spatial distribution of forest alliance groups in the Wellington Region. Most indigenous forests are in the Tararua mountains which form a north to south-west axis. Most of remaining indigenous forests are in Aorangi mountains in the southernmost part of region. Map projection is NZTM.

The tree mapping algorithm was applied to the CHM of the entire Wellington Region, taking several hours on a high-performance computer. Figure 4 shows a small extract of the Wellington canopy height model and identified tree tops greater than 30 m tall. The number of tall trees over 30 m occurring in indigenous forest in the entire Wellington Region is estimated to be 286,041. Table 3 breaks these down into different height ranges. There are 53,029 trees over 35 m tall and 7340 giant trees over 40 m tall. The giant trees are mainly in the beech-broadleaved-podocarp and broadleaved-podocarp forests, corresponding to the presence of podocarps in these group. The density of giant trees in broadleaved-podocarp forest is three times that of beech-broadleaved-podocarp forest (Table 4). The tall trees between 30 and 35 m tall occur at sites with relatively high mean elevation in the beech forest (638 m) and low mean elevation in the broadleaved-podocarp forest (332 m) (Table 5).

Table 3. Number of trees in forest alliance groups in selected tree-height ranges.

Forest Alliance Group	Area (ha)	>30 m	>35 m	>40 m	>45 m	>50 m
Beech forest	37,402	47,646	4911	410	44	0
Beech-broadleaved forest	7946	4911	598	108	9	0
Beech-broadleaved podocarp forest	81,155	176,518	28,903	3202	293	0
Broadleaved-podocarp forest	17,341	33,493	11,268	2103	226	0
Podocarp forest	71	129	4	0	0	0
Broadleaved forest	5172	3689	1357	302	43	0
Unspecified indigenous forest	9956	19,202	5944	1210	170	0
Subalpine shrubland	3841	453	44	5	0	0
Total	162,884	286,041	53,029	7340	785	0

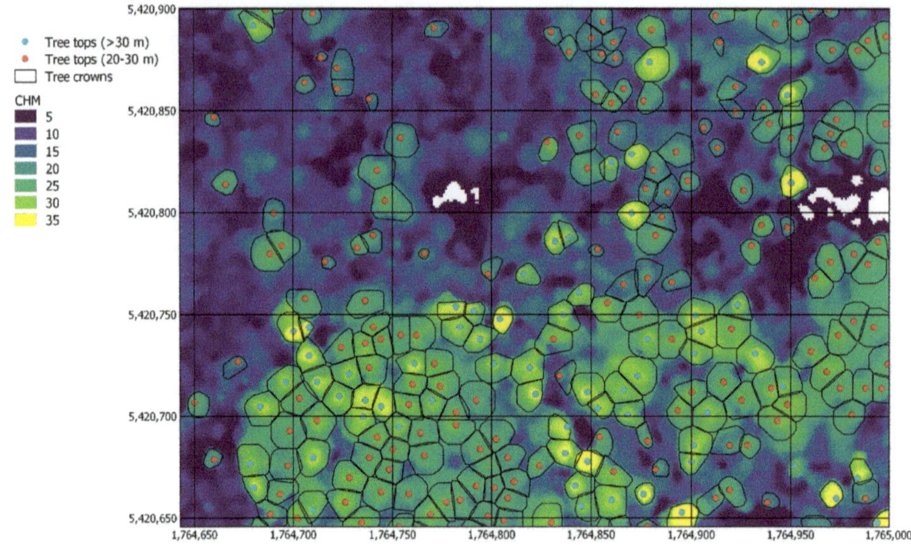

Figure 4. Extract of the canopy height model in the Wellington Region and identified tree tops greater than 30 m tall (cyan) and between 20 and 30 m tall (red). Map projection is NZTM.

Table 4. Density of trees (n/ha) in selected tree-height ranges.

Forest Alliance Group	>30 m	>35 m	>40 m	>45 m
Beech forest	1.27	0.13	0.01	0.001
Beech-broadleaved forest	0.62	0.08	0.01	0.001
Beech-broadleaved podocarp forest	2.18	0.36	0.04	0.004
Broadleaved-podocarp forest	1.93	0.65	0.12	0.013
Podocarp forest	1.82	0.06	0.00	0.000
Broadleaved forest	0.71	0.26	0.06	0.008
Unspecified indigenous forest	1.93	0.60	0.12	0.017
Subalpine shrubland	0.12	0.01	0.00	0.000

Table 5. Mean elevation of trees (m) in height classes.

Forest Alliance Group	30–35 m	35–40 m	40–45 m	45–50 m
Beech forest	638	565	478	332
Beech-broadleaved forest	385	338	224	181
Beech-broadleaved podocarp forest	470	418	373	332
Broadleaved-podocarp forest	332	307	291	286
Podocarp forest	461	464		
Broadleaved forest	278	245	218	217
Unspecified indigenous forest	191	191	172	154
Subalpine shrubland	731	742		

In the stereo models, the 3-dimensional points representing tree tops were mostly located on the very tops of tall trees. Occasionally, where trees were very close to each other but were distinct in the stereo model due to different foliage color, only one tree top was identified in the canopy height model (i.e., a tree top was missed). This happened for 0.8% of the actual trees. Occasionally, where tree branches from the same tree were large and appeared to be separate trees, more than one tree top was identified (i.e., a tree top was added incorrectly). There was 1.6% of estimated tall trees added in this way. Figure 5 shows the number of tall trees identified using stereo-models (actual) versus the number of tall trees identified using the automated mapping approach (estimated). It shows that our automated tree top identification compares well with a manual approach using stereo-imagery—in absence of ground-based measurements—and can be used to count trees over large areas with a systematic error of about one percent (i.e., $\approx$(0.8–1.6)%).

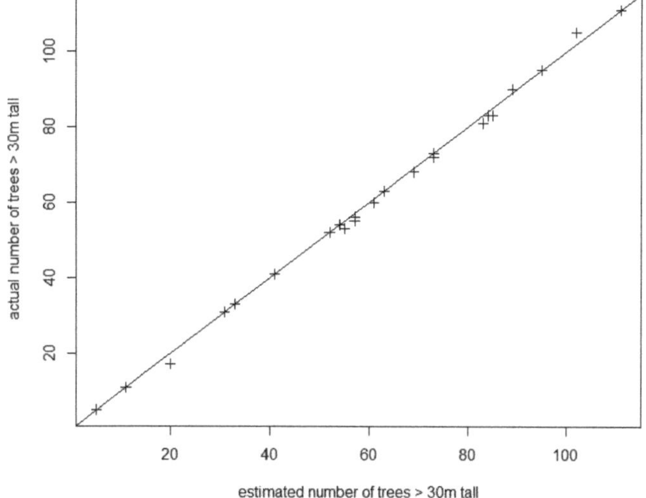

Figure 5. Number of tall trees within each land component as identified in stereo-imagery (actual) versus the number of tall trees as identified by the automated tree mapping algorithm (estimated).

In conjunction with the accuracy assessment of tree counting, we also tested the sensitivity of our automated mapping approach to varying inputs, that is, a CHM smoothed using different window sizes for the median filter and varying sizes of the local maximum filter to detect tall trees (Table 6). The number of trees detected is largest when a small window size is used, for example, 1081 tall trees using 3 m radius filters versus only 492 tall trees using 7 m radius filters. This reduction is caused by merging of trees and branches due to excessive smoothing of the CHM and by lower sensitivity to fine features in the CHM when using large filter sizes for the local peak detection.

Table 6. Sensitivity analysis of filters used for tree detection for one LiDAR tile (index row: 82, column: 72). Shown are the counts of tall trees (>30 m) detected using different combination of filter sizes for the CHM Median Filter (smoothing filter) and the Local Maximum Filter (peak detection in CHM). The unit of the window sizes (ws) is meter.

		CHM Median Filter		
		$ws = 3$	$ws = 5$	$ws = 7$
	$ws = 3$	1081	768	546
Local Maximum Filter	$ws = 5$	882	711	511
	$ws = 7$	777	646	492

Our tree mapping approach includes the automated correction of tree top positions in steep terrain. We noted for each individual tree whether its top position has been corrected or not and which correction method was used, that is, either the high point from the surface model within its crown or the center of mass of its crown. Analyzing the entire dataset of tall trees in Greater Wellington showed that about 3.1% of all tree tops have been corrected. Of these, 63.2% were corrected using the DSM and the remaining 36.8% using the COM method. We also calculated the reduction in tree height due to the correction procedure for one particular tile in mountainous terrain near the Totara Flats (index row: 82, column: 72, NZTM coordinates: 1,801,893:1,803,253 m and 5,466,814:5,468,104 m) (Figure 6). Of all trees detected in the 1 × 1 km tile, 145 had their height initially over-estimated, 13 of them by over 4 m. The average reduction in tree height following the correction method was −1.9 m and the maximum reduction was −9.4 m (for one tree on a cliff edge).

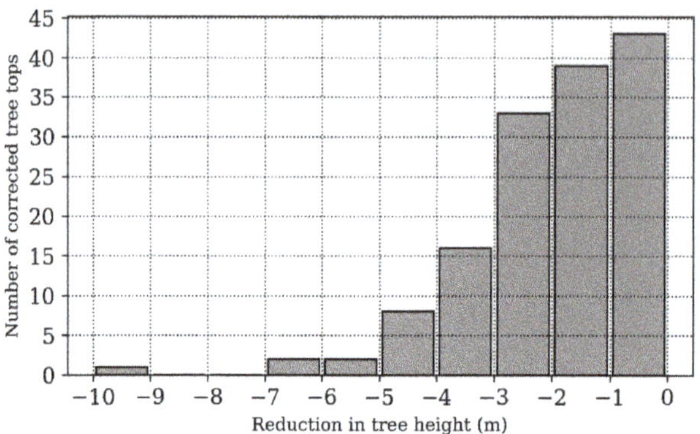

Figure 6. Effect of the automated tree top correction method on the estimated tree top height. Trees situated on steep slopes or overhanging a cliff edge typically have their top height overestimated due to the terrain normalization of the DSM. Most tree tops are reduced by 1 to 4.

4. Discussion

This is the first time in New Zealand that a spatial tree inventory, where individual trees are labelled and heights recorded, has been achieved for an area as large as a region. This has produced unprecedented characterization of indigenous forests for which tree height information has previously been scant (the New Zealand Indigenous Carbon Monitoring System collected height data for 1304 trees nationally [58]). Our tree mapping algorithm was applied to the entire Wellington Region, taking only several hours on a high-performance computer. It identified over one million tall trees (>30 m) in the region, of which over a quarter of a million were in indigenous forest. Through evaluation of stereo-models of aerial photography, the algorithm was found to be accurate to within plus or minus one percent. Taking this uncertainty into account, the number of tall trees (>30 m) in the Wellington Region may be reported as 1,174,000 (±1%), of which 286,000 (±1%) are in indigenous forest.

While there are many tall trees in the Wellington Region, their density in the three most common forest alliance groups—beech, beech-broadleaved podocarp and broadleaved-podocarp—is relatively low at ~2 tall trees per hectare. The density, however, varies markedly with elevation and can reach over 10 trees per hectare in lowlands, reflecting the more equable climates including reduced wind strength. The relatively high density in three alliance groups containing podocarps is due to mature NZ conifers generally being more than twice as tall as mature broadleaved (angiosperm) trees [11]. In other alliance groups, tall angiosperm trees, such as *Metrosideros umbellata* and *Knightia excelsa* R.Br., have podocarp-like traits including thick, tough leaves and slow growth [11].

The spatial distribution of forest alliance groups is controlled by climate and soil [59] and by disturbance history, both natural, such as earthquakes, landslides and high winds and anthropogenic, such as fire and logging [60]. Broadleaved-podocarp forest is common at low altitudes on the western side of the Tararuas, where there is hill country with mild climate and consistent rainfall. Further into the mountainous Tararuas, hardy beech trees become more common due to the cold climate and strong prevailing westerly winds. Near the tree line, at about 1000 m, the forest is nearly all beech. On the eastern side of the Tararuas there is a narrower band of hill country between the mountains and the plains and so the elevation progression starts with beech-broadleaved podocarp forest, rather than broadleaved-podocarp forest. In the Aorangi mountains in the south-east of the region, beech-broadleaved forest is more common, than in the Tararuas, due to a more recent history of logging [61].

The accuracy assessment of the tree mapping algorithm for counting tall trees showed an overestimation by 1.6% due to incorrect separation of tree crowns and an underestimation by 0.8% due to incorrect merging of tree crowns. This implies a net systematic over-estimation of tree numbers by 0.8%, that is, 1.6%–0.8%. Rather than make a systematic downwards adjustment of 0.8% to all estimated tree numbers and estimating the uncertainty to which the adjustment can be determined (as in estimation of land cover areas from remote sensing [62]), we took a conservative approach and set the uncertainty to the required adjustment, rounded up to the nearest percent, that is, ±1%, giving a conservative range of 2% uncertainty, without actually making the adjustment. Not making a systematic adjustment to estimated tree numbers achieves the desirable property of being consistent with the individual tree inventory. Using stereoscopy of aerial photographs to check delineation of tall trees assumes that heights of the canopy in the CHM are accurate and that all uncertainty is due to finding tree top positions in the CHM. This avoided the difficulty of field measurement of tree heights in thick mountainous forests. For some lone tall trees on flat land we found that the CHM was indeed accurate to plus or minus 0.5 m, similar to that reported by Gatzioli et al. [63].

Coomes et al. [12] also identified biases in the number of segmented trees compared with ground-based field measurements (mainly of trees smaller than 30 m). They found that the CHM-based tree mapping algorithm was sensitive to the tree height—strongly under-estimating the number of short understory trees (<12 m) and over-estimating the number of taller trees (>12 m) by 16% (subdivision of single overstory trees). However, identifying tree tops from the ground, especially of overstory trees, is challenging in thick forest canopies. Based on these previous findings we carefully evaluated settings to minimize uncertainties arising from both the incorrect separation and merging of tree crowns.

The choice of parameters of the tree mapping algorithm not only affects the number of trees detected but also the identified tree-top height. Three parameters most affect the results: the pixel resolution of the raster data sets (CHM, DSM and DTM); the window size of the smoothing filter for the CHM; and the maximum filter for the tree top detection. Our accuracy assessment revealed that, on the one hand, in 0.8% of the cases two trees were too close together to be detected as two separate trees by a 5 × 5 m median filter and 5 × 5 m maximum filter. This suggests a smaller window size (e.g., 3 × 3 m) would be beneficial in detecting more trees. On the other hand, about 1.6% of tall trees appeared to be separate trees because tree branches from the same tree were very large. This suggests a larger window size (e.g., 7 × 7 m) would be beneficial in detecting fewer trees. This is supported by our sensitivity analysis which showed both the effects of merging of trees due to excessive smoothing of the CHM and a lower sensitivity for tree top detection using too large window sizes.

We also tested the effect of a higher resolution CHM (0.5 × 0.5 m pixels) on the identified number of tall trees but found that almost identical results could be achieved with a 1 × 1 m pixel resolution and small window sizes for the smoothing and local maxima identification. We conclude that the overall number of trees using our set of parameters is accurate and the two contrasting effects balance each other out on a large-scale data set. However, we note that for an accurate local analysis, the set of parameters should be adjusted to the local canopy structure and properties.

Application of the algorithm in a region with steep landscapes, such as the Wellington Region, is made possible by the check and correction for over-estimated heights of trees on steep terrain. In a 1 × 1 km LiDAR tile containing typically mountainous terrain, 145 trees had their height initially over-estimated by 1.9 m on average; of these, 13 were overestimated by more than four meters. This would have introduced errors for many trees if the check and correction had not been applied. Previous studies reported similar reductions in tree top height following the correction of the terrain-effect. Khosravipour et al. [38] observed an average reduction of about 0.4 m and maximum reductions of 1.8 m of pine trees in the French Alps. Alexander et al. [40] conducted a research study in the tropical forest in Sumatra and found differences of 16.6 m in tree heights estimated from the CHM and DSM. Our findings are consistent with these previous efforts and we present a flexible and fast method to correct for the terrain-effect on a large-scale LiDAR dataset, which has not been done before and consider tree specific parameters such as the crown shape.

A source of error for estimated tree height that we did not consider correcting is tree lean. On steep slopes it is possible that trees may lean preferentially downhill. We did not see any evidence in the stereo models of tree lean in the tall indigenous trees on steep slopes, however, researchers have found systematic tree lean elsewhere. Gatziolis et al. [63] found that trees on steep slopes in the US Pacific Northwest had a mean offset of about 1 m horizontal between base and top of tree. A 1 m offset for a tall tree on a 20 degree slope would result in a small overestimation of tree height by 0.4 m, so should not have a significant impact on the statistics of tree heights. In extreme cases when the offset is 5 m, the tree height would be overestimated by about 1.5 m. This represents a possible, or maximum, error of plus 5% for tree heights on steep slopes.

While the tree-mapping algorithm took several hours for the Wellington Region, the processing of the raw LiDAR point cloud data to DTMs/DSMs/CHMs took much longer. The LiDAR data were divided into 9480 tiles of 1 km^2, with each one taking several hours to process on the high-performance computer in 2014—when we first processed the dataset. Now, the processing time has reduced to less than 30 min per tile and can be conveniently parallelized by taking advantage of modern cluster computing systems. While the preparation of the data represents a significant investment in time and compute resources, it only has to be done once and there are many other uses of the LiDAR-derived data in addition to forest characterization, such as ecological modelling [64], flood control [65] and roading. Indeed, several more regional authorities in New Zealand are currently acquiring or planning regional LiDAR surveys, including the Gisborne district, Northland Regional Council and Hawke's Bay Regional Council. (The Bay of Plenty already has regional LiDAR coverage, as does Wellington.) To coordinate a national approach to LiDAR acquisition, Land Information New Zealand has brought out a base specification [66] providing a foundation for public sector LiDAR data procurement.

The map of forest alliance groups is based on a previously existing map of indigenous forests, EcoSat Forests. This 1:50,000 scale database was derived by combining the vegetation layer in New Zealand Land Resource Inventory (NZLRI [67]) with the Land Cover Database (LCDB). In the 1:50,000 scale NZLRI, the area of indigenous forest types within landform polygons was estimated from a combination of fieldwork and photointerpretation of stereo aerial photographs. The nominal date of the mapping is 1980. The LCDB has a higher spatial resolution with a minimum mapping unit of 1 ha and a later currency with mapping dates of 1995, 2001–2002, 2008 and 2012; however, it has lower thematic resolution with only two indigenous forest classes. EcoSat Forests combines the two databases in a way that takes advantage of the high thematic resolution of NZLRI with the high spatial resolution of LCDB but this assumes that there has been little change in the indigenous forest classes since 1980. Indeed, the main forest alliance groups are unlikely to have changed since 1980 but as the accuracy of the NZLRI indigenous forest class mapping is not well characterized, the map of forest alliance groups used here should be considered a preliminary version only. Nevertheless, we believe the general spatial patterns of forest alliance groups as reported in this paper are well represented. Work is underway to produce a more current national map of forest alliance groups by integrating recent satellite imagery with recently surveyed forest plots.

Our tree inventory provides a useful baseline and starting point for monitoring the condition of indigenous forests in New Zealand. If diseases, such as kauri dieback and myrtle rust, spread through indigenous forests in New Zealand, then the impact on individual trees could be assessed and recorded. There is also the expectation of an increasing frequency and intensity of natural disturbances under climate change and continuing damage from mammalian pests. Certainly, tree mortality can be determined through repeat LiDAR surveys and possibly deteriorating condition could be assessed through repeated hyperspectral surveys [68]. Furthermore, the tree inventory lays the foundation for future work such as modelling the spatial distribution of tall trees in the landscape and investigating the underlying drivers governing their distribution.

5. Conclusions

Region-wide tree inventories can significantly improve characterization of natural forests and provide a useful baseline for monitoring conditions. It is possible to derive region-wide models of forest canopy height from LiDAR at reasonable cost. From the CHM it is further possible to derive a region-wide inventory of tall trees. Our tree top detection and correction scheme, which identifies tree tops from local maxima in the CHM and uses the DSM and DTM to correct for terrain-effects, provides a simple and fast method to accurately map overstory trees in flat as well as mountainous areas. It can be directly applied to improve existing and build new tree inventories in regions where LiDAR data is available.

The number of tall trees over 30 m in the Wellington Region was estimated to be 286,041 ($\pm$1%) and the number of giant trees over 40 m was estimated to be 7340 ($\pm$1%). The giant trees were found mainly in broadleaved-podocarp forest (density = 0.12 trees/ha) and beech-broadleaved-podocarp forest (density = 0.04 trees/ha).

Author Contributions: J.Z., conceptualization, methodology, formal analysis, software implementation, writing and editing; J.R.D., conceptualization, methodology, validation, writing and editing; J.D.S., conceptualization and methodology; S.K.W. reviewing and editing; B.J., formal analysis.

Funding: The Ministry of Business Innovation & Employment funded this research under contract C09X1709.

Acknowledgments: We would like to thank David Pairman from Landcare Research for the fruitful discussions and internal review.

Conflicts of Interest: The authors declare no conflict of interest.

References

1. Dymond, J.R.; Shepherd, J.D.; Newsome, P.F.; Belliss, S. Estimating change in areas of indigenous vegetation cover in New Zealand from the New Zealand Land Cover Database (LCDB). *N. Z. J. Ecol.* **2017**, *41*. [CrossRef]
2. Dymond, J.R.; Ausseil, A.-G.E.; Peltzer, D.A.; Herzig, A. Conditions and trends of ecosystem services in New Zealand—A synopsis. *Solut. J.* **2014**, *5*, 38–45.
3. Wardle, P. *Vegetation of New Zealand*; Cambridge University Press: Cambridge, UK, 1991; ISBN 9780521258739.
4. Díaz, I.A.; Sieving, K.E.; Peña-Foxon, M.E.; Larraín, J.; Armesto, J.J. Epiphyte diversity and biomass loads of canopy emergent trees in Chilean temperate rain forests: A neglected functional component. *For. Ecol. Manag.* **2010**, *259*, 1490–1501. [CrossRef]
5. Dislich, R.; Mantovani, W. Vascular epiphyte assemblages in a Brazilian Atlantic Forest fragment: Investigating the effect of host tree features. *Plant Ecol.* **2016**, *217*. [CrossRef]
6. Li, S.; Liu, S.; Shi, X.-M.; Liu, W.-Y.; Song, L.; Lu, H.-Z.; Chen, X.; Wu, C.-S. Forest Type and Tree Characteristics Determine the Vertical Distribution of Epiphytic Lichen Biomass in Subtropical Forests. *Forests* **2017**, *8*, 436. [CrossRef]
7. Lesak, A.A.; Radeloff, V.C.; Hawbaker, T.J.; Pidgeon, A.M.; Gobakken, T.; Contrucci, K. Modeling forest songbird species richness using LiDAR-derived measures of forest structure. *Remote Sens. Environ.* **2011**, *115*, 2823–2835. [CrossRef]
8. Flaspohler, D.J.; Giardina, C.P.; Asner, G.P.; Hart, P.; Price, J.; Lyons, C.K.; Castaneda, X. Long-term effects of fragmentation and fragment properties on bird species richness in Hawaiian forests. *Biol. Conserv.* **2010**, *143*, 280–288. [CrossRef]
9. Greene, D.F.; Johnson, E.A. A Model of Wind Dispersal of Winged or Plumed Seeds. *Ecology* **1989**, *70*, 339–347. [CrossRef]
10. Loehle, C. Strategy Space and the Disturbance Spectrum: A Life-History Model for Tree Species Coexistence. *Am. Nat.* **2000**, *156*, 14–33. [CrossRef] [PubMed]
11. McGlone, M.S.; Richardson, S.J.; Jordan, G.J. Comparative biogeography of New Zealand trees: Species richness, height, leaf traits and range sizes. *N. Z. J. Ecol.* **2010**, *34*, 137–151.
12. Coomes, D.A.; Šafka, D.; Shepherd, J.; Dalponte, M.; Holdaway, R. Airborne laser scanning of natural forests in New Zealand reveals the influences of wind on forest carbon. *For. Ecosyst.* **2018**, *5*, 10. [CrossRef]

13. Koukoulas, S.; Blackburn, G.A. Mapping individual tree location, height and species in broadleaved deciduous forest using airborne LIDAR and multi-spectral remotely sensed data. *Int. J. Remote Sens.* **2005**, *26*, 431–455. [CrossRef]

14. Latifi, H.; Nothdurft, A.; Koch, B. Non-parametric prediction and mapping of standing timber volume and biomass in a temperate forest: Application of multiple optical/LiDAR-derived predictors. *Forestry* **2010**, *83*, 395–407. [CrossRef]

15. O'Neil-Dunne, J.; MacFaden, S.; Royar, A. A Versatile, Production-Oriented Approach to High-Resolution Tree-Canopy Mapping in Urban and Suburban Landscapes Using GEOBIA and Data Fusion. *Remote Sens.* **2014**, *6*, 12837–12865. [CrossRef]

16. McPherson, E.G.; Simpson, J.R.; Xiao, Q.; Wu, C. Million trees Los Angeles canopy cover and benefit assessment. *Landsc. Urban Plan.* **2011**, *99*, 40–50. [CrossRef]

17. Crowther, T.W.; Glick, H.B.; Covey, K.R.; Bettigole, C.; Maynard, D.S.; Thomas, S.M.; Smith, J.R.; Hintler, G.; Duguid, M.C.; Amatulli, G.; et al. Mapping tree density at a global scale. *Nature* **2015**, *525*, 201–205. [CrossRef] [PubMed]

18. Lefsky, M.A.; Cohen, W.B.; Acker, S.A.; Parker, G.G.; Spies, T.A.; Harding, D. Lidar Remote Sensing of the Canopy Structure and Biophysical Properties of Douglas-Fir Western Hemlock Forests. *Remote Sens. Environ.* **1999**, *70*, 339–361. [CrossRef]

19. Hyyppä, J.; Kelle, O.; Lehikoinen, M.; Inkinen, M. A segmentation-based method to retrieve stem volume estimates from 3-D tree height models produced by laser scanners. *IEEE Trans. Geosci. Remote Sens.* **2001**, *39*, 969–975. [CrossRef]

20. Zhen, Z.; Quackenbush, L.; Zhang, L. Trends in Automatic Individual Tree Crown Detection and Delineation—Evolution of LiDAR Data. *Remote Sens.* **2016**, *8*, 333. [CrossRef]

21. Ayrey, E.; Fraver, S.; Kershaw, J.A.; Kenefic, L.S.; Hayes, D.; Weiskittel, A.R.; Roth, B.E. Layer Stacking: A Novel Algorithm for Individual Forest Tree Segmentation from LiDAR Point Clouds. *Can. J. Remote Sens.* **2017**, *43*, 16–27. [CrossRef]

22. Hamraz, H.; Contreras, M.A.; Zhang, J. A robust approach for tree segmentation in deciduous forests using small-footprint airborne LiDAR data. *Int. J. Appl. Earth Obs. Geoinf.* **2016**, *52*, 532–541. [CrossRef]

23. Duncanson, L.I.; Cook, B.D.; Hurtt, G.C.; Dubayah, R.O. An efficient, multi-layered crown delineation algorithm for mapping individual tree structure across multiple ecosystems. *Remote Sens. Environ.* **2014**, *154*, 378–386. [CrossRef]

24. Bunting, P.; Lucas, R. The delineation of tree crowns in Australian mixed species forests using hyperspectral Compact Airborne Spectrographic Imager (CASI) data. *Remote Sens. Environ.* **2006**, *101*, 230–248. [CrossRef]

25. Wang, Y.; Hyyppa, J.; Liang, X.; Kaartinen, H.; Yu, X.; Lindberg, E.; Holmgren, J.; Qin, Y.; Mallet, C.; Ferraz, A.; et al. International Benchmarking of the Individual Tree Detection Methods for Modeling 3-D Canopy Structure for Silviculture and Forest Ecology Using Airborne Laser Scanning. *IEEE Trans. Geosci. Remote Sens.* **2016**, *54*, 5011–5027. [CrossRef]

26. Li, W.; Guo, Q.; Jakubowski, M.K.; Kelly, M. A New Method for Segmenting Individual Trees from the Lidar Point Cloud. *Photogramm. Eng. Remote Sens.* **2012**, *78*, 75–84. [CrossRef]

27. Pirotti, F.; Kobal, M.; Roussel, J.R. A comparison of tree segmentation methods using very high density airborne laser scanner data. *Int. Arch. Photogramm. Remote Sens. Spat. Inf. Sci.* **2017**, *42*, 285–290. [CrossRef]

28. Dalponte, M.; Coomes, D.A. Tree-centric mapping of forest carbon density from airborne laser scanning and hyperspectral data. *Methods Ecol. Evol.* **2016**, *7*, 1236–1245. [CrossRef] [PubMed]

29. Fonstad, M.A.; Dietrich, J.T.; Courville, B.C.; Jensen, J.L.; Carbonneau, P.E. Topographic structure from motion: A new development in photogrammetric measurement. *Earth Surf. Process. Landf.* **2013**, *38*, 421–430. [CrossRef]

30. Baltsavias, E.; Gruen, A.; Eisenbeiss, H.; Zhang, L.; Waser, L.T. High-quality image matching and automated generation of 3D tree models. *Int. J. Remote Sens.* **2008**, *29*, 1243–1259. [CrossRef]

31. White, J.; Wulder, M.; Vastaranta, M.; Coops, N.; Pitt, D.; Woods, M. The Utility of Image-Based Point Clouds for Forest Inventory: A Comparison with Airborne Laser Scanning. *Forests* **2013**, *4*, 518–536. [CrossRef]

32. St-Onge, B.; Jumelet, J.; Cobello, M.; Véga, C. Measuring individual tree height using a combination of stereophotogrammetry and lidar. *Can. J. For. Res.* **2004**, *34*, 2122–2130. [CrossRef]

33. Dandois, J.P.; Ellis, E.C. Remote Sensing of Vegetation Structure Using Computer Vision. *Remote Sens.* **2010**, *2*, 1157–1176. [CrossRef]

34. Westoby, M.J.; Brasington, J.; Glasser, N.F.; Hambrey, M.J.; Reynolds, J.M. 'Structure-from-Motion' photogrammetry: A low-cost, effective tool for geoscience applications. *Geomorphology* **2012**, *179*, 300–314. [CrossRef]

35. Brandtberg, T.; Warner, T.A.; Landenberger, R.E.; McGraw, J.B. Detection and analysis of individual leaf-off tree crowns in small footprint, high sampling density lidar data from the eastern deciduous forest in North America. *Remote Sens. Environ.* **2003**, *85*, 290–303. [CrossRef]

36. Hirata, Y. The effects of footprint size and sampling density in airborne laser scanning to extract individual trees in mountainous terrain. *Int. Arch. Photogramm. Remote Sens. Spat. Inf. Sci.* **2004**, *36*, 102–107.

37. Hansen, E.; Ene, L.; Gobakken, T.; Ørka, H.; Bollandsås, O.; Næsset, E. Countering Negative Effects of Terrain Slope on Airborne Laser Scanner Data Using Procrustean Transformation and Histogram Matching. *Forests* **2017**, *8*, 401. [CrossRef]

38. Khosravipour, A.; Skidmore, A.K.; Wang, T.; Isenburg, M.; Khoshelham, K. Effect of slope on treetop detection using a LiDAR Canopy Height Model. *ISPRS J. Photogramm. Remote Sens.* **2015**, *104*, 44–52. [CrossRef]

39. Duan, Z.; Zhao, D.; Zeng, Y.; Zhao, Y.; Wu, B.; Zhu, J. Assessing and Correcting Topographic Effects on Forest Canopy Height Retrieval Using Airborne LiDAR Data. *Sensors* **2015**, *15*, 12133–12155. [CrossRef] [PubMed]

40. Alexander, C.; Korstjens, A.H.; Hill, R.A. Influence of micro-topography and crown characteristics on tree height estimations in tropical forests based on LiDAR canopy height models. *Int. J. Appl. Earth Obs. Geoinf.* **2018**, *65*, 105–113. [CrossRef]

41. Regional High-Resolution DEM Now on LDS, Wellington Regional Government GIS Group (2016). Available online: http://mapping.gw.govt.nz/News06.htm (accessed on 3 October 2018).

42. Coomes, D.A.; Dalponte, M.; Jucker, T.; Asner, G.P.; Banin, L.F.; Burslem, D.F.R.P.; Lewis, S.L.; Nilus, R.; Phillips, O.L.; Phua, M.-H.; et al. Area-based vs tree-centric approaches to mapping forest carbon in Southeast Asian forests from airborne laser scanning data. *Remote Sens. Environ.* **2017**, *194*, 77–88. [CrossRef]

43. Kandare, K.; Dalponte, M.; Ørka, H.; Frizzera, L.; Næsset, E. Prediction of Species-Specific Volume Using Different Inventory Approaches by Fusing Airborne Laser Scanning and Hyperspectral Data. *Remote Sens.* **2017**, *9*, 400. [CrossRef]

44. Allen, R.B.; Bellingham, P.J.; Holdaway, R.J.; Wiser, S.K. New Zealand's indigenous forests and shrublands. In *Ecosystem Services in New Zealand—Conditions and Trends*; Dymond, J.R., Ed.; Manaaki Whenua Press: Lincoln, New Zealand, 2013; pp. 34–48. ISBN 9780478347364.

45. Müller, M.U.; Shepherd, J.D.; Dymond, J.R. Support vector machine classification of woody patches in New Zealand from synthetic aperture radar and optical data, with LiDAR training. *J. Appl. Remote Sens.* **2015**, *9*, 095984. [CrossRef]

46. Bunting, P.; Armston, J.; Clewley, D.; Lucas, R. The sorted pulse data software library (SPDLib): Open source tools for processing LiDAR data. In Proceedings of the SilviLaser 2011, 11th International Conference on LiDAR Applications for Assessing Forest Ecosystems, University of Tasmania, Hobart, Australia, 16–20 October 2011; pp. 1–11.

47. Zhang, K.; Chen, S.; Whitman, D.; Shyu, M.; Yan, J.; Zhang, C. A progressive morphological filter for removing nonground measurements from airborne LiDAR data. *IEEE Trans. Geosci. Remote Sens.* **2003**, *41*, 872–882. [CrossRef]

48. Evans, J.S.; Hudak, A.T. A Multiscale Curvature Algorithm for Classifying Discrete Return LiDAR in Forested Environments. *IEEE Trans. Geosci. Remote Sens.* **2007**, *45*, 1029–1038. [CrossRef]

49. Dalponte, M. itcSegment: Individual Tree Crowns Segmentation. R Package Version 0.8. Available online: https://CRAN.R-project.org/package=itcSegment (accessed on 3 October 2018).

50. Zörner, J.; Dymond, J.; Shepherd, J.; Jolly, B. *PyCrown-Fast Raster-Based Individual Tree Segmentation for LiDAR Data*; Landcare Research Ltd.: Lincoln, New Zealand, 2018. [CrossRef]

51. Roussel, J.-R.; Auty, D. lidR: Airborne LiDAR Data Manipulation and Visualization for Forestry Applications. R Package Version 1.5.1. Available online: https://CRAN.R-project.org/package=lidR (accessed on 3 October 2018).

52. NeSI Pan Cluster, Centre for eResearch. Available online: https://wiki.auckland.ac.nz/display/CER/NeSI+Pan+Cluster (accessed on 3 October 2018).

53. Wiser, S.; Thomson, F.; de Cáceres, M. Expanding an existing classification of New Zealand vegetation to include non-forested vegetation. *N. Z. J. Ecol.* **2016**, *40*, 160–178. [CrossRef]

54. Wiser, S.K.; de Cáceres, M. New Zealand's plot-based classification of vegetation. *Phytocoenologia* **2018**, *48*, 153–161. [CrossRef]

55. Shepherd, J.D.; Ausseil, A.-G.; Dymond, J.R. *EcoSat Forests: A 1:750,000 Scale Map of Indigenous Forest Classes in New Zealand*; Manaaki Whenua Press: Lincoln, New Zealand, 2005; Available online: https://lris.scinfo.org.nz/ (accessed on 3 October 2018).

56. Erdas Imagine, Hexagon Geospatial. Available online: https://www.hexagongeospatial.com (accessed on 3 October 2018).

57. Dymond, J.R.; Derose, R.C.; Harmsworth, G.R. Automated mapping of land components from digital elevation data. *Earth Surf. Process. Landf.* **1995**, *20*, 131–137. [CrossRef]

58. Bellingham, P.; Wiser, S.; Coomes, D.; Dunningham, A. *Review of Permanent Plots for Long-Term Monitoring of New Zealand's Indigenous Forests*; Science for Conservation; Department of Conservation: Wellington, New Zealand, 2000; ISBN 9780478219586.

59. Leathwick, J.; Wilson, G.; Rutledge, D.; Wardle, P.; Morgan, F.; Johnston, K.; McLeod, M.; Kirkpatrick, R. *Land Environments of New Zealand*; David Bateman Ltd.: Auckland, New Zealand, 2003; ISBN 9781869535223.

60. Wardle, P. Facets of the distribution of forest vegetation in New Zealand. *N. Z. J. Bot.* **1964**, *2*, 352–366. [CrossRef]

61. Ministry for Culture and Heritage. Te Ara: The Encyclopaedia of New Zealand. Available online: https://teara.govt.nz/en/logging-native-forests/page-1 (accessed on 3 October 2018).

62. Dymond, J.R.; Shepherd, J.D.; Newsome, P.F.; Gapare, N.; Burgess, D.W.; Watt, P. Remote sensing of land-use change for Kyoto Protocol reporting: The New Zealand case. *Environ. Sci. Policy* **2012**, *16*, 1–8. [CrossRef]

63. Gatziolis, D.; Fried, J.S.; Monleon, V.S. Challenges to estimating tree height via LiDAR in closed-canopy forests: A parable from Western Oregon. *For. Sci.* **2010**, *56*, 139–155.

64. Vierling, K.T.; Vierling, L.A.; Gould, W.A.; Martinuzzi, S.; Clawges, R.M. Lidar: Shedding new light on habitat characterization and modeling. *Front. Ecol. Environ.* **2008**, *6*, 90–98. [CrossRef]

65. Turner, A.; Colby, J.; Csontos, R.; Batten, M. Flood Modeling Using a Synthesis of Multi-Platform LiDAR Data. *Water* **2013**, *5*, 1533–1560. [CrossRef]

66. New Zealand National Aerial LiDAR Base Specification, Version 1.1. 2018. Available online: https://www.linz.govt.nz/system/files_force/media/doc/loci_nz-national-aerial-lidar-base-specification_20180629.pdf (accessed on 3 October 2018).

67. The New Zealand Land Resource Inventory. Available online: https://lris.scinfo.org.nz/layer/48055-nzlri-vegetation/ (accessed on 3 October 2018).

68. Asner, G.P.; Martin, R.E.; Keith, L.M.; Heller, W.P.; Hughes, M.A.; Vaughn, N.R.; Hughes, R.F.; Balzotti, C. A Spectral Mapping Signature for the Rapid Ohia Death (ROD) Pathogen in Hawaiian Forests. *Remote Sens.* **2018**, *10*, 404. [CrossRef]

© 2018 by the authors. Licensee MDPI, Basel, Switzerland. This article is an open access article distributed under the terms and conditions of the Creative Commons Attribution (CC BY) license (http://creativecommons.org/licenses/by/4.0/).

 *forests*

Article

Spatiotemporal Variations of Aboveground Biomass under Different Terrain Conditions

Aihua Shen [1], Chaofan Wu [2,*], Bo Jiang [1], Jinsong Deng [3], Weigao Yuan [1], Ke Wang [3], Shan He [3], Enyan Zhu [3], Yue Lin [3] and Chuping Wu [1]

[1] Zhejiang Academy of Forestry, Hangzhou 310023, China; mailahshen@126.com (A.S.); jiangbof@126.com (B.J.); Zfaywg@126.com (W.Y.); wcp1117@hotmail.com (C.W.)
[2] College of Geography and Environmental Sciences, Zhejiang Normal University, Jinhua 321004, China
[3] College of Environmental and Resource Sciences, Zhejiang University, Hangzhou 310058, China; jsong_deng@zju.edu.cn (J.D.); kwang@zju.edu.cn (K.W.); heshan33@zju.edu.cn (S.H.); eyzhu@zju.edu.cn (E.Z.); joyelin2017@163.com (Y.L.)
* Correspondence: cfwdh@zjnu.edu.cn; Tel.: +86-579-8228-2273

Received: 30 October 2018; Accepted: 13 December 2018; Published: 17 December 2018

Abstract: Biomass is a key biophysical parameter used to estimate carbon storage and forest productivity. Spatially-explicit estimation of biomass provides invaluable information for carbon stock calculation and scientific forest management. Nevertheless, there still exists large uncertainty concerning the relationship between biomass and influential factors. In this study, aboveground biomass (AGB) was estimated using the random forest algorithm based on remote sensing imagery (Landsat) and field data for three regions with different topographic conditions in Zhejiang Province, China. AGB distribution and change combined with stratified terrain classifications were analyzed to investigate the relations between AGB and topography conditions. The results indicated that AGB in three regions increased from 2010 to 2015 and the magnitude of growth varied with elevation, slope, and aspect. In the basin region, slope had a greater influence on AGB, and we attributed this negative AGB-elevation relationship to ecological forest construction. In the mountain area, terrain features, especially elevation, showed significant relations with AGB. Moreover, AGB and its growth showed positive relations with elevation and slope. In the island region, slope also played a relatively more important role in explaining the relationship. These results demonstrate that AGB varies with terrain conditions and its change is a consequence of interactions between the natural environment and anthropogenic behavior, implying that biomass retrieval based on Landsat imagery could provide considerable important information related to regional heterogeneity investigations.

Keywords: aboveground biomass; Landsat; random forest; topography; human activity

---

1. Introduction

Biomass is an important biophysical parameter used to understand carbon dynamics on the background of global climate change, and the spatiotemporal estimation of biomass will provide invaluable information for carbon calculation and scientific forest management [1,2]. In the past few decades, remote sensing has been increasingly used to estimate aboveground biomass because of its macroscopical, nondestructive, and efficiency advantages compared to time- and space-limited field survey methods [3,4]. Historically, the field-measured biomass has been usually calculated by establishing species-specific allometric equations based on height and diameter at breast height (DBH) gauged within standardized plots, which provide the basic samples for remote sensing-based biomass simulation [5,6]. Spatiotemporal biomass retrieval based on remote sensing has been increasingly implemented, because when compared to single-period biomass distribution, it provides more details for biomass change detection and further exploration of the influencing mechanisms [7]. Among various remote sensing data sources, Landsat imagery has acquired wide applications in biomass estimation due to its open-access data availability, appropriate spatial resolution, and abundant history archive [3,8]. Multiple potential features can be derived from the existing Landsat imagery, including multispectral bands, vegetation indices, texture bands, and time-sequence data, which provide abundant information for biomass retrieval [9–11].

Statistical methods and radiative transfer models are two common methods used to quantify biomass spatial distributions [12,13]. Radiative transfer models usually describe mechanisms using combinations of complicated parameters, which makes the process difficult to implement. Comparatively speaking, statistical methods realize this prediction processes by establishing relationships more directly. Among the most advanced approaches, machine learning methods have received considerable attention in recent years [14,15]. Compared to traditional regression algorithms such as multiple linear regression, machine learning algorithms have no strict assumptions on input variables or relationships between response variables and explanatory variables [16]. Support vector machine, random forest, and k-nearest neighbor are the most often implemented algorithms that result in satisfactory prediction [17–19]. Random forest (RF) is frequently selected as the regression method for biomass retrieval because of its outstanding performance, for example with higher prediction accuracy [20,21]. The difference between RF and other state-of-the art machine learning methods is that RF requires fewer parameters, but provides more accurate predictions [22].

Varied natural environments and human activities greatly influence changes in AGB (aboveground biomass). In the anthropogenic era, forest management has a profound impact on forest ecosystem dynamics. China's Natural Forest Conservation Program has increased the total biomass in China by persistent afforestation and reforestation [7]. At the same time, the influence of natural conditions on biomass change has also been investigated. Sattler et al. found that biomass could get different accumulations in sloped and flat regions after afforestation [23]. However, Lee et al. stated that there were no significant relationships between biomass and topographical factors in that intact lowland forest [24]. Therefore, the biomass distribution and change influenced by topography should be further explored. Although Du et al. pointed out that the biomass spatial distribution in Zhejiang Province was related to topographical factors including altitude and slope [25], biomass heterogeneity caused by topography across different districts has not been investigated, especially when combined with remote sensing techniques.

The impact of hierarchical elevation and ecological forests on biomass spatiotemporal change has been explored previously [9]. In this study, the objectives are: (1) based on the field measurements in 2010 and 2015, to map the distribution of aboveground biomass under different topographic conditions in Zhejiang Province in both years; (2) to inspect further topographical factors including slope and aspect on AGB and its change; and (3) to reveal the regular pattern within different regions under discrepant natural environment and human conditions. The flowchart of this research was displayed in Figure 1.

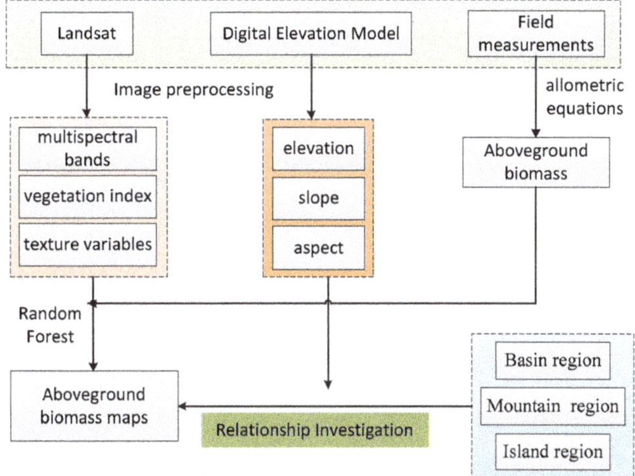

Figure 1. The flowchart of this research.

2. Materials and Methods

2.1. Study Area

Zhejiang Province is located in the southeastern region of China, ranging from 118°02′ E–123°08′ E, 27°03′ N–31°11′ N, with a subtropical monsoon climate. The annual average temperature is between 15 and 18 °C and the annual precipitation is between 1100 and 2000 mm. Being one of the most developed provinces with regard to the economy in China, it has consumed a large number of wood resources in the past several decades, and now, a majority of the land is covered with secondary forests. The local government has made great efforts to protect forests. As a result, the total forest coverage in the province has reached 60.91%. To investigate the influence of different topographic factors on AGB and its change, three counties named Wuyi County (administered by Jinhua City), Xianju County (belonging to Taizhou City), and Dinghai District (governed by Zhoushan City) were selected as the study areas (Figure 2). They are representative of basin, mountain, and island regions, which are located in the middle, southeast, and northeast of Zhejiang Province, respectively.

2.2. Field-Measured Data

Field investigations were carried out in 2010 and 2015 by Zhejiang Academy of Forestry. Sample plot design and selection were completed by taking into account the local geographical environment factors across the whole province. The size of each plot was 20 m × 20 m for trees, with three 2 m × 2 m subplots set in the diagonal line of each plot for shrubs and grasses [14]. The total biomass in each plot was calculated by summing the biomass of all trees, shrubs, and herbs, which was further defined as the final aboveground biomass (AGB) with a unit of Mg/ha. AGB values of broadleaved forests, coniferous, and broadleaved mixed forests, shrubs, bamboo forests, pine forests, and Chinese fir forests were calculated using measured DBH and height values embedded in specific allometric equations developed by Yuan et al. [26].

Quadrats outside the forest region and administrative boundary were deleted after checking their positions on remote sensing images and Google Earth based on visual interpretation. The outliers were selected and removed using the Pauta method, also named 3σ (standard deviation) measurement, by calculating the mean and variance values [20]. The statistics of the final dataset for subsequent analysis are shown in Table 1.

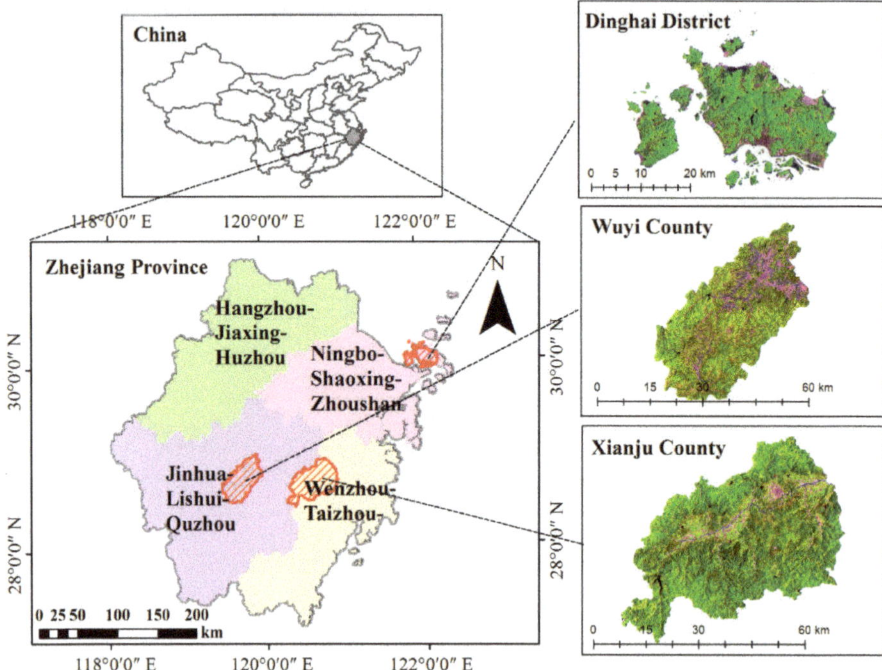

Figure 2. Locations of the study areas in Zhejiang Province, China.

Table 1. Statistics of field-based aboveground biomass (Mg/ha).

Statistic	Wuyi County		Xianju County		Dinghai District	
	2010	2015	2010	2015	2010	2015
Number of plots	130	130	49	49	43	43
Mean	91.22	99.37	79.07	93.78	87.96	112.81
Max	188.29	212.01	131.92	144.51	184.64	299.81
Min	20.32	16.85	7.71	13.28	3.38	5.38
SD	38.61	41.68	27.90	31.51	50.74	66.74

2.3. Remote Sensing Data

Landsat-5 Thematic Mapper (TM) and Landsat-8 Operational Land Imager (OLI) images (L1T) were downloaded from the United States Geological Survey (USGS) website [27]. Compared to the selected multispectral bands with a spatial resolution of 30 m, the thermal infrared channels for TM (120 m) and OLI (100 m) were abandoned for their coarser resolution. The first blue band (0.43–0.45 μm) of Landsat 8 OLI was removed to keep the bands consistent with Landsat 5 TM. The dates of the acquired images were almost in the same season by considering the phenology. Nevertheless, limited by the availability of cloudless images, the selections were based on the hypothesis that there was no significant biomass difference between imagery acquisition and field investigation. Detailed information about the Landsat images is listed in Table 2.

Digital Elevation Model (DEM) data were acquired from the Advanced Spaceborne Thermal Emission and Reflection Radiometer Global Digital Elevation Model (ASTER GDEM) V2 product with a spatial resolution of 30 m. Based on the DEM data, elevation, slope, and aspect were generated as the three basic geomorphology features. Administrative boundaries, present land use maps, forest maps, Google earth images, and socioeconomic statistics were collected as additional datasets.

To reduce the impact of the atmosphere, radiometric calibration was completed by inputting gain and offset information from the attached files, and an atmospheric correction using fast line of sight atmospheric analysis of spectral hypercubes (FLAASH) was executed. C correction was adopted as the topographic correction method to reduce the impact of terrain effects, especially for regions with shady slopes.

Table 2. Landsat imagery acquisition information.

Study Area	Path/Row	Sensor	Imagery Acquisition Time
Dinghai District	118/39	TM5 OLI	17 July 2009 3 August 2015
Wuyi County	119/40	TM5 OLI	24 May 2010 13 October 2015
Xianju County	118/40	TM5 OLI	28 July 2007 20 July 2016

2.4. Biomass Estimation from Remote Sensing

2.4.1. Feature Derivation

Consequently, the pixels containing corresponding sample plots were selected to link the spectral information of Landsat imagery with the biomass density of quadrats based on the hypothesis that there should be no significant difference between the biomass per area in the 20 m × 20 m field plots and their position-homologous 30 m × 30 m Landsat pixels. Candidate predictor variables were extracted from the remote sensing imagery specific to previous studies, including multispectral bands, vegetation indices, and texture information. The corrected Normalized Difference Vegetation Index (NDVIc), incorporating shortwave infrared bands (SWIR), was calculated [28]. Three components, including brightness, greenness, and wetness, were also derived through the tasseled cap (TC) transformation [29]. Texture variables were also extracted using the gray-level co-occurrence (GLCM) method.

2.4.2. Machine Learning Method

Biomass data selected from field plots were considered as the response variable, and derivatives from remote sensing imagery were treated as predictive variables. Random forest was selected as the prediction method to establish the relationship between AGB and derivatives because numerous researchers have testified to this algorithm's outstanding performance in biomass estimation [5,20,30]. The algorithm randomly selects variables at each node in the regression and classification tree and uses the bootstrap method to construct training samples without pruning. During this process, 2/3 samples are usually selected as training data, and the others are treated as validation data, which is also called "out-of-bag" [22]. The random selection of samples and variables makes the prediction results variable, but efficient. There are two important parameters, named *mtry* and *ntree*, that should be adjusted during the modeling process [31]. *ntree* controls the number of trees and is usually set to 500, while *mtry* determines the number of features and is usually set to 1/3 of the total number of input features. In addition, random forest is capable of estimating the relative importance of input features, which can be indicated by two built-in indices named %IncMSE and IncNodePurity. %IncMSE refers to Mean Decrease Accuracy and is calculated by constructing each tree of an ensemble with and without the specific variable. For all trees, the differences in error of these two variants are recorded, averaged, and normalized by their standard deviation [32,33]. It has been used in many previous studies [22] and was adopted in the current study.

2.4.3. Precision Evaluation

Ten-fold cross-validation was selected as the accuracy assessment approach. It divided the dataset into 10 groups. Once one of the groups was selected as the validation set, the other groups were treated as the training set each time. The process was repeated 10 times until all the groups had been traversed. Random forest modeling and accuracy assessment were implemented in the R 3.5.1 © open source software through the "caret" package [34]. R^2 values indicating the variance in the response variable explained by the predictor variables were computed to evaluate modeling accuracy. Furthermore, scatter diagrams of predicted and field-measured biomass values were plotted.

2.5. AGB Mapping and Spatio-Temporal Characteristic Analysis

To improve the accuracy of identification, a hierarchical system with six categories (0–30, 30–60, 60–90, 90–120, 120–150, >150 Mg/ha) used in Zhao et al.'s study was applied to the estimated AGB maps in all regions [6]. Simultaneously, topographic variables including elevation, slope, and aspect were classified into different levels by referring to Du et al.'s work [25]. Elevation was reclassified with an interval of 200 m for Wuyi and Xianju. Considering the relative lower elevation in Dinghai District, an interval of 50 m was set by consulting the work of Pan et al. [35]. In terms of slope and aspect, six grades of slope (0–5°, 5–15°, 15–25°, 25–35°, 35–45°, and >45°) and eight categories of aspect (north, northeast, east, southeast, south, southwest, west, northwest) were adopted to investigate the influence of slope and aspect on AGB distribution and its change. Meanwhile, mean values for each category were calculated from the estimated AGB maps for comparison. Besides, regression methods were used to check whether there was significant correlation between AGB/change and corresponding topography variables.

3. Results

3.1. The Importance Rank of Variables

Due to the randomness of the random forest algorithm, all the modeling process had to be repeated 20 times [20,36]. To investigate the performance of different variables involved in biomass estimation, %IncMSE were normalized for more convenient comparison [36], and the top 10 variables participated in the modeling process in three regions were selected to form the rank of the most important variables shown in Figure 3. Among all the selected variables, NDVIc was always selected as the most significant predictor in 2010, but the SWIR band (OLI Band 6) and derivatives from tassel transformation were among the best in 2015.

3.2. Accuracy Assessment

All the samples were used to predict the AGB as a previous study had found that using all samples, when compared to a smaller sample size, was propitious to biomass estimation [5]. R^2 were calculated to inspect the modeling accuracy (Figure 4). It can be seen that R^2 values fluctuated with different magnitudes, where the estimation in 2010 in Wuyi County (blue, solid line) obtained the best performance, while the prediction in Dinghai (green) showed relatively lower accuracy. Besides, Figure 5 shows the scatter plots by using the simple linear regression method for estimated AGB and field measurements. Limited by the total number of samples, Xianju and Dinghai had lower R^2 values than Wuyi, but all values of R^2 were beyond 0.8. Meanwhile, underestimation of the values of the highest biomass and overestimation on the values of the lowest biomass cannot be ignored.

3.3. Bitemporal Distribution and Change of Aboveground Biomass

When inspecting the spatial change in different periods, Figure 6 showed that the overall AGB in three regions increased from 2010–2015. AGB values below 30 Mg/ha accounted for a relatively small proportion in all the regions, which were mostly distributed in water areas and construction lands (when compared to the land use map, but not shown here). Simultaneously, AGB values beyond 150 Mg/ha also occupied a low percentage, and AGB in most regions changed in the range of 30–120 Mg/ha. Moreover, AGB in Dinghai District increased evidently, almost covering the whole region.

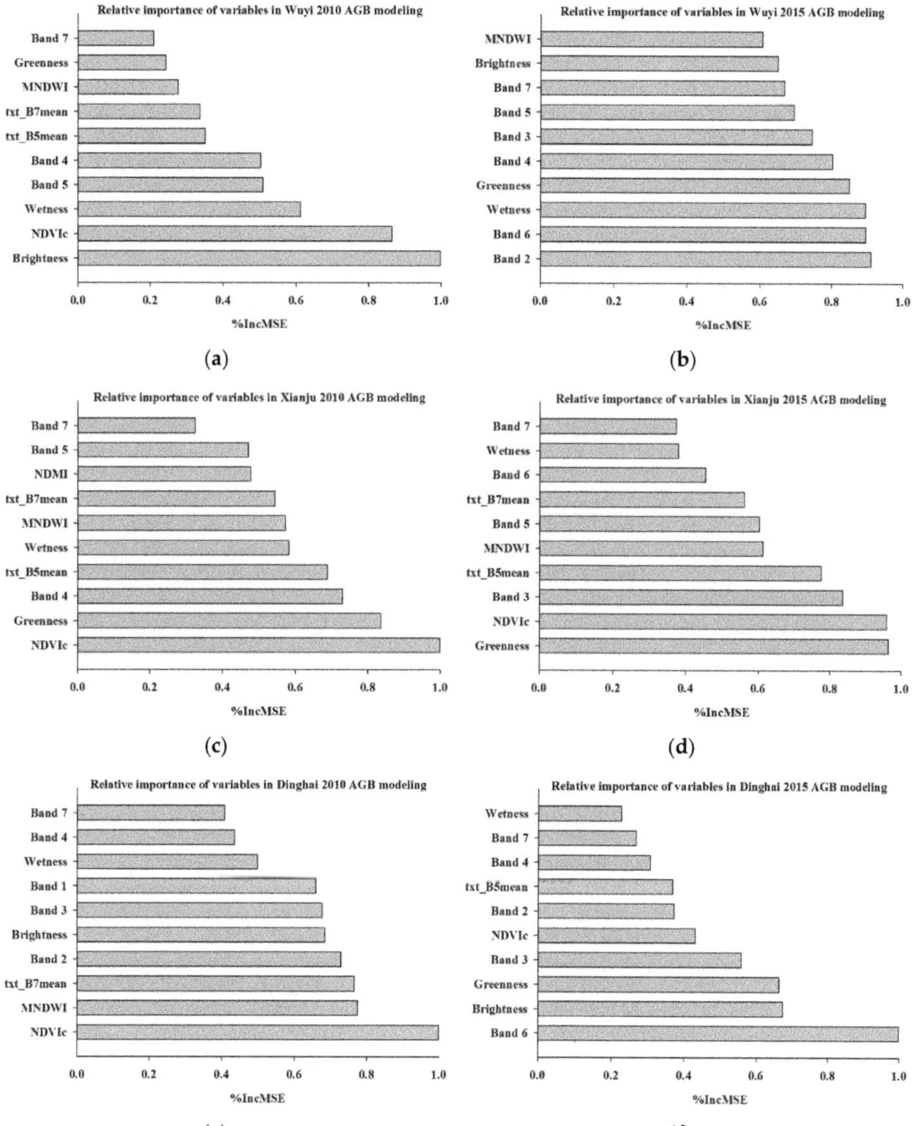

Figure 3. The relative importance of the top 10 variables in three regions: (**a**) relative importance of variables in Wuyi 2010 AGB modeling; (**b**) relative importance of variables in Wuyi 2015 AGB modeling; (**c**) relative importance of variables in Xianju 2010 AGB modeling; (**d**) relative importance of variables in Xianju 2015 AGB modeling; (**e**) relative importance of variables in Dinghai 2010 AGB modeling; (**f**) relative importance of variables in Dinghai 2015 AGB modeling. AGB: aboveground biomass.

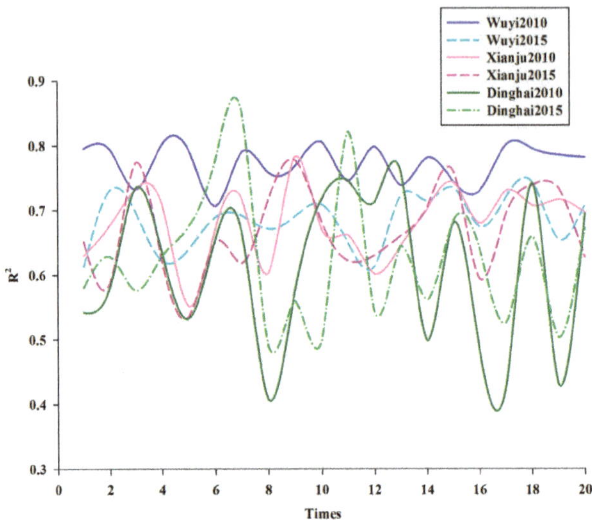

Figure 4. Prediction accuracy (R^2) of modeling in different regions.

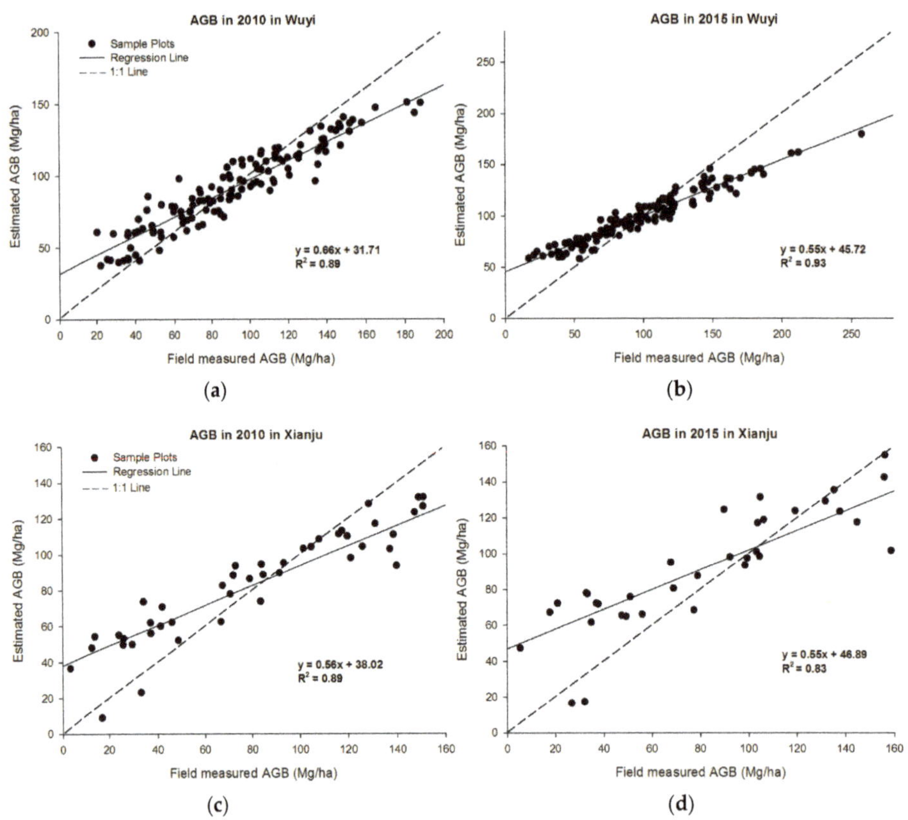

(**a**)

(**b**)

(**c**)

(**d**)

Figure 5. *Cont.*

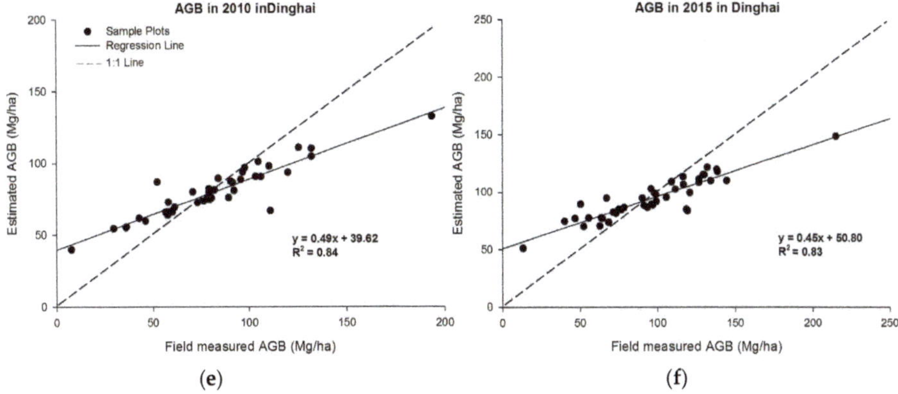

(e) (f)

Figure 5. The relationship between estimated and field-measured AGB in three regions: (**a**) AGB in 2010 in Wuyi; (**b**) AGB in 2015 in Wuyi; (**c**) AGB in 2010 in Xianju; (**d**) AGB in 2015 in Xianju; (**e**) AGB in 2010 in Dinghai; (**f**) AGB in 2015 in Dinghai.

The topographic features in the three regions in Table 3 indicated that Dinghai District, occupying the smallest area, had the lowest mean values of elevation, slope, and aspect, while Xianju County, covering the largest area, owned the highest mean values. Meanwhile, Wuyi and Xianju Counties had higher mean values of AGB than Dinghai District had both in 2010 and 2015. However, the AGB in Xianju County acquired the least increase with the minimum increase rate, and Dinghai had the highest increase rate of AGB.

Table 3. Basic information and AGB change in three regions.

Regions	Area (km²)	Elevation (m)	Slope °	Aspect °	AGB (Mg/ha) 2010	AGB (Mg/ha) 2015	Increase (Mg/ha)	Increase Rate (%)
Wuyi County	1583.13	383.83	16.01	177.23	91.18	106.23	15.05	16.51
Xianju County	1999.78	408.71	19.18	179.83	79.55	88.10	8.55	10.75
Dinghai District	534.40	62.97	10.87	169.44	70.59	83.23	12.64	17.91

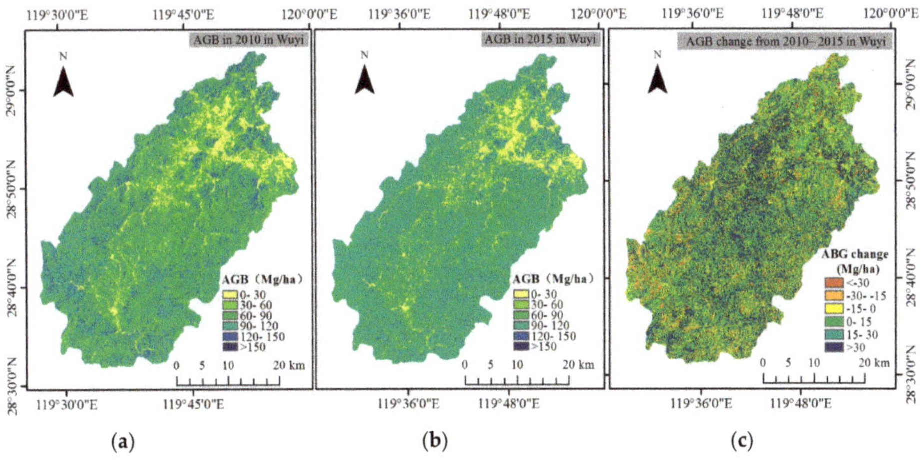

(a) (b) (c)

Figure 6. *Cont.*

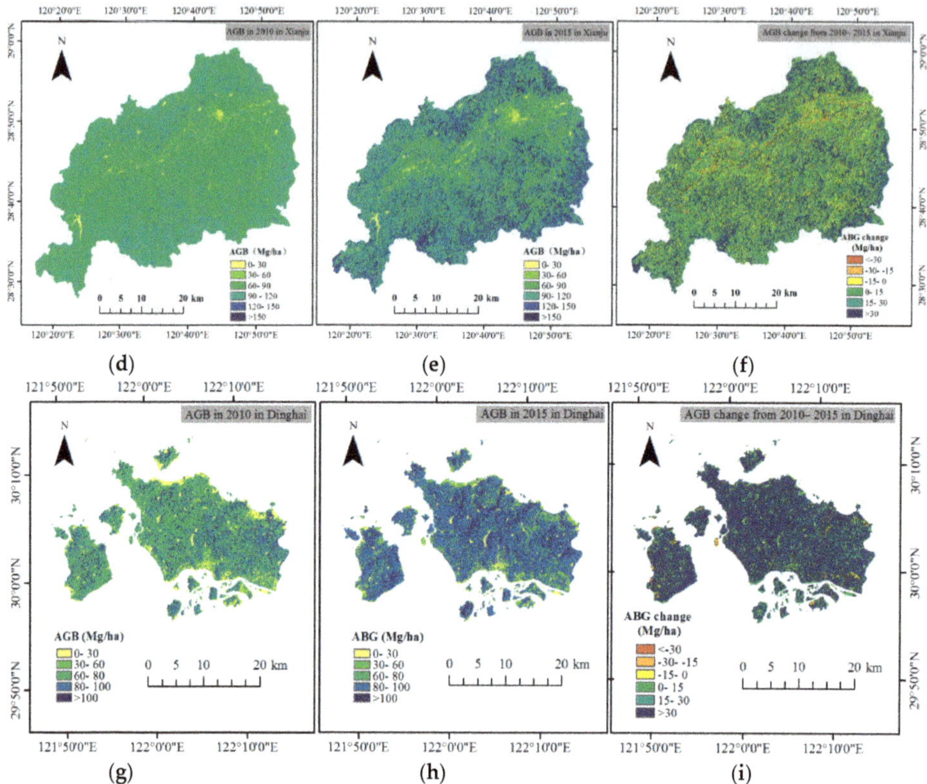

Figure 6. AGB and change maps in the three regions: (**a**) AGB in 2010 in Wuyi; (**b**) AGB in 2015 in Wuyi; (**c**) AGB change from 2010–2015 in Wuyi; (**d**) AGB in 2010 in Xianju; (**e**) AGB in 2015 in Xianju; (**f**) AGB change from 2010–2015 in Xianju; (**g**) AGB in 2010 in Dinghai; (**h**) AGB in 2015 in Dinghai; (**i**) AGB from 2010–2015 in Dinghai.

3.4. Spatiotemporal Biomass Change in the Three Regions

3.4.1. AGB Change in Wuyi County

Figure 7 showed that the relations between AGB/change and terrain features in Wuyi County had different trends. When the elevation became higher, AGB gradually grew until the elevation reached 900 m, then it started to decrease with higher elevation. This situation took place both in 2010 and 2015. Although in general, AGB increased with higher elevation, the magnitude of biomass increase was getting smaller (Figure 7a). In terms of slope (Figure 7b), a higher slope possessed higher biomass values, but a steeper slope made lower AGB increase, especially when the slope was higher than 45°; AGB in 2015 was obviously smaller than that in 2010. As for the aspect, Figure 7c shows that the mean values of AGB were almost similar in both years; even the change during this period in different aspects was almost the same. The radar chart (Figure 7d) provides more information to understand the AGB distribution characteristics for eight aspects, where all the octagons had nearly equal angles. No significant AGB difference could be found for different aspects, which was verified by the regression methods displayed in Table 4. It was also demonstrated that the magnitude of AGB increase had significant negative correlations with elevation and slope.

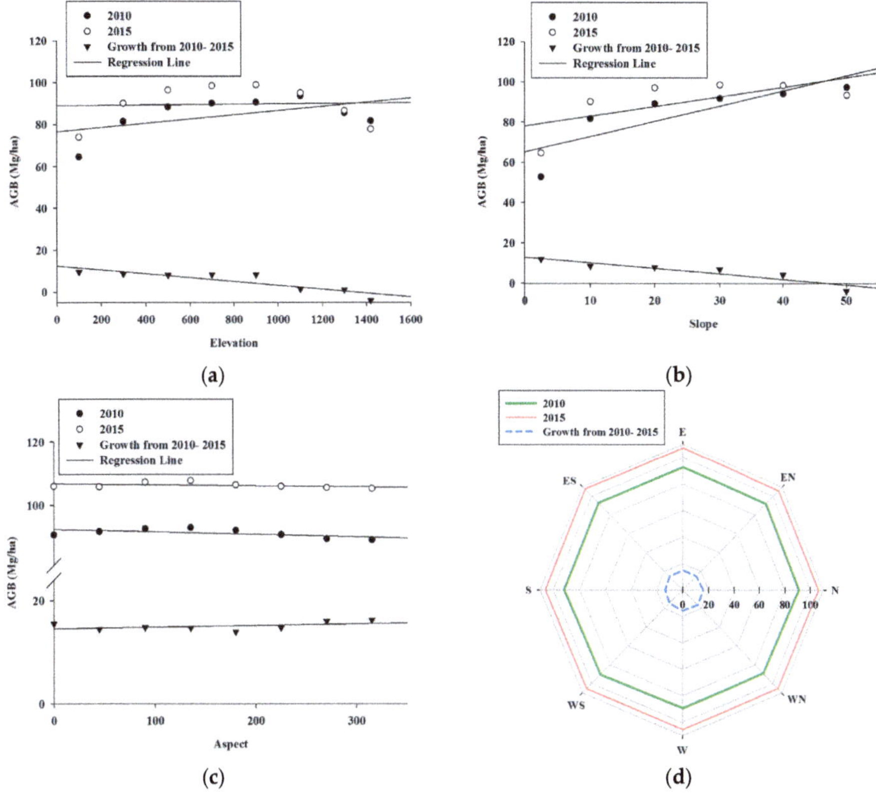

Figure 7. AGB/change within the stratified topography in Wuyi County: (**a**) AGB/change with elevation; (**b**) AGB/change with slope; (**c**) AGB/change with aspect; (**d**) radar chart of AGB/change with aspect.

Table 4. Regression results between AGB/change and terrain features in Wuyi County.

Year	Elevation			Slope			Aspect		
	R^2	*p*-Value	Sig.	R^2	*p*-Value	Sig.	R^2	*p*-Value	Sig.
2010	0.28	0.1817	–	0.69	0.0393	*	0.35	0.1243	–
2015	0.00	0.1470	–	0.45	0.1470	–	0.16	0.3330	–
2010–2015	0.77	0.0044	**	0.85	0.0085	**	0.20	0.2658	–

Note: * significant at the 0.05 level, ** significant at the 0.01 level, – insignificant.

3.4.2. AGB Change in Xianju County

The relation between AGB/change and terrain features in Xianju County is displayed in Figure 8. Figure 8a,b indicates that biomass increased with higher elevation and a steeper slope both in 2010 and 2015, but Figure 8c states that aspect had different relationships with biomass in these two years. Combined with the results in Table 5, AGB and its increase had a positive relation with elevation and slope. Meanwhile, AGB had a significant negative relation with aspect in 2010, but the relationship became insignificant in 2015 (Figure 8c). On the basis of the result in the radar chart (Figure 8d), AGB acquired a higher increase in magnitude in the western aspects.

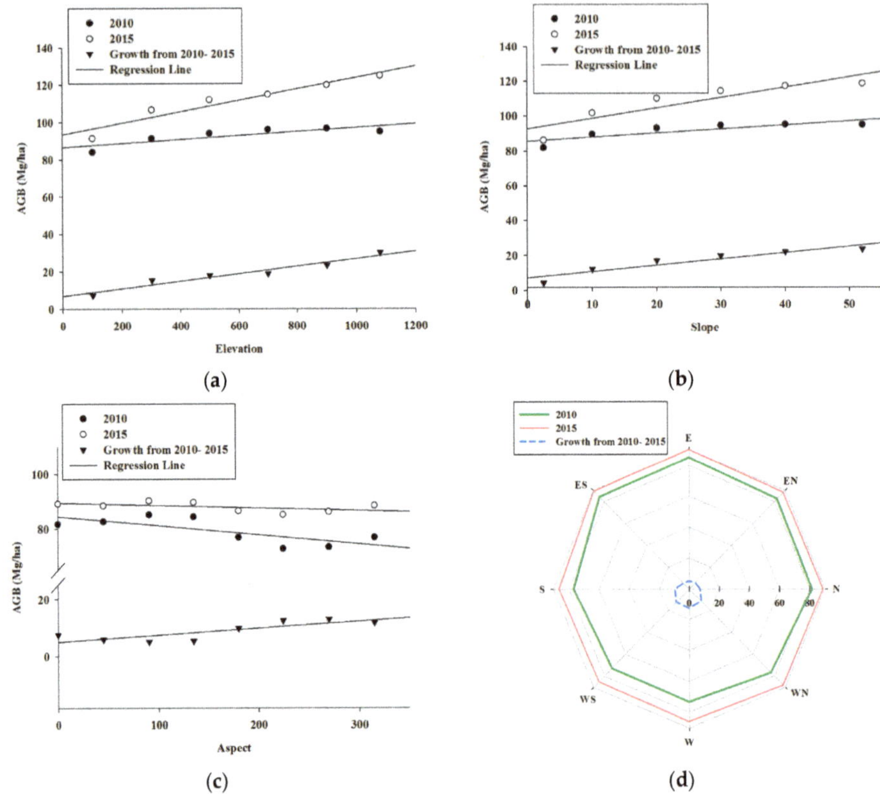

Figure 8. AGB/change within the stratified topography in Xianju County: (**a**) AGB/change with elevation; (**b**) AGB/change with slope; (**c**) AGB/change with aspect; (**d**) radar chart of AGB/change with aspect.

Table 5. Regression results between AGB/change and terrain features in Xianju County.

Year	Elevation			Slope			Aspect		
	R^2	*p*-Value	Sig.	R^2	*p*-Value	Sig.	R^2	*p*-Value	Sig.
2010	0.69	0.0410	*	0.68	0.0444	*	0.56	0.0324	*
2015	0.92	0.0026	**	0.79	0.0171	*	0.31	0.1493	–
2010–2015	0.94	0.0016	**	0.86	0.0077	**	0.65	0.0152	*

Note: * significant at the 0.05 level, ** significant at the 0.01 level, – insignificant.

3.4.3. AGB Change in Dinghai District

From Figure 9, different relations can be found between AGB/change and three topographic factors. Connecting the results of Figure 9a to Table 6, it can be seen that AGB and its change increased with high elevation, but in 2010 and from 2010–2015, the linear correlations were not significant. When inspecting the slope, AGB generally increased with higher slope in each year, but when the slope became steeper, AGB changed with a lower increase un magnitude, especially when the slope was higher than 50°; AGB in 2015 was smaller than 2010 (Figure 9b). Although Figure 9c revealed that in the south aspect, AGB had the lowest value both in 2010 and 2015, the regression results (Table 6) affirmed that aspect had no significant relations with AGB in two years, and AGB kept at a relatively steady level during this period.

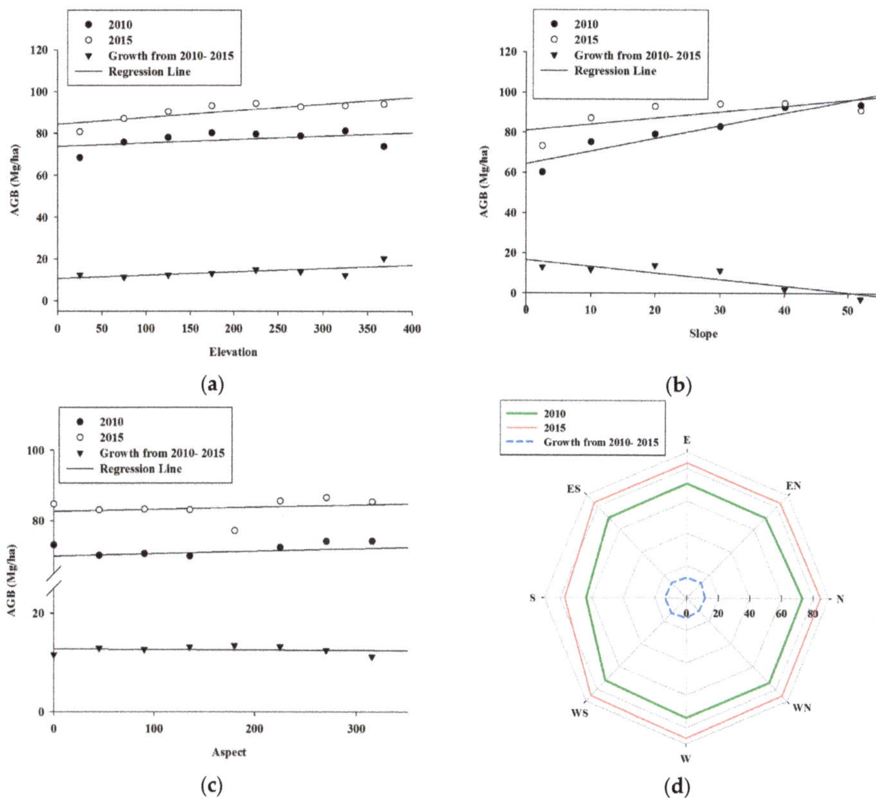

Figure 9. AGB/change within the stratified topography in Dinghai District: (**a**) AGB/change with elevation; (**b**) AGB/change with slope; (**c**) AGB/change with aspect; (**d**) radar chart of AGB/change with aspect.

Table 6. Regression results between AGB/change and terrain features in Dinghai District.

Year	Elevation			Slope			Aspect		
	R^2	*p*-Value	Sig.	R^2	*p*-Value	Sig.	R^2	*p*-Value	Sig.
2010	0.22	0.2385	–	0.90	0.0041	**	0.05	0.5934	–
2015	0.69	0.0109	*	0.49	0.1217	–	0.05	0.5822	–
2010–2015	0.45	0.0693	–	0.78	0.0202	*	0.01	0.7868	–

Note: * significant at the 0.05 level, ** significant at the 0.01 level, – insignificant.

3.4.4. Comparison of AGB/Change in Three Regions

To investigate the AGB difference between the three terrain regions scientifically, the values of AGB/change were compared under the same assessment system using a unified stratified classification (Figure 10). Figure 10a shows that the mountains in Wuyi County had the highest elevation beyond 1000 m and Dinghai District held the lowest elevation. However, Xianju County always possessed the highest AGB in each subclass with a considerable magnitude of increase. AGB change in Wuyi in the period from 2010–2015 showed a decrease tendency when the elevation was higher, while in the other two regions, AGB obtained the opposite trend. In terms of slope, Xianju stood out for its notably higher AGB in 2015 among all the stratifications. Besides, when the slope became steeper, AGB increased with less magnitude in Wuyi and Dinghai, and it even became a negative number when the slope was larger than 45 degrees. However, the increase of AGB in Xianju kept growing during this period, independent of the change of slope (Figure 10b). As for the aspect, it can be observed from Figure 10c that Wuyi had the highest AGB and Dinghai had the lowest AGB in each aspect, both in 2010 and 2015. Furthermore, three regions all achieved increased AGB in all aspects, with Wuyi having the highest magnitude, but Xianju holding the least growth in each aspect.

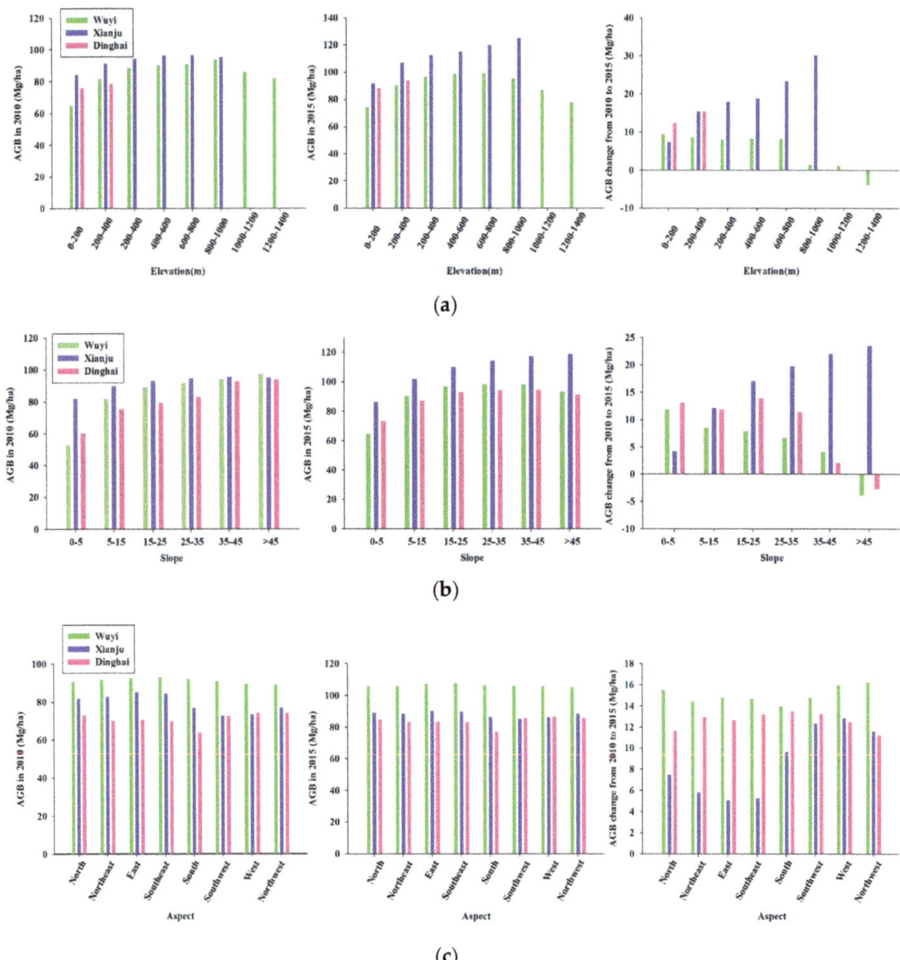

Figure 10. AGB/change within the stratified topography in the three regions: (**a**) AGB/change with stratified elevation; (**b**) AGB/change with stratified slope; (**c**) AGB/change with stratified aspect.

4. Discussion

4.1. Comparison of Variable Importance

The relative importance of different variables involved in the modeling process is normalized and compared in Figure 3. Among the most important variables in 2015, except for that in Xianju, SWIR (OLI Band 6) held relatively higher score values of %IncMSE, which can be explained by the reason that SWIR is more sensitive to moisture and shade components inherent in the forest stand structure and less impacted by atmospheric conditions [6,37]. Moreover, NDVIc incorporating SWIR acquired better performance in 2010 than it did in 2015 for all regions, which suggested that it would be apt to support the hypothesis that this variable was more suitable for open forest stands [28,38]. The preceding presentation showed that AGB generally increased from 2010–2015, while here, the relative importance of NDVIc descended on the whole. In terms of tasseled cap, its components have been widely used in biomass estimation [14,21,39]. Brightness, greenness, and wetness were successively selected as important predictor variables, similar to other Landsat-related biomass estimation research [40]. However, it has been stated that Landsat 8 has a refined near-infrared spectral band for more accurate spectral acquisition when compared to the Landsat former series [11], but the advantage in the near-infrared band of OLI Band 5 over TM Band 4 remains to be further investigated.

4.2. The Effect of Forest Policy on Biomass Spatiotemporal Variations

The total afforested area in Zhejiang Province increased from 2010–2012, but slightly decreased from 2013–2015 (Table 7). In addition, the total forestry production value increased from 2010–2015 (Figure 11). This implies that afforestation in the future will be limited by the finite area of land resources combined with rapid socioeconomic consumption, which indicates that the configuration characteristic of Zhejiang forests will change from quantity augmentation to quality improvement. Biomass is defined as the total amount of organic matter present at a given time per unit area and is the foundation of energy and nutrient exchange for forest ecosystem. Therefore, it is usually treated as an important indicator of forest quality, and the results of our previous study showed that forest policy implementation of ecological forests was beneficial for the increase of biomass [9]. Thus, proper management such as establishing nature reserves, implementing forest protection policies, and enhancing public awareness of forest ecological benefit would have a powerful effect on the spatial change of biomass.

In this context, forest protection campaigns such as plain greening and ecological forests, which promote afforestation and prohibit deforestation, will provide a better environment for biomass accumulation and in some way explain the general increasing tendency of AGB in our study area. The government of Wuyi County delimited the ecological forest in 2001, and up to September 2010, the provincial ecological forest reached 42,894.89 ha, which accounted for 44.72% of the total forested land area in Wuyi [41]. It holds a larger proportion of mountain areas with relatively high forest age that promote higher biomass density. Comparably, there has been 18,996.21 ha of ecological forest in Dinghai District until 2015, holding a proportion of 64.2% [42]. Additionally, Xianju is one of the earliest pilot counties in Zhejiang province to implement the ecological forest program. Therefore, all of these measures would facilitate biomass increase.

Table 7. The total afforested area in Zhejiang Province from 2010–2015.

Year	2010	2011	2012	2013	2014	2015
Total Afforested Area (km^2)	15.21	40.47	43.92	42.36	39.40	32.02

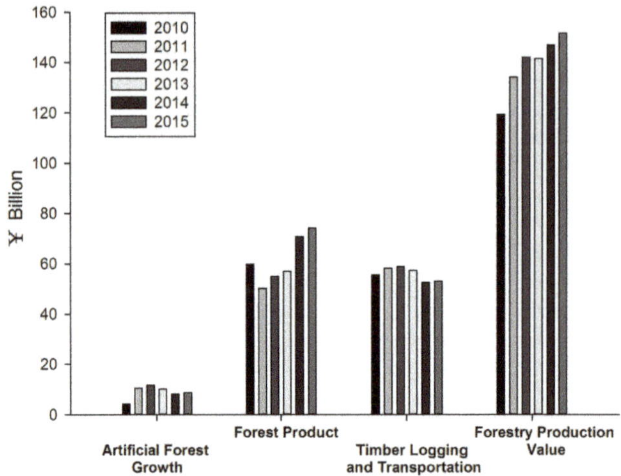

Figure 11. Forestry production value change in Zhejiang Province from 2010–2015.

4.3. The Terrain Impact on Biomass Distribution and Change

Biomass distribution is effected by many factors. In this study, we focus on the relationship between AGB and three important topographic factors including elevation, slope, and aspect among different regions. By summing up the above results, some interesting points have been found.

For the elevation feature, AGB in the three regions had a positive correlation with elevation more or less, especially for Xianju County. At the same time, it should be pointed out that when the elevation was higher, the AGB change obtained quite different trends in the three regions, where Wuyi, Xianju, and Dinghai had negative, positive, and insignificant correlations with elevation, respectively. As stated in the study of Zhang et al. [43], the property of main local tree species largely determined the characteristics of biomass spatial and temporal variation. There is a great percentage of broad forest in Xianju County, especially for the locations in the ecological forest, whose community composition containing large broadleaf species is mainly distributed at higher altitudes with better hydrothermal conditions. This may be the reason why in this region, AGB increases faster in higher mountains than that in flat areas. However, in Wuyi, a possible reason to explain why lower regions have a larger magnitude of AGB increase is that forest management activities like plain greening and ecological forest construction have been implemented [41]. Furthermore, the relatively limited elevation with a highest elevation below 400 m and the surrounding sea and ocean environment lead AGB in Dinghai to have a weakened relation with elevation.

With regard to slope, AGB/change in the three regions had significant relations with slope to varying degrees. Both in 2010 and 2015, AGB in Wuyi, Xianju, and Dinghai had a positive relation with slope. Du et al. assumed that vegetation distributed in higher slopes avoided the frequent intervention of human activity, and could be better preserved, contributing to more plentiful forest growth that promoted biomass accumulation [25]. However, in Wuyi and Dinghai, AGB acquired a negative increase when the slope was larger than 45°.

When it comes to aspect, it seems that aspect only had a significant correlation with AGB/change to some extent. Aspects in the south, southwest, west, and northwest are called sunny slopes [44,45], and AGB in these aspects achieved a larger magnitude of increase in Xianju. However, AGB at different aspects in Wuyi and Dinghai had no distinct difference. Comparably, aspect had the lowest relation with AGB and its change among the three topographic factors.

When comparing the relative importance of the three terrain parameters, elevation and slope both played significant roles in temporal AGB change. Although all features influenced AGB in

Xianju, a typical mountainous area, elevation was at the top of the list. By comparison, slope was inclined to be the most important determinant for the island region of Dinghai. In Wuyi, slope also had a relatively more significant relation with AGB. Despite this, at the provincial level, Du et al. found that forest carbon density increased with higher altitude in Zhejiang Province [25]. The situation becomes complicated when the provincial scale changed to smaller regional scales. In our study, unambiguous principles to explain the AGB-terrain relationship in the three regions are still needed in further investigations, as a number of studies has found that topography is closely related to solar radiation, temperature, moisture, and soil condition, all these interrelate with vegetation growth and human activities [9,46,47]. The objective of this study was to reveal the relationship between AGB/change and topography factors, while the spatiotemporal characteristics of AGB should be explained by the combination of natural condition and anthropogenic behavior.

4.4. Future Works

Although the spatial AGB maps in different periods were produced and the characteristics of AGB/change were analyzed, further studies should be conducted in the future. First, TM images for Wuyi County (in May and October) and Xianju County (in 2007) were an expedient selection restricted by the coverage of clouds, as mentioned above. Strictly speaking, remote sensing data should be selected by referring to field data to ensure that both were collected from the same period. Second, total biomass should include tree trunks, branches, and foliage, but the optical imagery we used captured only the signals from the vegetation canopy, and even contained noise from soil and other environmental backgrounds, which resulted in great uncertainty in the biomass estimation. Lidar has been widely used as it has the ability to provide vertical information that is closely related to AGB [48,49]. Third, the random forest algorithm was used to produce the AGB map and rank the relative importance of selected variables, but it is still a black box in which the interaction mechanism between remote sensing data and forest biochemical parameters is unrevealed. Approaches with more distinct mechanisms can be explored. Fourth, limited by our available forest investigation dataset, a typical region representing plain terrain in the north of Zhejiang has not be included to form a more comprehensive comparison. Fifth, time series biomass estimation should be used instead of bi-temporal biomass change to find more valuable and detailed information [50]. It should be noted that in this study, the mean aboveground biomass density in Dinghai district was the lowest, but when investigating the industrial structure in three regions, Dinghai had the highest GDP (gross domestic product) during 2010–2015 (Figure 12). Therefore, more frequent remote sensing imagery can be explored to find out whether a correlation exists between biomass and socioeconomic factors.

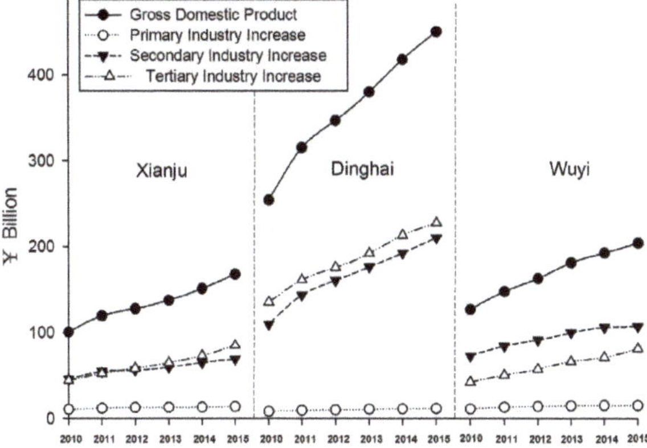

Figure 12. Industrial structure change in study area from 2010–2015.

5. Conclusions

The distribution of aboveground biomass changes with various natural and anthropogenic environment conditions. Remote sensing provides a nondestructive and efficient way to describe the temporal and spatial characteristics of this information and has earned increased attention in recent years. In this study, Landsat imagery covering three regions in Zhejiang Province, China, was collected, representative of different topographic regions including basin, island, and mountain areas. Combined with field-measured plots, bi-temporal aboveground maps (2010 and 2015) were produced based on the random forest algorithm. The spatial distributions and changes in AGB were investigated and analyzed through establishing stratified topographic categories based on DEM data. As a result, the biomass in all regions increased from 2010–2015. In the basin region that had more frequent human activities, forests in lower-altitude regions had higher biomass increases. In the mountain region, AGB was uppermost influenced by elevation, and forests at higher elevations acquired both a higher value and an increased magnitude. For the islands, with limited elevation and water surrounding environment, the dominant influence for AGB/change was inclined to be the slope. Comparatively speaking, aspect had the weakest relation with AGB. More works should be done to clarify the complex relationships between AGB and diverse terrain conditions.

The local government of Zhejiang Province has taken such actions as plain greening and ecological forest implementation to promote the increase of forest coverage and prevent deforestation, which is beneficial to regional forest ecosystems. In the future, with the excavation of limited potential forests, the improvement of forest quality should be the focus of forest management. Biomass is an important parameter for forest ecosystem assessment. As shown by this study, the interaction between natural conditions and human activities, which cannot be completely separated, exerts considerable influence on local forests. Remote sensing-based biomass estimation, with especial attention paid to regional heterogeneity, can provide scientific guidance for natural resource management.

Author Contributions: Conceptualization, A.S. and C.W.; methodology, C.W.; investigation, A.S., B.J., W.Y. and C.W.; data curation, A.S., C.W., S.H., E.Z. and Y.L.; writing, original draft preparation, A.S. and C.W.; writing, review and editing, J.D. and K.W.; funding acquisition, A.S., C.W., B.J., W.Y. and C.W.

Funding: This research was funded by the Key Research & Development Program of Zhejiang Province, China (No. 2017C02028) and the Open Fund of the Institute of Agricultural Remote Sensing and Information Technology of Zhejiang Province (No. ZJRS-2017004).

Acknowledgments: The authors are appreciative of the USGS and NASA for the open archives of Landsat imagery, and we would like to acknowledge the R Development Team for the open-source package used for the statistical analysis. The authors thank Qiming Zheng at Zhejiang University for his constructive comments, suggestions, and help in enhancing this manuscript. The authors thank the Editor and two anonymous reviewers for their constructive comments, suggestions and help in enhancing the manuscript.

Conflicts of Interest: The authors declare no conflict of interest.

References

1. Pan, Y.; Birdsey, R.A.; Fang, J.; Houghton, R.; Kauppi, P.E.; Kurz, W.A.; Phillips, O.L.; Shvidenko, A.; Lewis, S.L.; Canadell, J.G. A large and persistent carbon sink in the world's forests. *Science* **2011**, *333*, 988–993. [CrossRef]

2. Shuman, J.K.; Shugart, H.H.; Krankina, O.N. Assessment of carbon stores in tree biomass for two management scenarios in Russia. *Environ. Res. Lett.* **2013**, *8*, 045019. [CrossRef]

3. Lu, D.; Chen, Q.; Wang, G.; Liu, L.; Li, G.; Moran, E. A survey of remote sensing-based aboveground biomass estimation methods in forest ecosystems. *Int. J. Digit. Earth* **2014**, 1–43. [CrossRef]

4. Pflugmacher, D.; Cohen, W.B.; Kennedy, R.E.; Yang, Z. Using Landsat-derived disturbance and recovery history and lidar to map forest biomass dynamics. *Remote Sens. Environ.* **2014**, *151*, 124–137. [CrossRef]

5. Fassnacht, F.E.; Hartig, F.; Latifi, H.; Berger, C.; Hernández, J.; Corvalán, P.; Koch, B. Importance of sample size, data type and prediction method for remote sensing-based estimations of aboveground forest biomass. *Remote Sens. Environ.* **2014**, *154*, 102–114. [CrossRef]

6. Zhao, P.; Lu, D.; Wang, G.; Wu, C.; Huang, Y.; Yu, S. Examining Spectral Reflectance Saturation in Landsat Imagery and Corresponding Solutions to Improve Forest Aboveground Biomass Estimation. *Remote Sens.* **2016**, *8*, 469. [CrossRef]

7. Wang, X.; Shao, G.; Chen, H.; Lewis, B.J.; Qi, G.; Yu, D.; Zhou, L.; Dai, L. An Application of Remote Sensing Data in Mapping Landscape-Level Forest Biomass for Monitoring the Effectiveness of Forest Policies in Northeastern China. *Environ. Manag.* **2013**, *52*, 612–620. [CrossRef]

8. Roy, D.P.; Wulder, M.A.; Loveland, T.R.; Woodcock, C.E.; Allen, R.G.; Anderson, M.C.; Helder, D.; Irons, J.R.; Johnson, D.M.; Kennedy, R.; et al. Landsat-8: Science and product vision for terrestrial global change research. *Remote Sens. Environ.* **2014**, *145*, 154–172. [CrossRef]

9. Wu, C.; Shen, H.; Wang, K.; Shen, A.; Deng, J.; Gan, M. Landsat Imagery-Based Above Ground Biomass Estimation and Change Investigation Related to Human Activities. *Sustainability* **2016**, *8*, 159. [CrossRef]

10. Zhu, X.; Liu, D. Improving forest aboveground biomass estimation using seasonal Landsat NDVI time-series. *ISPRS J. Photogramm.* **2015**, *102*, 222–231. [CrossRef]

11. Dube, T.; Mutanga, O. Evaluating the utility of the medium-spatial resolution Landsat 8 multispectral sensor in quantifying aboveground biomass in uMgeni catchment, South Africa. *ISPRS J. Photogramm.* **2015**, *101*, 36–46. [CrossRef]

12. Baret, F.; Buis, S. Estimating Canopy Characteristics from Remote Sensing Observations: Review of Methods and Associated Problems. In *Advances in Land Remote Sensing*; Springer: Dordrecht, The Netherlands, 2008.

13. Verrelst, J.; Camps-Valls, G.; Muñoz-Marí, J.; Rivera, J.P.; Veroustraete, F.; Clevers, J.G.P.W.; Moreno, J. Optical remote sensing and the retrieval of terrestrial vegetation bio-geophysical properties—A review. *ISPRS J. Photogramm. Remote Sens.* **2015**, *108*, 273–290. [CrossRef]

14. Wu, C.; Tao, H.; Zhai, M.; Lin, Y.; Wang, K.; Deng, J.; Shen, A.; Gan, M.; Li, J.; Yang, H. Using nonparametric modeling approaches and remote sensing imagery to estimate ecological welfare forest biomass. *J. For. Res.* **2017**, 151–161. [CrossRef]

15. Jachowski, N.R.A.; Quak, M.S.Y.; Friess, D.A.; Duangnamon, D.; Webb, E.L.; Ziegler, A.D. Mangrove biomass estimation in Southwest Thailand using machine learning. *Appl. Geogr.* **2013**, *45*, 311–321. [CrossRef]

16. Latifi, H.; Fassnacht, F.E.; Hartig, F.; Berger, C.; Hernández, J.; Corvalán, P.; Koch, B. Stratified aboveground forest biomass estimation by remote sensing data. *Int. J. Appl. Earth Obs.* **2015**, *38*, 229–241. [CrossRef]

17. Li, M.; Im, J.; Beier, C. Machine learning approaches for forest classification and change analysis using multi-temporal landsat tm images over huntington wildlife forest. *Mapp. Sci. Remote Sens.* **2013**, *50*, 361–384. [CrossRef]

18. Dube, T.; Mutanga, O.; Elhadi, A.; Ismail, R. Intra-and-Inter Species Biomass Prediction in a Plantation Forest: Testing the Utility of High Spatial Resolution Spaceborne Multispectral RapidEye Sensor and Advanced Machine Learning Algorithms. *Sensors* **2014**, *14*, 15348–15370. [CrossRef] [PubMed]

19. Guo, Y.; Li, Z.; Zhang, X.; Chen, E.; Bai, L.; Tian, X.; He, Q.; Feng, Q.; Li, W. Optimal Support Vector Machines for Forest Above-ground Biomass Estimation from Multisource Remote Sensing Data. In Proceedings of the IEEE International Symposium on Geoscience and Remote Sensing IGARSS, Munich, Germany, 22–27 July 2012; pp. 6388–6391.

20. Wu, C.; Shen, H.; Shen, A.; Deng, J.; Gan, M.; Zhu, J.; Xu, H.; Wang, K. Comparison of machine-learning methods for above-ground biomass estimation based on Landsat imagery. *J. Appl. Remote Sens.* **2016**, *10*, 035010. [CrossRef]

21. Powell, S.L.; Cohen, W.B.; Healey, S.P.; Kennedy, R.E.; Moisen, G.G.; Pierce, K.B.; Ohmann, J.L. Quantification of live aboveground forest biomass dynamics with Landsat time-series and field inventory data: A comparison of empirical modeling approaches. *Remote Sens. Environ.* **2010**, *114*, 1053–1068. [CrossRef]

22. Belgiu, M.; Drăguţ, L. Random forest in remote sensing: A review of applications and future directions. *ISPRS J. Photogramm.* **2016**, *114*, 24–31. [CrossRef]

23. Sattler, D.; Murray, L.T.; Kirchner, A.; Lindner, A. Influence of soil and topography on aboveground biomass accumulation and carbon stocks of afforested pastures in South East Brazil. *Ecol. Eng.* **2014**, *73*, 126–131. [CrossRef]

24. Lee, S.; Lee, D.; Yoon, T.K.; Salim, K.A.; Han, S.; Yun, H.M.; Yoon, M.; Kim, E.; Lee, W.K.; Davies, S.J. Carbon stocks and its variations with topography in an intact lowland mixed dipterocarp forest in Brunei. *J. Ecol. Environ.* **2015**, *38*, 75–84. [CrossRef]

25. Du, Q.; Xu, J.; Wang, J.; Zhang, F.; Ji, B. Correlation between forest carbon distribution and terrain elements of altitude and slope. *J. Zhejiang A F Univ.* **2013**, *30*, 330–335.

26. Yuan, W.; Jiang, B.; Ge, Y.; Zhu, J.; Shen, A. Study on Biomass Model of Key Ecological Forest in Zhejiang Province. *J. Zhejiang For. Sci. Technol.* **2009**, *29*, 1–5.

27. U.S. Geological Survey. Available online: http://glovis.usgs.gov (accessed on 20 July 2016).

28. Nemani, R.; Pierce, L.; Running, S.; Band, L. Forest ecosystem processes at the watershed scale: Sensitivity to remotely-sensed Leaf Area Index estimates. *Int. J. Remote Sens.* **1993**, *14*, 2519–2534. [CrossRef]

29. Crist, E.P.; Cicone, R.C. A Physically-Based Transformation of Thematic Mapper Data—The TM Tasseled Cap. *IEEE Trans. Geosci. Remote* **1984**, *GE-22*, 256–263. [CrossRef]

30. Zhang, J.; Huang, S.; Hogg, E.H.; Lieffers, V.; Qin, Y.; He, F. Estimating spatial variation in Alberta forest biomass from a combination of forest inventory and remote sensing data. *Biogeosciences* **2014**, *11*, 2793–2808. [CrossRef]

31. Mutanga, O.; Adam, E.; Cho, M.A. High density biomass estimation for wetland vegetation using WorldView-2 imagery and random forest regression algorithm. *Int. J. Appl. Earth Obs.* **2012**, *18*, 399–406. [CrossRef]

32. Löw, F.; Knöfel, P.; Conrad, C. Analysis of uncertainty in multi-temporal object-based classification. *ISPRS J. Photogramm. Remote Sens.* **2015**, *105*, 91–106. [CrossRef]

33. Gessner, U.; Machwitz, M.; Conrad, C.; Dech, S. Estimating the fractional cover of growth forms and bare surface in savannas. A multi-resolution approach based on regression tree ensembles. *Remote Sens. Environ.* **2013**, *129*, 90–102. [CrossRef]

34. Kuhn, M. Building Predictive Models in R Using the caret Package. *J. Stat. Softw.* **2008**, *28*, 1–26. [CrossRef]

35. Pan, Y. Spatiotemporal Dynamics of Island Urbanization in Response to Integrated Ocean and Coastal Development. Ph.D. Thesis, Zhejiang University, Hangzhou, China, 2016.

36. Zhang, Q.; Gao, W.; Su, S.; Weng, M.; Cai, Z. Biophysical and socioeconomic determinants of tea expansion: Apportioning their relative importance for sustainable land use policy. *Land Use Policy Int. J. Cover. All Asp. Land Use* **2017**, *68*, 438–447. [CrossRef]

37. Gao, Y.; Lu, D.; Li, G.; Wang, G.; Chen, Q.; Liu, L.; Li, D. Comparative analysis of modeling algorithms for forest aboveground biomass estimation in a subtropical region. *Remote Sens.* **2018**, *10*, 627. [CrossRef]

38. Zheng, D.; Rademacher, J.; Chen, J.; Crow, T.; Bresee, M.; Le Moine, J.; Ryu, S.-R. Estimating aboveground biomass using landsat 7 etm+ data across a managed landscape in northern wisconsin, USA. *Remote Sens. Environ.* **2004**, *93*, 402–411. [CrossRef]

39. Sadeghi, Y.; St-Onge, B.; Leblon, B.; Prieur, J.-F.; Simard, M. Mapping boreal forest biomass from a srtm and tandem-x based on canopy height model and landsat spectral indices. *Int. J. Appl. Earth Obs. Geoinf.* **2018**, *68*, 202–213. [CrossRef]

40. Nguyen, T.; Jones, S.; Soto-Berelov, M.; Haywood, A.; Hislop, S. A comparison of imputation approaches for estimating forest biomass using landsat time-series and inventory data. *Remote Sens.* **2018**, *10*, 1825. [CrossRef]

41. Hu, X. Study on the Ecological & Social Benefits of Non-Commercial Forest in Wuyi County. Master's Thesis, Zhejiang A & Fu University, Hangzhou, China, 2011.

42. Xu, C. First exploration of public welfare forest construction and management in Dinghai Distinct. *China For. Ind.* **2016**, 274.

43. Zhang, J.; Gao, H.; Ying, B.; Wang, J.; Yuan, W.; Zhu, J.; Yi, L.; Jiang, B. The biomass dynamic analysis of public waifare forest in Xianju county of Zhejiang province. *J. Nanjing For. Univ. (Nat. Sci. Ed.)* **2011**, *35*, 147–150.

44. Fan, Y.; Zhou, G.; Shi, Y.; Du, H.; Zhou, Y.; Xu, X. Effects of terrain on stand structure and vegetation carbon storage of phyllostachys edulis forest. *Sci. Silvae Sin.* **2013**, *49*, 177–182.

45. Li, P.; Wei, X.; Tang, M. Forest site classification based on nfi and dem in zhejiang province. *J. Southwest For. Univ.* **2018**, *38*, 137–144.

46. Flores, A.N.; Ivanov, V.Y.; Entekhabi, D.; Bras, R.L. Impact of hillslope-scale organization of topography, soil moisture, soil temperature, and vegetation on modeling surface microwave radiation emission. *IEEE Trans. Geosci. Remote Sens.* **2009**, *47*, 2557–2571. [CrossRef]

47. Ai, Z.; He, L.; Xin, Q.; Yang, T.; Liu, G.; Xue, S. Slope aspect affects the non-structural carbohydrates and c:N:P stoichiometry of artemisia sacrorum on the loess plateau in china. *Catena* **2017**, *152*, 9–17. [CrossRef]

48. Laurin, G.V.; Puletti, N.; Chen, Q.; Corona, P.; Papale, D.; Valentini, R. Above ground biomass and tree species richness estimation with airborne lidar in tropical ghana forests. *Int. J. Appl. Earth Obs. Geéoinf.* **2016**, *52*, 371–379. [CrossRef]

49. Ahmed, O.S.; Franklin, S.E.; Wulder, M.A.; White, J.C. Characterizing stand-level forest canopy cover and height using Landsat time series, samples of airborne LiDAR, and the Random Forest algorithm. *ISPRS J. Photogramm.* **2015**, *101*, 89–101. [CrossRef]

50. Gómez, C.; White, J.C.; Wulder, M.A.; Alejandro, P. Historical forest biomass dynamics modelled with Landsat spectral trajectories. *ISPRS J. Photogramm.* **2014**, *93*, 14–28. [CrossRef]

© 2018 by the authors. Licensee MDPI, Basel, Switzerland. This article is an open access article distributed under the terms and conditions of the Creative Commons Attribution (CC BY) license (http://creativecommons.org/licenses/by/4.0/).

 *forests*

Article

Improving Forest Aboveground Biomass (AGB) Estimation by Incorporating Crown Density and Using Landsat 8 OLI Images of a Subtropical Forest in Western Hunan in Central China

Chao Li [1,2], Yingchang Li [1,2] and Mingyang Li [1,2,*]

[1] College of Forestry, Nanjing Forestry University, Nanjing 210037, Jiangsu, China;
    gislichao@hotmail.com (C.L.); lychang312@126.com (Y.L.)
[2] Co-Innovation Center for Sustainable Forestry in Southern China, Nanjing Forestry University,
    Nanjing 210037, Jiangsu, China
* Correspondence: lmy196727@njfu.edu.cn; Tel.: +86-025-8542-7327

Received: 17 December 2018; Accepted: 25 January 2019; Published: 29 January 2019

Abstract: Forest aboveground biomass (AGB) estimation modeling based on remote sensing is an important method for large-scale biomass estimation; the accuracy of the estimation models has been a topic of broad and current interest. In this study, we used permanent sample plot data and Landsat 8 Operational Land Imager (OLI) images of western Hunan. Remote-sensing-based models were developed for different vegetation types, and different crown density classes were incorporated. The linear model, linear dummy variable model, and linear mixed-effects model were used to determine the most effective and accurate method for remote-sensing-based AGB estimation. The results show that the adjusted coefficient of determination (R^2_{adj}) and root mean square error (RMSE) of the linear dummy model and linear mixed-effects model were significantly better than those of the linear model; the R^2_{adj} increased more than 0.16 and the RMSE decreased more than 2.12 for each vegetation type, and the F-test also showed significant differences between the linear model and linear dummy variable model and between the linear model and linear mixed-effects model. The accuracies of the AGB estimations of the linear dummy variable model and the linear mixed-effects model were significantly better than those of linear model in the thin and dense crown density classes. There were no significant differences in the AGB estimation performance between the linear dummy variable model and linear mixed-effects model; these two models were more flexible and more suitable than the linear model for remote-sensing-based AGB estimation. The results of this study provide a new approach for solving the low-accuracy estimations of linear models.

Keywords: aboveground biomass estimation; remote sensing; crown density; low-accuracy estimation; model comparison

---

1. Introduction

Forest ecosystems provide important ecosystem services and are an important component of the earth's energy cycle. Forest biomass is a fundamental parameter for describing the structure and function of forest ecosystems [1,2]. Many ecosystem processes are impacted by forest biomass and, in turn, forest biomass is impacted by these processes [3]. Forests provide important terrestrial carbon storage. Studies on forest biomass are essential for determining the carbon storage, carbon balance, and carbon cycling at the regional and global levels.

Due to difficulties in measuring forest belowground biomass, the majority of previous studies have mostly focused on forest aboveground biomass (AGB). The estimation of AGB is an essential task for assessing carbon stocks and carbon balance [4]. In past studies, three main approaches have been used to estimate forest AGB, namely: process-based ecosystem models, field measurements, and a combination of forest inventory plots and remotely sensed data [5,6]. The remote-sensing-based method has been commonly used in the last decades for several reasons: (1) Remote sensing data covers large areas, allowing for the assessment of the spatial variation of vegetation and making it possible to determine the spatial distribution and pattern of biomass in large areas and complex forest landscapes; (2) multiple sensors and multiple spatial resolutions can be used for forest biomass research at different scales; and (3) multi-temporal remote sensing images provide long-term, dynamic, and continuous AGB observations [7,8].

The rapid development of remote sensing technology has provided a wide variety of remotely sensed imagery data for AGB estimation. The data can be divided into three categories: (1) optical remote sensing data such as Landsat, Systeme Probatoire d'Observation de la Terre (SPOT), moderate-resolution imaging spectroradiometer (MODIS), QuickBird, ASTER, Advanced Very High-Resolution Radiometer (AVHRR), and China-brazil earth resource satellite (CBERS); (2) active remote sensing data including Radar and Lidar; and (3) the integration of multisource remote sensing data [5,9–13]. In particular, Landsat has been commonly used for forest biomass estimation in combination with sample plots because the images can be freely downloaded, have medium spatial (30 m × 30 m) and temporal (16 days) resolutions, and have wide coverage [14,15]. In many countries, the spatial resolution of Landsat is similar to the size of sample plots in national forest inventories, thus reducing the spatial errors in matching the pixels and the sample plots [8].

Generally, forest stands with different biomass have different forest structures and different biophysical parameters. These features are reflected in remote sensing images as different colors, structures, and textures. Using feature extraction methods, the image parameters that are closely related to forest biomass can be extracted from the remote sensing images, and forest biomass can be estimated. Vegetation information in remote sensing images is mainly reflected by the spectral characteristics. The spectral differences in leaves and vegetation canopies and their changes over time differ in different spectral bands [9,16,17]. Vegetation parameters derived from optical remote sensing include vegetation indices, leaf area index, absorbed photosynthetically active radiation (APAR), and various image transformations [18–20]. Landsat images can be used to derive spectral information that can be correlated with forest inventory AGB data [21]. The remote sensing information is strongly related to several forest parameters and the use of spectral variables in modeling forest biomass has a long history. The Landsat variables that have been commonly used include spectral bands, vegetation indices (e.g., normalized differential vegetation index (NDVI), Enhanced Vegetation Index (EVI)), image transformations (e.g., principal component analysis (PCA) and tasseled cap transformation (TCT)), and texture images [5,15,22–26].

Parametric algorithms and nonparametric algorithms have been applied for AGB estimation [27]. In parametric algorithms, it is assumed that the direct or indirect relationships between the remotely sensed parameters and the forest AGB can be expressed using regression models. The application of parametric algorithms over large areas requires the assumption of spatially homogeneous relationships between the ground-based information and remote-sensing data. Parametric algorithms are easy to apply but are weak in terms of describing the complex relationship between AGB and remote sensing data. In addition, the accuracy of the algorithms largely relies on the statistical robustness. In contrast to parametric algorithms, nonparametric algorithms do not have explicit equations [28] and do not assume a normal distribution of the independent and dependent variables. Nonparametric algorithms are more flexible to describe the nonlinear relationship between AGB and image data, but the physical mechanisms of the models are not clear and there are risks of over-fitting.

The linear model was frequently used in forest biomass estimation based on remote sensing. In previous studies, when the linear models were built for estimating AGB, the remote sensing factors were directly considered as fixed effect variables. The linear models did not consider the effects of forest characteristics, effects which may influence the independent variables and the model fitting, which in turn affect the fitting accuracy of the models. In this study, based on the analysis of the differences between the independent variables and AGB of different vegetation types in different crown density classes, the basic AGB linear models using remote sensing were built. The crown density classes which were considered as the influencing factor (random effect or dummy variable) were introduced into the model, and the linear dummy variable model and linear mixed-effects model were fitted to estimate AGB. The accuracies of the linear model, linear dummy variable model, and linear mixed-effects model were compared.

2. Materials and Methods

2.1. Study Area

The study area is located in "Greater Xiangxi", an area that borders on the Hubei, Chongqing, and Guizhou provinces in the west of Hunan Province, including Xiangxi Tujia and Miao, Zhangjiajie, and Huaihua City (Figure 1). The study area is located in a transition zone between the Yunnan-Guizhou Plateau and the Jiangnan hills where medium and low mountains account for more than 70% of the area. The climate of this region is a typical subtropical monsoon humid climate with an average annual temperature of about 16 °C and an annual precipitation of about 1400 mm. The natural conditions of this region are complex with a sensitive ecological environment, and the area is underdeveloped in terms of socioeconomic development. The area is an important forestry area in Hunan Province with abundant tree biodiversity. The forest area in the region covers more than 49,000 square kilometers and the tree harvest volume is 156,000,000 m^3. However, the distribution of the forest resources in this region is extremely uneven, the forest biomass in different stand ages is heterogeneous, and forest productivity is low [29,30].

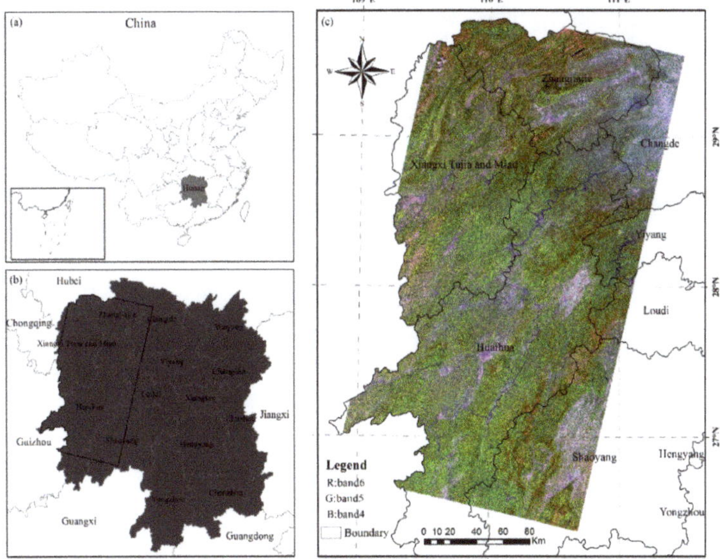

Figure 1. The location of study area: (**a**) The study area location in China; (**b**) the western Hunan in Hunan province; and (**c**) a false color composite of Landsat 8 OLI band 6 in red, band 5 in green, and band 4 in blue.

2.2. Field Survey Data

In this study, 377 fixed sample plots of typical forests were used including 125 fixed sample plots of pine forests (Pure or *Pinus massoniana* dominant forests with a small mixture of broadleaf trees and shrubs), 162 fixed sample plots of Chinese fir forests (Pure or *Cunnigjamia lanceolate* (*Lamb.*) *Hook* dominant forests with very small mixture of *Pinus massoniana* and shrubs), and 90 fixed sample plots of mixed forests (dominant species including *Pinus massoniana, Cunnigjamia lanceolate (Lamb.) Hook, Cinnamomum camphora (L.) Presl., Cupressus funebris Endl.*, and shrubs) (Figure 2). The fixed sample plots were surveyed in 2014 and the plots were systematically laid out in a grid of 4 × 8 km with a plot size of 0.067 ha (China National Forest Continuous Inventory (NFCI) Technical Regulations). The biomass conversion factor method was used to convert the stand volume into forest AGB [31,32]. The sample plots were divided into three vegetation types including pine, Chinese fir (fir), and mixed forest (mixed). The statistics of the sample plots of the crown density classes are summarized in Table 1. All of the plots had a mean AGB of 47.7 Mg/ha with a standard deviation of 30.06 Mg/ha. The mixed forest had the highest mean AGB and standard deviation, and pine forest had the lowest mean AGB and minimum AGB value (Table 1). The mean AGB values were lowest for the pine forest in each crown density class. The differences in the AGB were determined for the crown density classes: There were significant differences in the AGB of each vegetation type between the thin, medium, and dense crown density classes. The AGB of the medium plots was not significantly different from that of the average for four vegetation types.

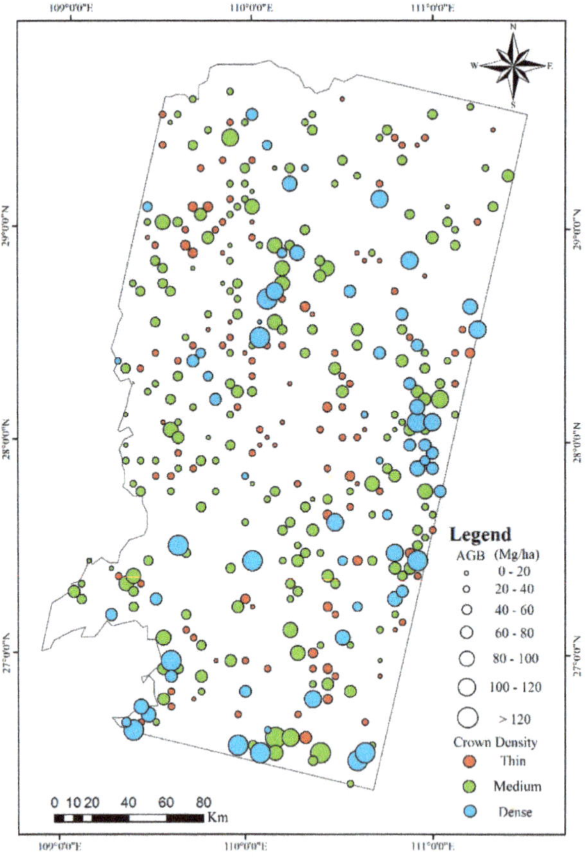

Figure 2. Spatial distribution of sampling plots corresponding to plots of aboveground biomass (AGB) values and crown density class across the western Hunan.

Table 1. The basic statistics of the sample plots by crown density classes and vegetation types.

Vegetation Type	Crown Density	AGB (Mg/ha)				
		No.	Minimum	Mean	Maximum	Standard Deviation
Pine	Thin	41	1.05	16.40	47.33	10.61
	Middle	70	3.61	33.60	83.58	17.06
	Dense	14	6.16	51.17	118.07	35.57
	Total	125	1.05	29.65	118.07	20.94
Fir	Thin	54	22.76	31.11	57.46	8.70
	Middle	77	24.72	51.18	130.87	18.37
	Dense	31	55.55	92.55	154.48	30.65
	Total	162	22.76	52.41	154.48	28.68
Mixed	Thin	18	24.52	37.70	65.42	10.53
	Middle	53	31.74	62.13	131.03	24.02
	Dense	19	41.28	92.57	171.53	37.06
	Total	90	24.56	63.67	171.53	30.86
Total	Thin	113	1.05	26.65	65.42	12.77
	Middle	200	3.61	48.11	131.03	22.68
	Dense	64	6.16	83.26	171.53	37.39
	Total	377	1.05	47.70	171.53	30.06

2.3. Remote Sensing Data

In this research, two Landsat 8 Operational Land Imager (OLI) L1T product images (path/rows: 119/39 and 119/40, cloud cover <10%) acquired on 24 December 2013 were used. The first seven bands of the images were used in this study, including the Coastal band, Blue band, Green band, Red band, near-infrared (NIR) band, and two shortwave infrared (SWIR) bands. The coordinate system of the images was the Universal Transverse Mercator coordinate system with zone 49 north. The dark object subtraction method was used for atmospheric calibration [33]. ASTER global digital elevation model (GDEM) data with the same coordinate system and same spatial resolution as the OLI images were used for the topographic correction of the Landsat 8 OLI images using a C-correction approach [34]. The images were mosaicked into one image (Figure 1).

The vegetation information in remote sensing imagery were reflected by the spectral characteristics, spectral differences, and spectral changes of the vegetation canopy in different bands. Vegetation indices were used to reflect the existence, quantity, quality, state, and spatial and temporal distribution characteristics of vegetation, and biophysical properties had already been estimated by vegetation indices. The most widely used vegetation indices were based on remotely sensed data measured in visible-red and near-infrared spectral wavebands such as the normalized difference vegetation index (NDVI) [35]. Atmospherically resistant vegetation index (ARVI), soil adjusted vegetation index (SAVI), atmospherically resistant vegetation index (ARVI), and enhance vegetation index (EVI) were derived from NDVI. The results of image transformations, such as the first principal component from the Principal Component Analysis, showed stronger relationships with biomass than individual spectral bands [5]. Texture information referred to the pattern of intensity variations in the remote sensing image, and the texture based on gray level co-occurrence matrix was effective and important in describing the spatial distribution and structure information of forest.

A total of 340 spectral variables were calculated from the OLI images to fully exploit the remote sensing information, including the original image bands, vegetation indices, image transform algorithms, and grey-level co-occurrence matrix-based texture measures (Table 2) [8]. The Pearson product-moment correlation coefficient was used to analyze the relationships between AGB and the spectral variables; the spectral variables which had significant correlations with AGB were used as independent variables. A stepwise regression was used to develop the AGB linear regression models.

Table 2. Spectral variables derived from a total of seven bands for the Landsat 8 OLI image.

Spectral Variables	Definitions of Spectral Variables	No.
Original Band	b_1—coastal, b_2—blue, b_3—green (GRN), b_4—red (RED), b_5—near infrared (NIR), b_6—shortwave infrared1 (SWIR1), b_7—shortwave infrared2 (SWIR2)	7
Inversions of band$_i$	$IB_i = 1/b_i, i = 1, \ldots ,7$	7
Simple two-band ratios ($SR_{i,j}$)	$SR_{i,j} = b_i/b_j, i, j = 1, \ldots 7; i \neq j$	42
Three-band ratios	$SR_{i,j,k} = b_i/\left(b_j + b_k\right), i, j, k = 1, \ldots ,7; i \neq j \neq k, j < k$	106
Vegetation indices	Normalized difference vegetation index (*NDVI*), atmospherically resistant vegetation index (*ARVI*), soil adjusted vegetation index ($SAVI_l = (b_5 - b_4)(1 + l)/(b_5 + b_4 + l), l = 0.1$), atmospherically resistant vegetation index (*ARVI*), enhance vegetation index (*EVI*), albedo, sum of three visible bands ($VIS_{234}, VIS_{234} = b_2 + b_3 + b_4$)	7
Principal component analysis	The first 3 PCs from principal component analysis (*PCA1, PCA2, PCA3*)	3
Texture measures	Grey-level co-occurrence matrix-based texture measures of original bands (b_i), including contrast (b_{iCONj}), correlation (b_{iCORj}), dissimilarity (b_{iDISj}), entropy (b_{iENj}), homogeneity (b_{iHOj}), angular second moment (b_{iSEMj}), mean (b_{iMEj}), and variançe(b_{iVAj}) with different window sizes j ($3 \times 3, 5 \times 5, 7 \times 7$)	168

2.4. Statistical Model

In forestry research, the variables are mostly continuous variables and can be directly used for model fitting. Sometimes, categorical and qualitative variables are also needed in some studies because they may influence the model results. In modeling, these variables are considered mixed-effects or dummy variables when they are added to regression models. The sample plots were divided into three crown density classes based on the inventory data, i.e., thin (<0.4), medium (0.4 ~ 0.7), and dense ($\geq$0.7) (Figure 2). The crown density classes represented the dummy variable and mixed-effects variable in the linear regression models.

For the AGB estimation, a linear regression model (model 1) without the crown density, linear dummy variable model (model 2), and linear mixed-effects model (model 3) were fitted and compared in this study. Model 2 and model 3 were implemented by considering the dummy variables and random-effects in the linear regression model, respectively. The equations of these three models were introduced by Tang et al. and Fu et al. [36,37].

During the stepwise regression for model 1, the multi-collinearity, which creates highly sensitive parameter estimators with inflated variances and improper model selection, was assessed for each pair of the selected spectral variables using the variance inflation factor (VIF). For the linear dummy variable model and linear mixed-effects model, two methods exist to add dummy variables or random-effects to the linear model. One approach is to add them to the intercept, and another approach is to add them to all parameters (intercept and slope) of the linear model. In order to avoid multicollinearity in the linear dummy variable model and allow for the comparison of the two models, both the linear dummy variable model and linear mixed-effects model were fitted by adding dummy variables or random-effects to the intercept. Furthermore, two variance-covariance structures needed to be determined to fit the linear mixed-effects model: (1) Determine the variance-covariance structure (R matrix) of the fixed effect; and (2) determine the variance-covariance structure (D matrix) of the random effect [38,39]. In this study, the D matrix was a diagonal matrix (*pdDiag*), the R matrix was divided into two parts, the variance structure of R was a power function, and the covariance structure of R was a spherical function.

2.5. Model Fitting and Evaluating

The linear regression model, linear dummy variable model, and linear mixed-effects model were used to establish the AGB estimation models of the pine forest, fir forest, mixed forest, and all-vegetation. All models were fitted using the RStudio software.

The accuracies of the predicted AGB values for the models were evaluated using the adjusted coefficient of determination (R^2_{adj}) and the root mean square error (RMSE). The difference between model 1 and model 2 and between model 1 and model 3 were evaluated using the F-test. The residuals were analyzed to determine the AGB estimation performance of the three models in the different crown density classes. In order to compare the performance improvement of the linear model by the linear dummy variable model (model 2) and linear mixed-effects model (model 3) for AGB estimation, the accuracy of the model 1, model 2, and model 3 were assessed using the percentage root mean square error (RMSE%) and percentage mean residual deviation (Bias%) of the different crown density classes (thin, medium, dense, and total). The difference between model 2 and model 3 was also assessed.

$$R^2 = 1 - \sum_{i=1}^{n} (y_i - \hat{y}_i)^2 / \sum_{i=1}^{n} (y_i - \overline{y}_i)^2 \tag{1}$$

$$R^2_{adj} = 1 - \left(1 - R^2\right)\frac{n-1}{n-k} \tag{2}$$

$$RMSE = \sqrt{\sum_{i=1}^{n} \frac{(y_i - \hat{y}_i)^2}{n}} \tag{3}$$

$$RMSE\% = \frac{RMSE}{\overline{y}} \times 100 \tag{4}$$

$$Bias\% = \frac{\sum_{i=1}^{n} \frac{(y_i - \hat{y}_i)}{n}}{\overline{y}} \times 100 \tag{5}$$

where y_i is the observed biomass values, $\overline{y}$ is the arithmetic mean of all observed biomass values, $\hat{y}_i$ is the estimated biomass values based on models, n is the sample number, and k is the number of parameters of each model.

3. Results

The Pearson correlation coefficients between all spectral variables and the AGB were calculated and 30 variables had significant correlation with the AGB of four vegetation types. The correlation coefficients are listed in Table 3. The result showed that the correlation coefficients were not higher than 0.260 for all the 30 spectral variables, and 11 texture features had significant correlation with the AGB.

Table 3. Pearson correlation coefficients between remote sensing factors and aboveground biomass (AGB).

Variables	Correlation Coefficients	Variables	Correlation Coefficients	Variables	Correlation Coefficients	Variables	Correlation Coefficients
$b3$	-0.254 **	SR_{37}	-0.236 **	SR_{416}	-0.210 **	b_{4ME5}	-0.276 **
$b4$	-0.233 **	SR_{46}	-0.207 **	SR_{417}	-0.215 **	b_{7COR7}	0.258 **
VIS_{234}	-0.260 **	SR_{47}	-0.227 **	SR_{426}	-0.206 **	b_{3ME7}	-0.251 **
$ARVI$	0.162 *	SR_{64}	0.227 **	b_{3ME3}	-0.265 **	b_{4ME7}	-0.242 **
IB_4	0.247 **	SR_{73}	0.236 **	b_{4ME3}	-0.247 **	b_{2SEM5}	0.251 **
IB_2	0.232 **	SR_{124}	0.210 **	b_{5VA3}	0.272 **	b_{2SEM7}	0.230 **
SR_{14}	0.228 **	SR_{134}	0.204 *	b_{2COR5}	0.260 **	——	——
SR_{41}	-0.244 *	SR_{327}	-0.229 **	b_{3ME5}	-0.279 **	——	——

* Indicates a significance level of 0.05 and ** a significance level of 0.01.

Three types of models for each dependent variable (i.e., AGB of total vegetation, AGB of pine forest, AGB of fir forest, and AGB of mixed forest) were developed using the spectral variables which were selected by stepwise regression as the independent variable (Tables 4 and 5). Twelve models were obtained. Parameter estimates of models 1–3 for different vegetation types are presented in

Tables 4 and 5. The independent variables of the total vegetation AGB were dominated by the image texture information, and the independent variables of the pine, fir, and mixed forests were dominated by the image texture information and spectral features. The model standard coefficients of the linear models showed that the texture information contributed more to the AGB estimation than the spectral features, which indicated that the texture information was important for AGB estimation in this study.

Table 4. Parameter estimates of the linear model (model 1).

Vegetation Type	Parameter	Estimate	Std.coef	*p*-Value	Vegetation Type	Parameter	Estimate	Std.coef	*p*-Value
Pine	b_{2COR5}	24.14	0.33	<0.01	Fir	b_{5VA3}	1.14	0.25	<0.01
	SR_{327}	−165.54	−0.27	<0.01		IB_2	1061.00	0.61	<0.01
	b_{2SEM5}	19.47	0.17	<0.01		b_{2SEM7}	36.49	0.24	<0.01
						b_{3ME5}	6.90	0.40	<0.01
Mixed	b_{7COR7}	30.55	0.30	<0.01	Total vegetation	b_{3ME7}	−3.62	−0.25	<0.01
	b_{4ME7}	−10.05	−0.60	<0.01		b_{5VA3}	0.83	0.14	<0.01
	VIS_{234}	9.00	0.36	<0.01		b_{2SEM5}	15.36	0.09	<0.05

Table 5. Parameter estimates of the linear dummy variable model (model 2) and linear mixed-effects model (model 3).

Vegetation Type	Model 2			Vegetation Type	Model 3		
	Parameter	Estimate	S.D.	*p*-Value		Parameter	Estimate	S.D.	*p*-Value
Pine	b_{2COR5}	16.70	4.77	<0.01	Pine	b_{2COR5}	13.35	3.77	<0.01
	SR_{327}	−113.52	48.43	<0.05		SR_{327}	−118.00	21.87	<0.01
	b_{2SEM5}	16.85	6.86	<0.05		b_{2SEM5}	12.67	6.15	<0.05
Fir	b_{5VA3}	0.62	0.22	<0.01	Fir	b_{5VA3}	0.16	0.20	<0.05
	IB_2	732.08	208.43	<0.01		IB_2	515.01	172.84	<0.01
	b_{2SEM7}	17.18	7.68	<0.05		b_{2SEM7}	5.74	6.06	<0.05
	b_{3ME5}	5.29	2.08	<0.01		b_{3ME5}	4.44	1.69	<0.01
Mixed	b_{7COR7}	23.65	7.636	<0.01	Mixed	b_{7COR7}	10.55	6.79	<0.05
	b_{4ME7}	−4.80	3.665	<0.05		b_{4ME7}	−1.42	2.86	<0.05
	VIS_{234}	6.28	5.165	<0.05		VIS_{234}	0.74	4.09	<0.05
Total vegetation	b_{3ME7}	−2.20	0.66	<0.01	Total vegetation	b_{3ME7}	−1.60	0.50	<0.01
	b_{5VA3}	0.48	0.21	<0.05		b_{5VA3}	0.25	0.20	<0.05
	b_{2SEM5}	8.05	5.49	<0.05		b_{2SEM5}	1.07	4.56	<0.05

The fitting results of models 1–3 are summarized in Tables 6 and 7. For the different vegetation types, the R^2 and R^2_{adj} of model 2 and model 3 were larger than those of model 1, and the RMSE values were smaller than those of model 1. These results indicate that the performances of model 2 and model 3 were better than that of model 1. The R^2_{adj} of model 2 and model 3 for pine forest had the smallest increase compared with model 1; the value of R^2_{adj} increased by 0.16, and the RMSE values were smaller for model 2 and model 3 than for model 1. For the fir forest, model 2 and model 3 had the largest R^2_{adj} values, and compared with model 1, the values increased more than 0.39. For the mixed forest and total vegetation, the R^2_{adj} and RMSE values of model 2 and model 3 were better than those of model 1. These results show that model 2 and model 3, which were considered the crown density classes, had higher accuracies of AGB estimation than model 1.

Table 6. The model fitting results of model 1 for different vegetation types.

Vegetation Type	R^2	R^2_{adj}	RMSE	Predict Mean
Pine	0.23	0.21	18.41	29.64
Fir	0.22	0.22	25.57	52.43
Mixed	0.21	0.19	27.28	63.67
Total vegetation	0.11	0.10	28.47	47.40

Table 7. The model fitting results of model 2 (linear dummy variable model) and model 3 (linear mixed-effects model) for different vegetation types.

Vegetation Type	Model#	R^2	R^2_{adj}	RMSE	Predict Mean
Pine	2	0.41	0.40	16.05	29.65
	3	0.39	0.38	16.29	29.36
Fir	2	0.61	0.61	17.88	52.39
	3	0.61	0.61	17.92	51.95
Mixed	2	0.46	0.44	22.56	63.64
	3	0.43	0.42	23.12	62.41
Total vegetation	2	0.41	0.41	22.99	47.70
	3	0.41	0.41	23.07	47.51

To further test whether model 2 and model 3 significantly improved the accuracy of model 1, the F-test was used for determining the differences between model 1 and model 2 and between model 1 and model 3 (Table 8). The F-test results show that, except for model 3 of the mixed forest, there were significant differences between model 2 and model 1 and between model 3 and model 1. This indicated that the performances of model 2 and model 3 were significantly better than that of model 1. The fitting results of the model 2 and model 3 had no significant differences.

Table 8. The comparisons of linear models (model 1), linear dummy variable models (model 2), and linear mixed-effects models (model 3). p-Value is from the F-test used to compare the similarity of models 1–3 against the null hypothesis of no significant difference.

Vegetation Type	Model#	Models 1–3		Model 2 and Model 3	
		F-Value	p-Value	F-Value	p-Value
Pine	1				
	2	11.76	<0.01		
	3	4.12	<0.05	2.44	0.12
Fir	1				
	2	29.58	<0.01		
	3	17.31	<0.01	1.28	0.26
Mixed	1				
	2	9.37	<0.01		
	3	0.77	0.38	4.69	0.03
Total vegetation	1				
	2	111.48	<0.01		
	3	66.03	<0.01	2.95	0.09

The performance of the predictions could be explained with the scatterplots showing the relationships between the predicted AGB values and observed AGB values (Figure 3). It indicates that the overestimation and underestimation problems were obvious for the linear model (model 1) for each vegetation type. This situation, especially, became worse for all the vegetation types in thin and dense plots. For model 2 and model 3, the overestimations and underestimations in thin and dense crown density plots were alleviated for four vegetation types, and the estimates were more accurate than model 1 (Figure 3). A single-sample t-test was used to compare the model residuals of models 1, 2, and 3 (Figure 4). In model 1, there were no significant differences between the residuals and 0 for the total plots and medium crown density plots for each vegetation type (Figure 4). In the thin crown density plots, the residual values of model 1 were significantly smaller than 0, and in the dense crown density plots, the residual values of model 1 were significantly larger than 0 (Figure 4). These results indicate that there were significant inaccuracies in the AGB estimations of the thin and dense plots of model 1 (the former was overestimated and the latter was underestimated) (Figure 4). The residuals of model 2 were significantly different from 0 only in the thin and medium crown density plots for the fir forest, whereas the other three vegetation types exhibited no significant differences in each crown density class. The residuals of model 3 were not significantly different from 0 for all vegetation types for the different crown density classes (Figure 4). The residual results indicate that model 2 and model 3 had higher accuracies of AGB estimation than model 1 for the different crown density classes.

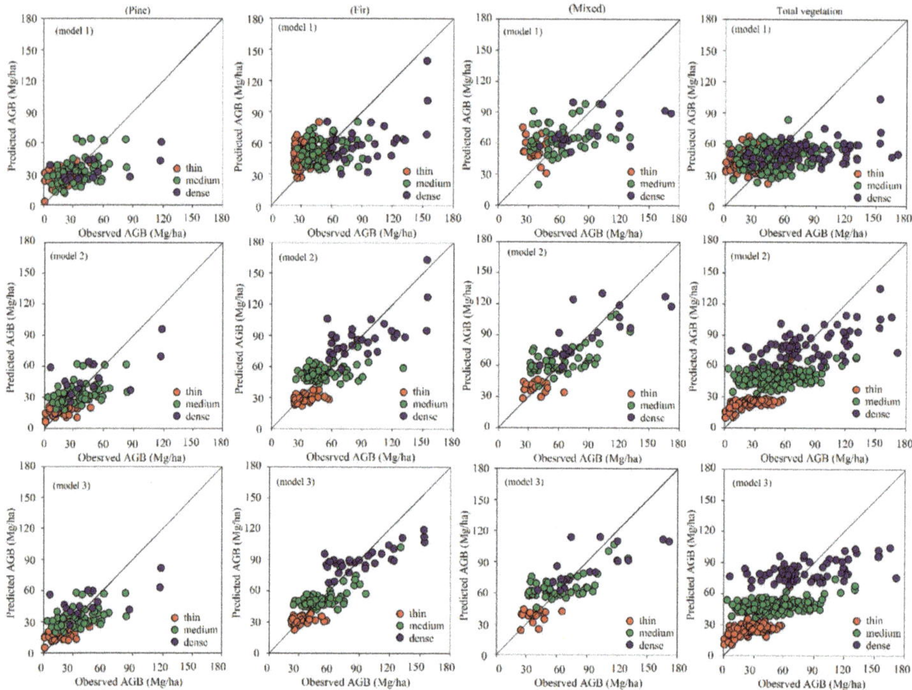

Figure 3. The relationships between predicted AGB from different models in different crown density against observed AGB for different vegetation types.

In this study, the RMSE% and Bias% of the three models of the different crown density classes were calculated for further comparison of the models (Figure 5). Generally, the RMSE% of model 2 and model 3 were lower than those of model 1 in the total plots for all vegetation types, and the differences in the RMSE% between model 1 and model 2 and between model 1 and model 3 were all significant. For the thin crown density plots, the differences in the RMSE% exceeded 27%, and both values were significantly different from the RMSE% of model 1. For the medium crown density plots, the RMSE% of model 2 and model 3 were smaller than those of model 1, but the differences between them were not significant. For the dense crown density plots, the differences in the RMSE% exceeded 5%, and the differences between model 2 and model 1 and between model 3 and model 1 were significant for the fir forest and total vegetation. In the thin and dense plots, the values of the Bias% for model 2 and model 3 were nearer to 0 than those of model 1, and the differences between model 2 and model 1 and between model 3 and model 1 were significant, indicating that model 2 and model 3 were more accurate than model 1 in these two crown density classes. In the medium crown density plots, the trends of the Bias% between model 1 and model 2 and between model 1 and model 3 were not clear, and significant decreases only existed in model 2 and model 3 of the pine forest. The total Bias% values were not significantly different between the three models for the different vegetation types, indicating that the overall estimated values obtained from models 1, 2, and 3 were not significantly different. The differences between model 2 and model 3 for the different vegetation types were compared. The overall RMSE% and Bias% values of model 2 and model 3 were not significantly different, and model 2 was slightly better than model 3, but the performances of model 2 and model 3 were different among the thin, medium, and dense crown density classes.

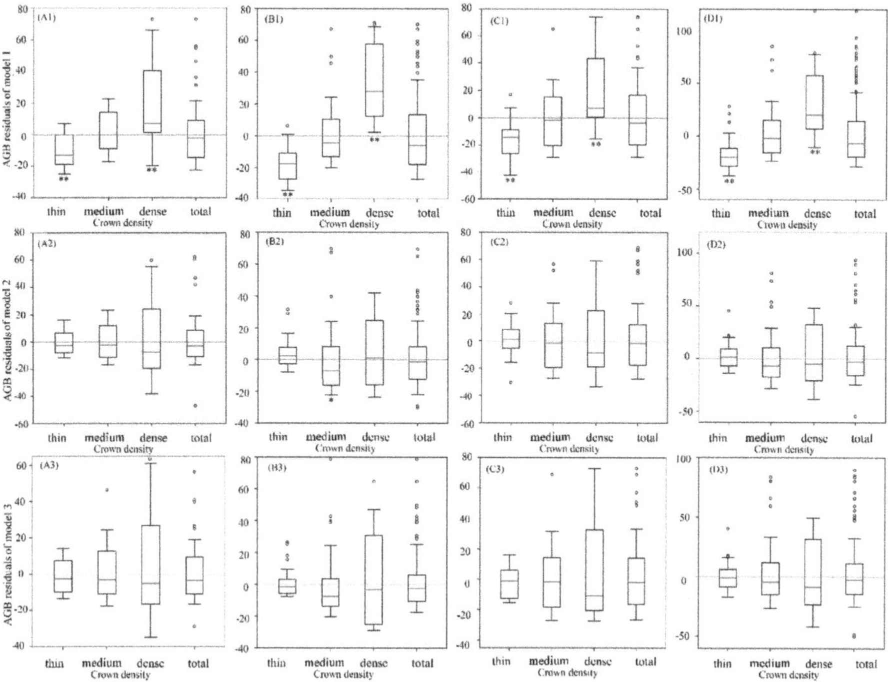

Figure 4. Residual boxplots of AGB of model 1, model 2, and model 3 for different vegetation types among different crown density classes: (**A–D**) represents pine forest, fir forest, mixed forest, and total vegetation, respectively (model 1—linear regression model; model 2—linear dummy variable model; model 3—linear mixed-effects model; ** indicates that the residuals were significantly different from 0 at the 0.01 level; * indicates that the residuals were significantly different from 0 at the 0.05 level).

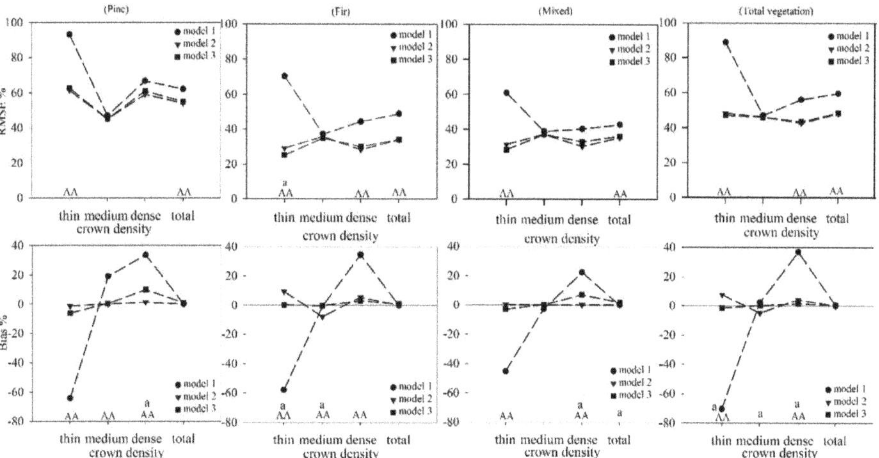

Figure 5. Comparison of root mean square error percent (RMSE%) and Bias percent (Bias%) results at different crown density classes of models 1–3 for pine forest, fir forest, mixed forest, and total vegetation. The significant differences between model 1 and model 2, and model 1 and model 3 for RMSE% and Bias% are expressed in capital letters (AA), and the lowercase letter (a) represents significant differences between model 2 and model 3.

4. Discussion

The choice of the independent variables is important for remote-sensing-based AGB estimation models, and potential variables from the images, such as single bands, vegetation indices, transformed images, textural information were applied because of the correlation with forest biomass. The correlation analysis results of over 300 spectral variables and the AGB of different vegetation types indicated that only 30 spectral variables simultaneously had significant correlation with AGB. This indicated that a large amount of remote sensing information does not fully reflect the forest characteristics. During the modelling process, stepwise regression was used to select the independent variables that were closely related to AGB. Although this variable selection method depended on the degree of linear correlation, the variables with low correlation coefficients may have been selected and thus affected the accuracy of the model.

Linear stepwise regression models have been widely used for AGB estimation using remote sensing [7,23]. In this study, the R^2 of the linear model (model 1) for the four vegetation types ranged from 0.1 to 0.3, indicating that the model had low accuracy. In addition, model 1 exhibited overestimation in the low crown density class and underestimation in the high crown density class of all vegetation types. The overestimations and underestimations of AGB were also investigated by Zhao et al., who determined that they were caused by the "global model (stepwise regression)" [40]. In addition, overestimations and underestimations have been observed when AGB was estimated using nonparametric models such as random forest, decision tree, and K-nearest neighbor methods [41–43]. In this study, the significant overestimations and underestimations of the linear model occurred in the thin (crown density < 0.4) and dense (crown density ≥ 0.7) plots, respectively. There were no significant overestimations or underestimations for model 2 and model 3 in the thin and dense plots. In addition, there were no significant differences between the linear dummy variable model (model 2) and linear mixed-effects model (model 3) except for the mixed forests (Table 8). However, in comparison with the model 1, model 2 and model 3 performed significantly better, and the results of the F-test and residuals verified the significant differences. The AGB estimation results of the three models were evaluated in the crown density classes and the results showed that the overestimation in the thin plots and underestimation in the dense plots of model 1 were not observed in model 2 and model 3.

The average AGB estimates of the sample plots for the total vegetation in the "Greater Xiangxi" varied from 47.4 Mg/ha to 47.7 Mg/ha, which were very close to the referenced value (47.7 Mg/ha) of the plots measured, and the average AGB estimates of pine forest, fir forest, and mixed forest were also very close to those of the referenced values. In Hunan province, the average AGB value of pine forest in 2011 was 31.61 Mg/ha, and the average AGB value of fir forest in the forest average AGB values obtained from the sample plots of the 4th and 8th national forest inventories in 1990 and 2009 were 31.76 Mg/ha [44] and 27.56 Mg/ha, respectively. This implied that the AGB values of forests in the "Greater Xiangxi" were larger than those of the whole Hunan mainly because the study area was a key forestry area and had various protected forests.

A comparison of the R^2_{adj} and RMSE of the three models indicated that the performances of model 2 and model 3 were better than that of model 1. The dummy variable model considered the group differences as special fixed parameters. The purpose of using the dummy variable model in this study was to introduce the parameter of crown density class into the intercept of the model so that the degree of freedom of the error was increased and the variance of the error was decreased, thereby improving the precision of the model [45]. The linear mixed-effects model considered the group differences as two parts: One part was the difference caused by different groups, and the other was the difference caused by random effects. Since the error and the random effect of the variance-covariance structures was considered, the model had high precision. Some studies compared dummy variable models with mixed-effects models for the estimation of large-scale forest growth models and the determination of biomass allometric growth equations. The linear mixed-effects model was a compromise between the dummy variable model and the linear model; in most cases, the dummy variable model was slightly better than the mixed-effects model, but this often depended on the sample size [45,46]. In this study, the sample plots were divided into the three categories of thin, medium, and dense crown density. The overall RMSE% and Bias% of model 2 were better than model 3, which supported the aforementioned results. In the past, the application of dummy variable models and mixed-effects models focused on the determination of allometric growth equations, whereas in this study, we considered whether the partition of the crown density classes improved the estimation accuracy of AGB using remote sensing data.

In statistics and biometrics, it is often debated whether the dummy variable model or mixed-effects model should be selected [46]. The choice often depends on the number of groups (random effects/dummy variables, crown density classes in this study) and the number of samples in each group. For a small group size (less than 10), the dummy variable model is commonly preferred; otherwise, the mixed-effects model is more appropriate [37,47]. Unlike in most other studies, we not only compared the overall differences between the linear dummy variable model and linear mixed-effects model but also the differences in model performance among different groups. Although the overall RMSE% and Bias% were better in model 2 than in model 3, this trend was not always the same for the different crown density classes. In the fir forest and the total vegetation, groups that had a large number of samples, the RMSE% and Bias% were smaller for model 3 than model 2 for all crown density classes. In pine and mixed forests, which had a small number of samples in each group, the RMSE% and Bias% were smaller for model 2 than model 3 for all crown density classes. Therefore, regardless of which of the models was chosen, we believe that if the overall differences between the two models are not significant, the fitting effects of the groups should be compared and the model with good performance in each group should be selected.

The climate of this region is a typical subtropical monsoon humid climate, and the typical forests are evergreen broad leaf forests and evergreen coniferous forests. In this study, the mixed forests were almost evergreen coniferous forests, and the seasonal variation of the vegetation types were not obvious. Many studies analyzed the variation of different vegetation types (NDVI) in the subtropical regions of China. They demonstrated that the NDVI of evergreen forests (evergreen broad leaf forest and evergreen coniferous forest) had no obviously seasonal variation [48]. Besides, the seasonal variations of leaf area index (LAI) and clumping index (CI) were very small because the canopy structure of evergreen forests were stable through the year [49,50], and texture information which referred the forest structure were relatively stable in the imagery. The spectral characteristics of remote sensing images are influenced by the soil, topography, vegetation type, forest structure, and other factors. It is important to choose appropriate spectral variables as independent variables in AGB estimation using remote-sensing-based methods [5,51]. Many studies have shown that when only spectral indices were used in AGB estimation, saturation occurred and caused inaccuracies of AGB estimation. Texture information calculated from a small neighborhood of pixels [26] may have a stronger correlation with AGB than spectral indices, and in some regions, AGB may only be closely correlated with texture information rather than spectral information. Texture information has been demonstrated to be an important factor in remote-sensing-based AGB estimation [52,53].

The independent variables of the linear model in this study illustrated that texture information had considerable influences on the accuracy of the AGB estimation in our study area. The linear models had low accuracy for the thin and dense crown density classes, and the linear dummy variable models and linear mixed-effects models had higher accuracy because the crown density classes were considered. The results indicate that the crown density class may be an important factor affecting the accuracy of AGB estimation. The sensitivity to the stand information decreased with increasing crown density in the dense stands; the spectral information may be affected by other non-forest characteristics in thin stands with low AGB, causing the low accuracy of the AGB estimation model. Many studies have demonstrated that a complex stand structure and high crown density caused saturation in remote sensing images and low crown density and sparse trees increased the occurrence of soil/vegetation mixed pixels [6,54,55]. The saturation and mixed pixels problems have attracted increased attention for remote-sensing-based AGB estimation. In this study, we demonstrated that the crown density classes influenced the accuracy of AGB estimation; however, the underlying mechanisms and relationships should be studied in more detail in the future.

In this study, the models for AGB estimation were explored combining sample plot data and remote sensing, and the results illustrated that the crown density was a factor that influences model accuracy. The crown density data incorporated in the linear dummy variable model and linear mixed-effects model were the most accurate. The aim of this study was to demonstrate that the crown density is an important factor that influences the accuracy of the models. A large amount of research has explored the potential of using satellite imagery for exploring remote-sensing-based methods of crown density, and there are more precise results [56]. This should be examined in future research for mapping large-scale AGB using our models when the crown density data were not available.

5. Conclusions

Permanent sample plot data of AGB of evergreen forests and Landsat 8 OLI images in the subtropical region of western Hunan province were used to develop remote-sensing-based AGB estimation models. The linear model, linear dummy variable model, and linear mixed-effects model were used to determine if the accuracy of the AGB linear estimation model could be improved by considering crown density classes. The forest AGB in our study exhibited significant differences between the thin, medium, and dense crown density classes for each vegetation type, and the AGB increased with increasing crown density. The results of the models indicate that the performance of the linear model was affected to a large extent by the crown density classes, resulting in the low accuracy of the linear model. The model-fitting results of the linear dummy variable models and linear mixed-effects models, which considered the crown density classes, were better than those of the linear models. The accuracy of the AGB estimation was significantly higher for the linear dummy variable models and linear mixed-effects models than the linear models, especially in the thin and dense crown density classes. There were no significant differences in the overall estimation accuracy between linear dummy models and linear mixed-effects models, but there were significant differences in some crown density classes of different vegetation types. The choice between the linear dummy variable model or linear mixed-effects model depended on the number of groups and sample size of the groups; when the sample size was large enough, each of the models met the accuracy requirements for AGB estimation.

Author Contributions: C.L. and M.L. conceived and designed the experiments. C.L. and Y.L. performed the experiments and analyzed the data. C.L. and M.L. wrote the paper.

Funding: This research was supported by the Doctorate Fellowship Foundation of Nanjing Forestry University and National Natural Science Foundation (NO. 31770679).

Acknowledgments: We would like to thank the Top-notch Academic Programs Project of Jiangsu Higher Education Institutions, China (TAPP, PPZY2015A062).

Conflicts of Interest: The authors declare no conflict of interest.

References

1. West, P.W. *Tree and Forest Measurement*, 2nd ed.; Springer: Berlin, Germany, 2009; ISBN 9783540959656.
2. Lieth, H.; Whittaker, R.H. *Primary Productivity of the Biosphere*; Springer-Verlag: Berlin, Germany, 1975; Volume 14, ISBN 978-3-642-80915-6.
3. Tian, X.; Li, Z.; Su, Z.; Chen, E.; van der Tol, C.; Li, X.; Guo, Y.; Li, L.; Ling, F. Estimating montane forest above-ground biomass in the upper reaches of the Heihe River Basin using Landsat-TM data. *Int. J. Remote Sens.* **2014**, *35*, 7339–7362. [CrossRef]
4. Ketterings, Q.M.; Coe, R.; van Noordwijk, M.; Ambagau, Y.; Palm, C.A. Reducing uncertainty in the use of allometric biomass equations for predicting above-ground tree biomass in mixed secondary forests. *For. Ecol. Manag.* **2001**, *146*, 199–209. [CrossRef]
5. Lu, D.; Chen, Q.; Wang, G.; Liu, L.; Li, G.; Moran, E. A survey of remote sensing-based aboveground biomass estimation methods in forest ecosystems. *Int. J. Digit. Earth* **2016**, *9*, 63–105. [CrossRef]

6. Safari, A.; Sohrabi, H.; Powell, S.; Shataee, S. A comparative assessment of multi-temporal Landsat 8 and machine learning algorithms for estimating aboveground carbon stock in coppice oak forests. *Int. J. Remote Sens.* **2017**, *38*, 6407–6432. [CrossRef]

7. Lu, D. Review Article The potential and challenge of remote sensing-based biomass estimation. *Int. J. Remote Sens.* **2006**, *27*, 1297–1328. [CrossRef]

8. Sun, H.; Qie, G.; Wang, G.; Tan, Y.; Li, J.; Peng, Y.; Ma, Z.; Luo, C. Increasing the accuracy of mapping urban forest carbon density by combining spatial modeling and spectral unmixing analysis. *Remote Sens.* **2015**, *7*, 15114–15139. [CrossRef]

9. Galidaki, G.; Zianis, D.; Gitas, I.; Radoglou, K.; Karathanassi, V.; Tsakiri–Strati, M.; Woodhouse, I.; Mallinis, G. Vegetation biomass estimation with remote sensing: Focus on forest and other wooded land over the Mediterranean ecosystem. *Int. J. Remote Sens.* **2017**, *38*, 1940–1966. [CrossRef]

10. Mitchard, E.T.A.; Saatchi, S.S.; Woodhouse, I.H.; Nangendo, G.; Ribeiro, N.S.; Williams, M.; Ryan, C.M.; Lewis, S.L.; Feldpausch, T.R.; Meir, P. Using satellite radar backscatter to predict above-ground woody biomass: A consistent relationship across four different African landscapes. *Geophys. Res. Lett.* **2009**, *36*. [CrossRef]

11. Sun, G.; Ranson, K.J.; Guo, Z.; Zhang, Z.; Montesano, P.; Kimes, D. Forest biomass mapping from lidar and radar synergies. *Remote Sens. Environ.* **2011**, *115*, 2906–2916. [CrossRef]

12. Qazi, W.A.; Baig, S.; Gilani, H.; Waqar, M.M.; Dhakal, A.; Ammar, A. Comparison of forest aboveground biomass estimates from passive and active remote sensing sensors over Kayar Khola watershed, Chitwan district, Nepal. *J. Appl. Remote Sens.* **2017**, *11*, 026038. [CrossRef]

13. Timothy, D.; Onisimo, M.; Cletah, S.; Adelabu, S.; Tsitsi, B. Remote sensing of aboveground forest biomass: A review. *Trop. Ecol.* **2016**, *57*, 125–132.

14. Zhu, C.; Lu, D.; Victoria, D.; Dutra, L.V. Mapping fractional cropland distribution in Mato Grosso, Brazil using time series MODIS enhanced vegetation index and Landsat Thematic Mapper data. *Remote Sens.* **2016**, *8*, 22. [CrossRef]

15. Zhu, X.; Liu, D. Improving forest aboveground biomass estimation using seasonal Landsat NDVI time-series. *ISPRS J. Photogramm. Remote Sens.* **2015**. [CrossRef]

16. Asner, G. Biophysical and Biochemical Sources of Variability in Canopy Reflectance. *Remote Sens. Environ.* **1998**, *64*, 234–253. [CrossRef]

17. Ollinger, S.V. Sources of Variability in Canopy Reflectance and the Convergent Properties of Plants. *New Phytol.* **2011**, *189*, 375–394. [CrossRef] [PubMed]

18. Ruimy, A.; Saugier, B.; Dedieu, G. Methodology for the estimation of terrestrial net primary production from remotely sensed data. *J. Geophys. Res.* **1994**, *99*, 5263–5283. [CrossRef]

19. Fensholt, R.; Sandholt, I.; Rasmussen, M.S.; Stisen, S.; Diouf, A. Evaluation of satellite based primary production modelling in the semi-arid Sahel. *Remote Sens. Environ.* **2006**, *105*, 173–188. [CrossRef]

20. Asner, G.P.; Levick, S.R.; Smit, I.P.J. Remote sensing of fractional cover and biochemistry in Savannas. In *Ecosystem Function in Savannas: Measurement and Modeling at Landscape to Global Scales*; CRC Press: Boca Raton, FL, USA, 2010; ISBN 9781439804711.

21. Kumar, L.; Sinha, P.; Taylor, S.; Alqurashi, A.F. Review of the use of remote sensing for biomass estimation to support renewable energy generation. *J. Appl. Remote Sens.* **2015**, *9*, 097696. [CrossRef]

22. Foody, G.M.; Boyd, D.S.; Cutler, M.E.J. Predictive relations of tropical forest biomass from Landsat TM data and their transferability between regions. *Remote Sens. Environ.* **2003**, *85*, 463–474. [CrossRef]

23. Lu, D.; Chen, Q.; Wang, G.; Moran, E.; Batistella, M.; Zhang, M.; Vaglio Laurin, G.; Saah, D. Aboveground Forest Biomass Estimation with Landsat and LiDAR Data and Uncertainty Analysis of the Estimates. *Int. J. For. Res.* **2012**, *2012*, 1–16. [CrossRef]

24. Lu, D.; Batistella, M. Exploring TM image texture and its relationships with biomass estimation in Rondônia, Brazilian Amazon. *Acta Amaz.* **2005**, *35*, 249–257. [CrossRef]

25. Powell, S.L.; Cohen, W.B.; Healey, S.P.; Kennedy, R.E.; Moisen, G.G.; Pierce, K.B.; Ohmann, J.L. Quantification of live aboveground forest biomass dynamics with Landsat time-series and field inventory data: A comparison of empirical modeling approaches. *Remote Sens. Environ. J.* **2010**, *114*, 1053–1068. [CrossRef]

26. Kelsey, K.C.; Neff, J.C. Estimates of aboveground biomass from texture analysis of landsat imagery. *Remote Sens.* **2014**, *6*, 6407–6422. [CrossRef]

27. Wang, G.; Zhang, M.; Gertner, G.Z.; Oyana, T.; McRoberts, R.E.; Ge, H. Uncertainties of mapping aboveground forest carbon due to plot locations using national forest inventory plot and remotely sensed data. *Scand. J. For. Res.* **2011**, *26*, 360–373. [CrossRef]

28. Fehrmann, L.; Lehtonen, A.; Kleinn, C.; Tomppo, E. Comparison of linear and mixed-effect regression models and a *k*-nearest neighbour approach for estimation of single-tree biomass. *Can. J. For. Res.* **2008**, *38*, 1–9. [CrossRef]

29. Yang, X.; Xue, S. Geographic Features of Forestry Resource in Hunan. *Econ. Geogr.* **2001**, *21*, 736–739. (In Chinese)

30. Xu, M.; Liu, C. Connotation and Evaluation of Regional Economic Transition Degree—A Case Study of the Western Hunan Area. *J. Nat. Resour.* **2015**, *30*, 1675–1685. (In Chinese)

31. Fang, J.; Chen, A.; Peng, C.; Zhao, S.; Ci, L. Changes in forest biomass carbon storage in China between 1949 and 1998. *Science* **2001**, *292*, 2320–2322. [CrossRef]

32. Zhang, M.; Wang, G. The forest biomass dynamics of Zhejiang Province. *Acta Ecol. Sin.* **2008**, *28*, 5665–5674. (In Chinese)

33. Chander, G.; Markham, B.L.; Helder, D.L. Summary of current radiometric calibration coefficients for Landsat MSS, TM, ETM+, and EO-1 ALI sensors. *Remote Sens. Environ.* **2009**, *113*, 893–903. [CrossRef]

34. Hantson, S.; Chuvieco, E. Evaluation of different topographic correction methods for landsat imagery. *Int. J. Appl. Earth Obs. Geoinf.* **2011**, *13*, 691–700. [CrossRef]

35. Boyd, D.S.; Foody, G.M.; Ripple, W.J. Evaluation of approaches for forest cover estimation in the Pacific Northwest, USA, using remote sensing. *Appl. Geogr.* **2002**, *22*, 375–392. [CrossRef]

36. Tang, S.Z.; Li, Y. *Statistical Foundation for Biomathematical Models*; Science Press: Beijing, China, 2002. (In Chinese)

37. Fu, L.Y.; Zeng, W.S.; Tang, S.Z.; Sharma, R.P.; Li, H.K. Using linear mixed model and dummy variable model approaches to construct compatible single-tree biomass equations at different scales—A case study for Masson pine in Southern China. *J. For. Sci.* **2012**, *58*, 101–115. [CrossRef]

38. Pinheiro, J.; Bates, D. *Mixed-Effects Models in S and S-PLUS*; Springer-Verlag: New York, NY, USA, 2000; ISBN 9780387989570.

39. Fang, Z.; Bailey, R.L. Nonlinear mixed effects modeling for slash pine dominant height growth following intensive silvicultural treatments. *For. Sci.* **2001**, *47*, 287–300.

40. Zhao, P.; Lu, D.; Wang, G.; Wu, C.; Huang, Y.; Yu, S. Examining spectral reflectance saturation in landsat imagery and corresponding solutions to improve forest aboveground biomass estimation. *Remote Sens.* **2016**, *8*, 469. [CrossRef]

41. Reese, H.; Nilsson, M.; Sandström, P.; Olsson, H. Applications using estimates of forest parameters derived from satellite and forest inventory data. *Comput. Electron. Agric.* **2002**, *37*, 37–55. [CrossRef]

42. Blackard, J.A.; Finco, M.V.; Helmer, E.H.; Holden, G.R.; Hoppus, M.L.; Jacobs, D.M.; Lister, A.J.; Moisen, G.G.; Nelson, M.D.; Riemann, R.; et al. Mapping U.S. forest biomass using nationwide forest inventory data and moderate resolution information. *Remote Sens. Environ.* **2008**, *112*, 1658–1677. [CrossRef]

43. Main-Knorn, M.; Moisen, G.G.; Healey, S.P.; Keeton, W.S.; Freeman, E.A.; Hostert, P. Evaluating the remote sensing and inventory-based estimation of biomass in the western carpathians. *Remote Sens.* **2011**, *3*, 1427–1446. [CrossRef]

44. Jiao, X.M.; Xiang, W.H.; Tian, D.L. Carbon Storage of Forest Vegetation and Its Geographical Distribution in Hunan Province. *J. Cent. South For. Univ.* **2005**, *25*, 4–8. (In Chinese)

45. Zeng, W.S.; Zhang, H.R.; Tang, S.Z. Using the dummy variable model approach to construct compatible single-tree biomass equations at different scales—A case study for Masson pine (*Pinus massoniana*) in southern China. *Can. J. For. Res.* **2011**, *41*, 1547–1554. [CrossRef]

46. Wang, M.; Borders, B.E.; Zhao, D. An empirical comparison of two subject-specific approaches to dominant heights modeling: The dummy variable method and the mixed model method. *For. Ecol. Manag.* **2008**, *255*, 2659–2669. [CrossRef]

47. Chen, D.; Huang, X.; Zhang, S.; Sun, X. Biomass modeling of larch (*Larix* spp.) plantations in China based on the mixed model, dummy variable model, and Bayesian hierarchical model. *Forests* **2017**, *8*, 268. [CrossRef]

48. Wang, Q.; Li, J. Seasonal variation of evergreen land coverage in poyang lake watershed using multi-temporal spot4-vegetation data. *Resour. Environ. Yangtze Basin* **2008**, *17*, 866–871. (In Chinese)

49. Gaolong, Z. Spatial-temporal characteristics of foliage clumping index in China during 2000–2013. *Chin. Sci. Bull.* **2016**, *61*, 1595–1603. (In Chinese) [CrossRef]

50. Liu, Y.B.; Ju, W.M.; Chen, J.M.; Zhu, G.L.; Xing, B.L.; Zhu, J.F.; He, M.Z. Spatial and temporal variations of forest LAI in China during 2000–2010. *Chin. Sci. Bull.* **2012**, *57*, 2846–2856. (In Chinese) [CrossRef]

51. Zhao, P.; Lu, D.; Wang, G.; Liu, L.; Li, D.; Zhu, J.; Yu, S. Forest aboveground biomass estimation in Zhejiang Province using the integration of Landsat TM and ALOS PALSAR data. *Int. J. Appl. Earth Obs. Geoinf.* **2016**, *53*, 1–15. [CrossRef]

52. Lu, D. Aboveground biomass estimation using Landsat TM data in the Brazilian Amazon. *Int. J. Remote Sens.* **2005**, *26*, 2509–2525. [CrossRef]

53. Eckert, S. Improved forest biomass and carbon estimations using texture measures from worldView-2 satellite data. *Remote Sens.* **2012**, *4*, 810–829. [CrossRef]

54. Du, H.; Cui, R.; Zhou, G.; Shi, Y.; Xu, X.; Fan, W.; Lü, Y. The responses of Moso bamboo (*Phyllostachys heterocycla* var. pubescens) forest aboveground biomass to Landsat TM spectral reflectance and NDVI. *Acta Ecol. Sin.* **2010**, *30*, 257–263. [CrossRef]

55. Wu, C.; Shen, H.; Wang, K.; Shen, A.; Deng, J.; Gan, M. Landsat imagery-based above ground biomass estimation and change investigation related to human activities. *Sustainibility* **2016**, *8*, 159. [CrossRef]

56. Karlson, M.; Ostwald, M.; Reese, H.; Sanou, J.; Tankoano, B.; Mattsson, E. Mapping tree canopy cover and aboveground biomass in Sudano-Sahelian woodlands using Landsat 8 and random forest. *Remote Sens.* **2015**, *7*, 10017–10041. [CrossRef]

© 2019 by the authors. Licensee MDPI, Basel, Switzerland. This article is an open access article distributed under the terms and conditions of the Creative Commons Attribution (CC BY) license (http://creativecommons.org/licenses/by/4.0/).

 *forests*

Article

Identifying European Old-Growth Forests using Remote Sensing: A Study in the Ukrainian Carpathians

Benedict D. Spracklen [1,*] **and Dominick V. Spracklen** [2]

[1]   The Rowans, Thomastown, Huntly AB54 6AJ, UK
[2]   School of Earth and Environment, University of Leeds, Leeds LS2 9JT, UK; D.V.Spracklen@leeds.ac.uk
*   Correspondence: b10spracklen@gmail.com; Tel.: +44-(0)-146-674-0307

Received: 4 January 2019; Accepted: 31 January 2019; Published: 5 February 2019

Abstract: Old-growth forests are an important, rare and endangered habitat in Europe. The ability to identify old-growth forests through remote sensing would be helpful for both conservation and forest management. We used data on beech, Norway spruce and mountain pine old-growth forests in the Ukrainian Carpathians to test whether Sentinel-2 satellite images could be used to correctly identify these forests. We used summer and autumn 2017 Sentinel-2 satellite images comprising 10 and 20 m resolution bands to create 6 vegetation indices and 9 textural features. We used a Random Forest classification model to discriminate between dominant tree species within old-growth forests and between old-growth and other forest types. Beech and Norway spruce were identified with an overall accuracy of around 90%, with a lower performance for mountain pine (70%) and mixed forest (40%). Old-growth forests were identified with an overall classification accuracy of 85%. Adding textural features, band standard deviations and elevation data improved accuracies by 3.3%, 2.1% and 1.8% respectively, while using combined summer and autumn images increased accuracy by 1.2%. We conclude that Random Forest classification combined with Sentinel-2 images can provide an effective option for identifying old-growth forests in Europe.

Keywords: old-growth forest; multispectral satellite imagery; random forest; forest classification

---

1. Introduction

Old-growth forest (OGF), also referred to as primary, virgin or ancient forest, are forests that have developed for a long period of time without significant human intervention and are characterised by the presence of old and large trees, multi-layered vertical structure and abundant standing and lying deadwood in different stages of decay [1–3]. OGF are important forest ecosystems, supporting significant biodiversity [4], storing and sequestering large amounts of carbon [5–9] and buffering microclimate [10].

In most European countries, centuries of exploitation have greatly reduced the extent of OGF. There are 1.4 Mha of primary forests remaining in Europe, equivalent to 0.7% of Europe's forest area [11]. Due to its scarcity and exceptional importance as a habitat for a wide variety of wildlife, conservation of OGF has become an important priority over the past few years. Despite this increased priority, continued loss of OGF from deforestation and conversion to managed plantations is occurring in Europe [12]. While there is no universally agreed definition of OGF, in most cases identification generally involves surveying indicators such as dead wood quantity and quality, forest structure and the degree of anthropogenic influence. This therefore requires time-intensive field surveys. Enabling the identification of such stands by remote sensing would therefore be highly useful. Even establishing the sites of potential OGF stands that could later be verified by field teams could help save time and expense.

While there have been a variety of studies using multispectral remote sensing to identify tree species in Europe [13–17], these are mostly not concerned with OGF. Variation in tree species, height, size and separation as well as the high number of shaded, dead and dying and spectrally unusual trees, mean that tree species in OGF are harder to classify than in other forest types [18]. At the same time, however, this spectral variability can potentially enable the distinction of OGF from younger forest stands.

There have been a number of previous investigations [19–23] into the effect of forest structure on satellite spectra in temperate zones using either Landsat or high resolution satellite imagery, mostly of closed canopy conifer stands (including OGF) in the western USA. Landsat and commercial satellite (10 m resolution) imagery was used to examine how tasselled cap vegetation indices varied with stand characteristics in closed canopy conifer forest in Oregon, USA [20]. Unsupervised classification of Landsat images (tasselled cap vegetation index) was also used in Oregon to map young, mature and old-growth stands [19]. Unsupervised classification of Landsat imagery was used to map mature and old-growth conifer stands in the Pacific Northwest, while Landsat imagery and a spectral mixing model was used to identify stand structural stages in Washington state, USA [23]. There have also been efforts to distinguish mature and old-growth forest using Lidar data and Random Forest classification [24] but Lidar data is usually both expensive and difficult to obtain. Satellite data has been used to identify potential OGF in Romania through manual inspection of images [25]. The recent European Space agency (ESA) Sentinel-2 (S2) mission provides freely available high spatial resolution (10 m) multispectral information and so offers great opportunities for such a forest classification study [26].

The Ukrainian Carpathians contain some of the largest remnants of old-growth fir-beech-spruce-pine forests remaining in Europe. The Carpathian Convention commits Ukraine to the protection of its virgin forests and in May 2017 the Ukrainian president signed an amendment to the Forest Code [27] protecting all OGF sites in Ukraine. An ongoing inventory of OGF in the Ukrainian Carpathians is being carried out by WWF Ukraine and can be viewed at gis-wwf.com.ua/.

In this paper we analyse the spectra of broadleaf, conifer and mixed forests in the Ukrainian Carpathians, using Sentinel-2 imagery and supervised classification to investigate the potential of machine learning to identify OGF, based on the hypothesis that there is a significant difference between the spectra of OGF and other forest types (non-Old-Growth Forest). To the best of our knowledge, this is the first such study employing Sentinel-2 imagery and a decision tree classifier to look at old-growth broadleaf, conifer and mixed broadleaf–conifer woodland in temperate regions. The key objectives of our study are to:

- Use machine learning (Random Forest classification) to identify different tree species in OGF.
- Determine if Random Forest classification can be used to identify and map potential OGF sites by differentiating between OGF and other forest types.
- Determine how combinations of spectral bands, multitemporal imagery and ancillary data affect map accuracy.

2. Material and Methods

2.1. Study Site

We analyse the ability of Sentinel-2 to identify OGF in the eastern Carpathian Mountains of SE Ukraine, a 42% forested region [28] covering about 24,000 km^2, ranging from 100–2060 m elevation and characterized in the upper elevations by dense forest stands on steep slopes [29]. Intensive land use and forest management has substantially affected the area's forests, with much of the lowlands being converted to agriculture. While over the past century forest cover has expanded in the region [30,31], forests are still subject to extensive logging, both legal and illegal [32–34] and there are large areas of intensively managed spruce plantations [35]. Nevertheless, the region still contains some of the largest areas of OGF remaining in Europe.

Species composition in OGF in our study site is dominated by beech (*Fagus sylvatica* L) (33% of area), Norway spruce (*Picea Abies* (L.) Karst.) (43% of area) and mountain pine (*Pinus mugo* Turra) (9% of area), with smaller areas of silver fir (*Abies alba* Mill.) and sessile oak (*Quercus petraea* (Matt.) Liebl.). Sycamore (*Acer psudeoplatanus* L), birch (*Betula verrucosa* Ehrh.), hornbeam (*Carpinus betulinus* L), rowan (*Sorbus aucuparia* L), aspen (*Populus tremula* L), Swiss pine (*Pinus cembra* L), Scots pine (*Pinus sylvestris* L), ash (*Fraxinus excelsior* L), wych elm (*Ulmus glabra* Huds.), hazel (*Corylus avellana* L), Norway maple (*Acer platanoides* L), green alder (*Alnus viridis* (Chaix.) D.C.) and grey alder (*Alnus incana* (L.) Moench) occur mixed in small quantities with these species. Tree species show a gradual transition with increased elevation, changing from oak and beech at lower elevations (300–500 m) to beech and Norway spruce dominated (500–1400 m) and to mountain pine and Norway spruce at the highest altitudes (1400–1800 m). Natural alpine meadows cover only the highest of the mountain peaks (>1800 m), though in most places the timberline has been artificially lowered through livestock grazing [36].

Mean annual precipitation varies by altitude from 600 mm in the lowlands to 1600 mm on the mountain peaks [28]. Natural disturbance regimes in the forest are dominated by small-scale loss, largely from low to moderate intensity windthrow damaging single or small groups of trees [37–39]. The study region covers three provinces (oblasts): Transcarpathian, Ivano-Frankivska and Chernivetska. Figure 1 (inset map) shows the location of the study area within Ukraine.

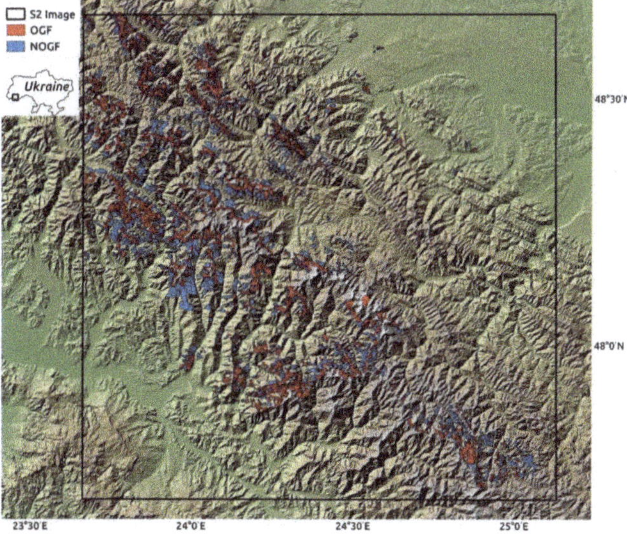

Figure 1. Relief map (elevation range 200–2060 m) of the study site showing Old-Growth Forest (OGF, shown as red polygons) and Non-Old Growth Forest (NOGF, shown as blue polygons). S2 image shows the extent of the Sentinel-2 image used in the study. Inset map shows the location of the study area within Ukraine.

2.2. OGF Survey Data

We use data on the spatial distribution of OGF from the ongoing survey (beginning in 2010) of OGF across the Ukrainian Carpathians. This data was provided by WWF Ukraine and covered the survey years 2010–2017 inclusive. This survey includes information on the location and spatial extent of OGF (shapefile polygons of identified OGF stands) as well as detailed information on tree species composition and age. The background to this WWF project and the criteria used for OGF identification can be found here [40]. A map [41] shows the areas surveyed for OGF up to 2017.

The main criteria used in the WWF study for classification of a forest area as OGF are as follows:

- standing and lying dead wood;

- complex structure (high variety of age groups and tree sizes);
- no non-native tree species;
- no visible traces of exploitation—i.e., logging.

While a minimum size criteria (of 20 ha) is given, in practise much smaller areas (down to 0.5 ha) of OGF were also recorded.

A visual inspection of the autumn 2017 Sentinel-2 image showed that since the WWF survey 188 OGF polygons in our study area (mostly Norway spruce forest along the border with Romania) had suffered disturbance through either clear felling, thinning or construction of tracks. These polygons were discarded and not used in our study. Details of the OGF polygons that were used can be seen in Table 1. There were a total of 4037 OGF polygons in our analysis, covering an area of 428 km^2. We defined the threshold for mixed forest as 20% and above.

Table 1. Number and area of polygons used in this study (not including polygons damaged by man) for different tree species in OGF polygons. We denoted mixed tree species polygons as "Dominant Tree Species Mix," while C and B stand for conifer and broadleaved respectively. Thus, Norway Spruce CBMix is Norway spruce dominant mixed with at least 20% broadleaved species, while Beech BMix is beech dominated mixed with at least 20% other broadleaved species.

Tree Species	Number of Polygons	Area (km^2)	Mean Elev (m)	Min Elev (m)	Max. Elev (m)	Mean Slope ($^{\circ}$)
Beech	1281	139.2	1055	394	1565	24.2
Oak	21	3.2	507	334	871	13.2
Mountain Pine	219	37.4	1477	1061	1982	22
Norway Spruce	1784	182.4	1343	519	1688	22.1
Silver Fir	20	1.3	598	481	946	12.5
Beech BCMix	189	16.0	1052	425	1443	24.5
Norway Spruce CBMix	226	19.2	1136	514	1620	29.2
Silver Fir CBMix	60	5.3	933	515	1286	24.8
Beech BMix	59	5.1	1039	454	1497	26.1
Other Bmix	15	1.5	618	342	1131	14
Other BCMix	6	0.7	1410	1030	1719	21.2
Norway Spruce CMix	98	10.6	1266	703	1568	22.3
Other CMix	48	6.0	1209	591	1953	23.1
Other CBMix	2	0.1	1598	1374	1722	24
Other B	2	0.07	1522	1422	1633	31.2
Other C	4	0.15	929	733	1373	24.5
Total Conifer	2173	237.6	1341	481	1982	22
Total Broadleaf	1378	149	1042	334	1633	24
Total Mixed	486	41.4	1084	342	1689	24.3
Total	4037	428	1208	334	1982	23

For comparison with OGF polygons, we created 4000 polygons randomly located within a buffer of 2 km of the OGF. This distance was chosen as it enabled the requisite number of appropriately sized NOGF polygons to fit in. To mimic the OGF polygons, polygon sizes were selected from a right-skewed distribution ranging in size from 0.05–200 ha. Polygons which comprised of open areas, non-surveyed areas or young forest were either eliminated or had their boundaries redrawn to exclude these areas. Open areas and young forest was identified either through the publicly available forest cover and forest loss data derived from Landsat timeseries [42] or through visual inspection since open ground and young forest shows up brightly in the images [43]. Since the remaining polygons were forest situated in areas that had been surveyed for OGF yet had not been identified as such, we were confident these polygons consisted of forest that was not OGF (NOGF). These NOGF polygons were then 'tidied up' by expansion to remove small gaps between polygons so that they shared a common border. Much of the OGF consisted of high altitude forest stretching up to the treeline. The neighbouring NOGF was therefore typically downhill from the OGF and consequently at a lower elevation and lacking high montane forest. To compensate, we therefore manually created a number of NOGF polygons along the treeline in areas that had been surveyed for OGF. Finally, all these NOGF polygons were classified through visual inspection of Sentinel-2 summer, autumn and winter imagery, as either broadleaved, evergreen or mixed forest. The end result was the creation of 4449 NOGF polygons (described in Table 2), of which a majority lay directly adjacent to the OGF polygons. The median NOGF polygon size was 0.08 km^2, compared to 0.07 km^2 for the OGF polygons. Figure 1 shows the study region with OGF and NOGF polygons overlaid.

Table 2. Number and area of polygons used in this study for different forest types in non-OGF polygons.

Forest Type	Number of Polygons	Area (km^2)	Mean Elev. (m)	Min Elev. (m)	Max Elev. (m)	Mean Slope ($^\circ$)
Conifer	2563	299.6	1238	457	1792	20.2
Broadleaved	1343	206.1	888	357	1456	23.6
Mixed	543	57.5	1045	438	1566	22.9
Total	4449	560.5	1108	357	1792	21.5

2.3. Sentinel-2 Images

Sentinel-2 (S2) features 13 spectral bands with 10, 20 and 60 m resolution [44]. We used the 10 and 20 m bands in our study (see Table 3). Two S2 images were downloaded (https://scihub.copernicus.eu/) as Level-1C Top-of-Atmosphere reflectance products: one for summer (2 August 2017) and one for autumn (16 October 2017), with codes:

"S2B_MSIL1C_20170802T092029_N0205_R093_T34UGU_20170802T092027.SAFE" and
"S2A_MSIL1C_20171016T092031_N0205_R093_T34UGU_20171016T092425.SAFE" respectively.

Table 3. Sentinel-2 bands with 10 m or 20 m resolution. Near IR is Near Infra-Red and SWIR is Short wave Infra-Red.

Sentinel-2 Bands	Central Wavelength (μm)	Resolution (m)
B2–Blue	0.490	10
B3–Green	0.560	10
B4-Red	0.665	10
B5–Red edge	0.705	20
B6–Red edge	0.740	20
B7–Red edge	0.783	20
B8–Near IR	0.842	10
B8A–Near IR	0.865	20
B11–SWIR	1.610	20
B12–SWIR	2.190	20

These particular images were chosen for their low cloud cover (5.2% and 0% respectively). The north-east and south-west corners of these images are 23°43′7.73″ E, 48°43′15.34″ N and 25°7′41.74″ E, 47°41′27.06″ N respectively. These images were then topographically and atmospherically corrected using the Sen2Cor module [45]. The 20 m resolution bands were resampled to 10 m spatial resolution. We investigated using spring or winter images but from December through to April most of the high altitude OGF was completely covered with snow, with the polygons completely white and providing limited useful information.

2.4. Sentinel-2 Image Evaluation

We used an object-based approach (as opposed to a pixel-based classification), where the mean and standard deviation of the pixel spectra and the mean of the associated vegetation indices and textural features within a forest polygon were used for the analysis. A number of studies have argued for the superiority of object-based over pixel-based approaches [15,18,46] and the WWF data included mixed forest polygons which suited an object-based approach. To further understand the distribution of the pixels within the polygons, we also calculated percentile values ranging from 5% to 99% for each polygon. T-tests of the band spectra mean values were calculated to test for significant differences between the OGF and NOGF polygons.

We calculated 6 vegetation indices from the S2 bands: the Normalized Vegetation Difference Index (NDVI) and the Enhanced Vegetation Index (EVI), probably the two most commonly used forest classification indices. Since the more heterogeneous structure of OGF compared to other forest types might help classification, we used two forest structure indices: Advanced Vegetation Index (AVI) and the Shadow Index (SI). A study [20] found the difference between SWIR and NIR bands most useful in distinguishing mature and OGF so we also used Normalised Difference Infrared Index (NDII). Lastly, the Red edge Normalized Difference Vegetation Index (RENDVI) was chosen to exploit information in the red edge bands.

$$NDVI = \frac{B8 - B4}{B8 + B4} \tag{1}$$

$$EVI = \frac{2.5(B8 - B4)}{(B8 + (6 \times B4) - (7.5 \times B2) + 1} \tag{2}$$

$$AVI = \sqrt[3]{(B8(1 - B4)(B8 - B4)} \tag{3}$$

$$SI = \sqrt[3]{(1 - B2)(1 - B3)(1 - B4)} \tag{4}$$

$$NDII = \frac{B8 - B11}{B8 + B11} \tag{5}$$

$$RENDVI = \frac{(B6 - B5)}{(B6 + B5)} \tag{6}$$

Spectral images vary not only in tone but also in texture (spatial variation). Texture measurements quantitively describe relationships of spectral values with neighbouring pixels, which information has been used previously to improve forest stand classification accuracy [47,48]. The most commonly used textural measure is the Grey Level Co-occurrence Matrix (GLCM) [49], essentially a description of how often different combinations of pixel brightness values (grey levels) occur in an image. A detailed overview of GLCM can be found here [50]. Generally, younger forest have a more uniform and low contrast image due to the trees' equal height and spatial distribution, whereas the heterogeneity of OGF, with a broader distribution of tree heights and ages, results in more shadows cast by emergent trees. OGF are therefore likely to have differences in texture compared to NOGF areas.

Use of GLCM requires choosing 6 parameters—textural features, pixel displacement and direction, the moving window size, quantisation level and spectral bands – giving rise to thousands of potential combinations. The textural features can be divided into contrast, orderliness and descriptive statistics groups [50]. We chose one textural feature from each of these groups that had been found useful in previous studies [51,52]: contrast, entropy and GLCM mean. Contrast is a weighted measure of the contrast between adjacent pixels—the greater the value the greater the contrast. Entropy corresponds to the orderliness of the image—larger entropy values indicate greater disorder. We calculate these features for a visual, near IR and shortwave IR band (B3, B8 and B12). A study [53] found that for spectrally homogenous classes, smaller window sizes improved classification accuracy. Combined with the coarse resolution of the S2 data, we therefore computed the selected textural variables with a relatively small 5×5 pixel window size over all directions, a pixel displacement of 1 and a 32 level quantization using the ESA Sentinel Application Platform (SNAP), available at http://step.esa.int/main/toolboxes/snap.

Mean and standard deviation for each polygon were extracted for each of the 10 bands, as well as the mean of the 6 vegetation indices and the 9 textural measures. Mean elevation and slope was also calculated (using 1 arc second resolution Shuttle Radar Topography Mission (SRTM) data [54]). The number of polygons and area for each forest type are given for OGF and NOGF in Tables 1 and 2 respectively.

2.5. Random Forest Method

Random Forest (RF) [55] is a non-parametric machine learning algorithm, selected for its high classification accuracy [56,57], ease of use [58,59] and its demonstrated ability in previous remote sensing forest classification studies [15–17,60].

The polygons were randomly divided into training and validation sets in a ratio of 75% and 25% respectively. The classification analysis was performed using the scikit-learn Python library [61]. The maximum number of features Random Forest was allowed to try in an individual tree was set as the square root of the total number of features. The number of trees built was set at 500. We found changing these parameters made little difference to model outcome. Feature importance was calculated by mean decrease impurity.

We report user's accuracy (how reliable is the map, that is, how often forest identified as, say, OGF in our model is actually present on the ground), producer's accuracy (how well is the situation on the ground mapped, that is, how often OGF on the ground is correctly identified as such by our model) and overall accuracy (how often all our forests were identified correctly). We report accuracy as the average across the relevant polygons.

The Random Forest classification between tree species was carried out using only the 1781 Norway spruce, 1281 beech, 219 mountain pine, 189 beech-conifer mixed (Beech BCMix) and 226 Norway spruce-broadleaved mixed (Norway spruce CBMix) OGF polygons. Due to their relative lack of polygons, no attempt was made to identify other tree species (such as oak and silver fir) and so these polygons were excluded from this Random Forest classification. Therefore, a total of 3696 OGF polygons covering about 92% of the total OGF area was used. No NOGF polygons were used for the Random Forest tree species classification.

In order to classify [62] the tree species we used 10 mean spectral band values (B), 10 standard deviation spectral band values (B_sd), mean elevation (Elev), 9 GLCM textural variables (TF) and 6 vegetation indices (VI). The classification was divided into 8 models: B, TF, VI, B+TF, B+VI, B+Elev, B+B_sd and B+B_sd+Elev+TF+VI. These models were conducted for summer, autumn and summer and autumn combined, resulting in 24 RF models.

A similar Random Forest classification was now made to distinguish OGF polygons from NOGF polygons, using all 4037 OGF and 4449 NOGF polygons, with the OGF and NOGF identified as either conifer, broadleaved or mixed. We used the same 8 models as for tree species classification, run for summer, autumn and summer and autumn combined. RF classification was carried out separately for conifer, broadleaf and mixed forest types and we therefore conducted a total of 72 RF models (3 × 24).

3. Results and Discussion

3.1. Distinguishing Old-Growth Forest Tree Species

We first explored whether S2 images could be used to identify different tree species within OGF polygons. Figure S1 shows boxplots of all the spectral signatures, the vegetation indices and the textural measures of the various tree species, including oak and silver fir, for both summer and autumn The impact of autumnal colours for beech results in a 140% and 40% increase in brightness in the autumn red (B4) and red edge (B5) bands respectively compared to the summer bands (see Figure S1a). Land class maps are shown in Figure 2.

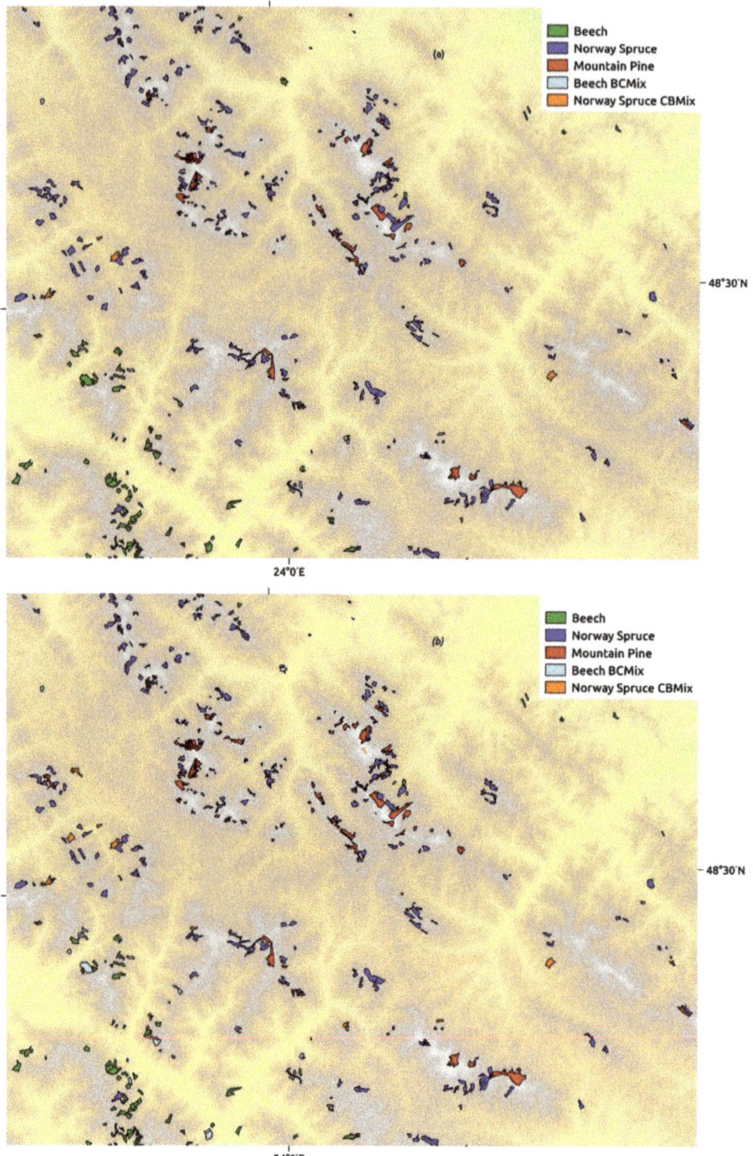

Figure 2. Map of Old-Growth Forest tree species based on (**a**) the Random Forest classifier using all features for combined summer and autumn images (**b**) WWF ground survey. Beech BCMix is dominant beech with at least 20% conifer species, Norway Spruce CBMix is dominant Norway spruce with at least 20% broadleaf species.

Figure S2 shows the ranking of features for importance. Figure 3 and Table S1 shows the classification accuracies for the tree species for summer and autumn images for the different Random Forest models. Beech and spruce consistently had the highest accuracies, with producer's accuracy of 95%–98% and user's accuracy of 85%–90%. Lower accuracy was achieved for mountain pine with producer's accuracy of 25%–60% and user's accuracy of 50%–90%. Classification was poorer for mixed forest with producer's accuracy ranging from 10%–30% and user's accuracy around 50%. For spruce and beech, producer's accuracy was consistently higher than user's accuracy, while for mixed and mountain pine the situation was reversed—a sign that the model was consistently misclassifying pine and mixed forest as spruce and beech. Similar remote sensing tree classification studies tend to obtain accuracies of between 70%–95% [15] and our study is generally in line with these.

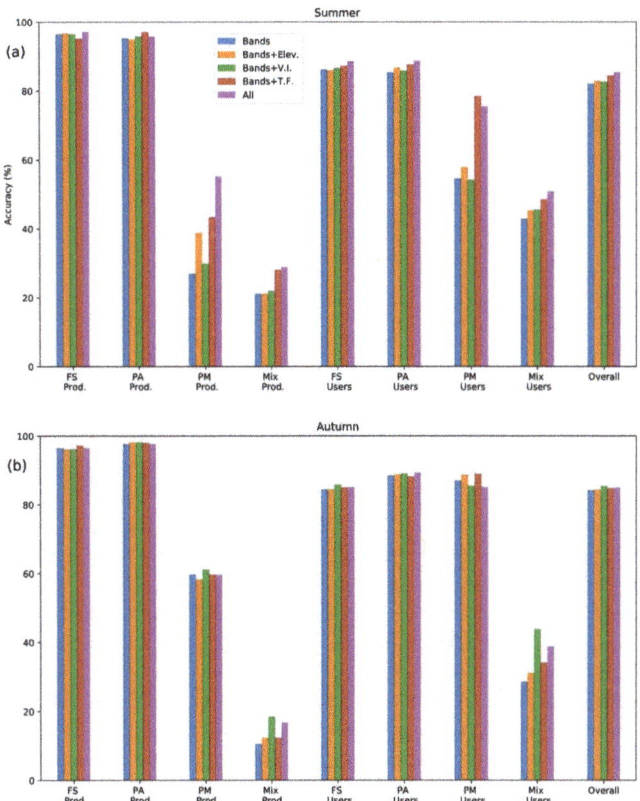

Figure 3. User's, producer's and overall accuracies resulting from Random Forest classification for Old-Growth Forest tree species for 5 selected models in (**a**) Summer (2 August 2017) and (**b**) Autumn (16 October 2017). Abbreviations: FS—Beech, PA—Norway spruce, PM—mountain pine.

In summer, SWIR, NIR and red edge bands (B5-7) were generally most important for classification (Table S1). Using only the bands, accuracy rates were higher using the autumn than the summer image by 2.2% (Figure 3)—the autumnal change in leaf colour was distinctive and consequently red and red edge bands were the best performing features for the autumn image (Table S1). Studies in eastern USA also found that mid-autumn was the best time for tree species classification [63,64]. For the summer image, adding elevation data improved overall accuracy. In particular, mountain pine, which only occurred at very high elevations in the study area, had its user's and producer's accuracy increased by 3.3% and 11.9% respectively. Previous studies have likewise found that topographic variables improved classification in studies in the USA [56,65] and Spain [56,65]. Vegetation indices performed better than the bands by 0.8% and 0.7% in summer and autumn respectively. Choice of features had a notable effect on accuracy for mountain pine, with user's accuracy varying from 50% to 90%. Using combined summer and autumn images increased accuracy by 1.5%, less than the 2%–7% found by another study [60].

The Confusion matrix for the most accurate classification is shown in Table 4, with the diagonal cells showing the number (in bold) of correct classifications and the off-diagonal cells indicating the mistakes. In distinguishing beech and spruce it performed well, making just a single mistake and producer's accuracies for these forest types was high (Figure 3). However, accuracy for mixed forests was poor, generally classing it as its pure tree species counterpart—i.e., beech mix was classed as pure beech and spruce mix as pure spruce. The model had trouble distinguishing between spruce and pine stands, classing 27 pine stands as spruce.

Table 4. Confusion matrix for three dominant tree species plus FS mix with conifers and PA mix with broadleaved, based on the most accurate Random Forest classification—summer and autumn mean band spectra with all features. FS—beech, PA—Norway spruce, PM—mountain pine.

		Predicted Species					
		FS	FS mix	PM	PA	PA mix	Sum
	FS	**300**	5	0	0	4	309
	FS mix	36	**9**	0	0	5	50
Actual	PM	0	0	**40**	27	0	67
species	PA	1	1	3	**460**	3	468
	PA mix	8	6	1	27	**22**	64
	Sum	345	21	44	514	34	958

3.2. Distinguishing between OGF and non-OGF

Land class maps are shown in Figure 4. Figure S3 shows the mean spectral signatures, vegetation indices and textural measures for OGF and NOGF. OGF had a lower mean brightness than NOGF for both broadleaf and mixed forests over all bands in both summer and autumn. For broadleaf, t-tests showed a significant difference between OGF and NOGF for all non-visible bands (B5-B8, B8A, B11, B12) and all bands for summer and autumn images respectively ($p < 0.05$). For mixed forest, t-tests showed a significant difference ($p < 0.05$) between OGF and NOGF for all bands and all bands except blue (B2) for summer and autumn images respectively. Younger forests tend to consist of small tree crowns packed tightly together with few gaps. As the forest ages, both mean crown size and the number of gaps increases. The increase in forest gap number and shadows cast by emergent trees results in a reduction in the reflected light, leading to a lower mean brightness in OGF. Due to the inverse relation between wavelength and atmospheric scattering, shadows will be illuminated more by visible light (skylight) than longer wavelength bands [66,67]. Structurally diverse OGF would likely result in more shadows compared to NOGF, resulting in a larger difference between OGF and NOGF in the red edge, NIR and SWIR bands than the visible bands.

There was less difference between conifer OGF and conifer NOGF in the summer mean band spectra, with significantly *higher* reflectance in OGF for all bands except B7 (t-test, $p < 0.05$). This is a surprising result, as it is contrary both to the result for broadleaf and mixed forest, as well as to a previous study [21] which found conifer OGF significantly darker than mature forest in summer in blue, green and NIR Landsat bands. In our analysis, conifer OGF was, on average, at higher elevations compared to conifer NOGF (1341 m and 1237 m respectively). Therefore, it is likely that a higher percentage of conifer OGF consisted of mountain pine than in conifer NOGF and mountain pine was significantly brighter than Norway spruce and silver fir across all bands (see Figure S1). Furthermore, OGF towards the treeline was more likely to contain open forest and clearings than NOGF. If this were the case then the OGF image would contain many more bright pixels comprised of ground vegetation and soils. Open areas were generally about 50%–100% brighter than conifer canopy for all bands. To test if this difference could explain our surprising result, we plotted mean percentile values for OGF and NOGF conifers split into subsets of mean elevation above and below 1250 m (Figure 5). OGF conifer contained significantly more bright pixels (percentile > 80%) than NOGF and for OGF below 1250 m more dark pixels (percentile < 20%). In other words, the higher mean brightness of OGF was due to the presence of more bright pixels (open areas), while the wooded areas are darker than NOGF. (It is worth noticing that this pattern also holds true for broadleaf and mixed forest, as shown in Figure S4). OGF conifer with a mean elevation below 1250 m was significantly darker than NOGF for Bands B6-B8A. As conifer OGF increased in elevation, it contains more open ground (bright pixels), as can be seen from comparing Figure 5a,b. An alternative explanation we considered for this surprising result is that OGF conifer stands contain a higher percentage of broadleaved species (which are brighter across all bands) than NOGF conifer stands. However, in autumn the red (B4) band in conifer OGF and NOGF polygons brightens by about the same percentage relative to the summer image, so this is unlikely to be a factor.

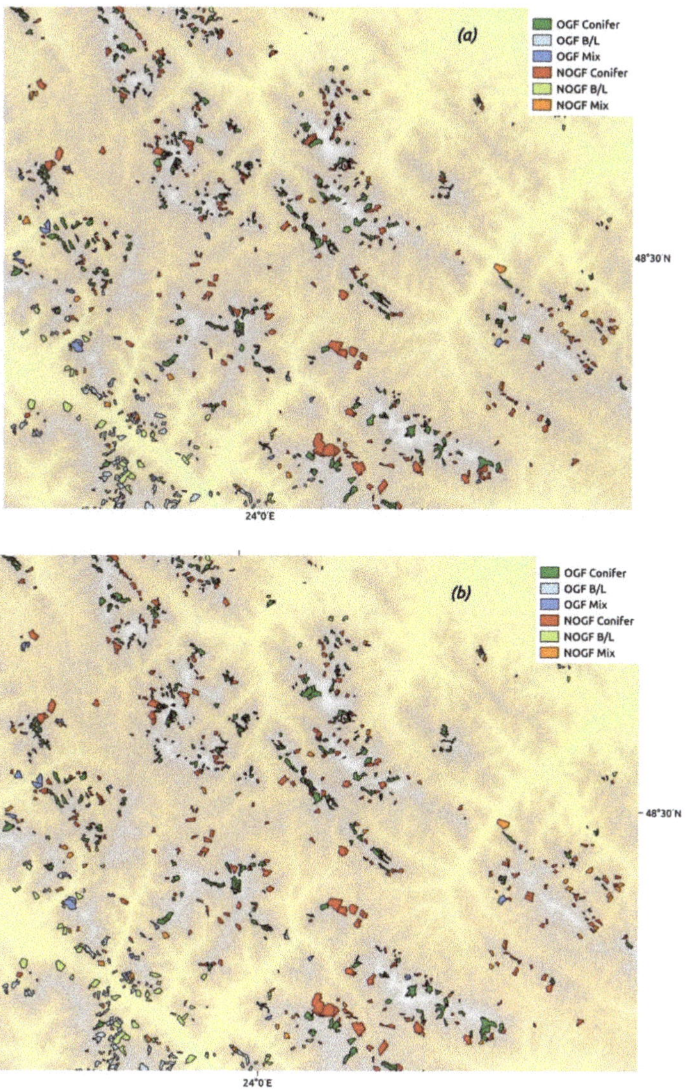

Figure 4. Map of Old-Growth Forest and Non-Old-Growth Forest based on (**a**) the Random Forest classifier using all features for combined summer and autumn images (**b**) the WWF ground survey.

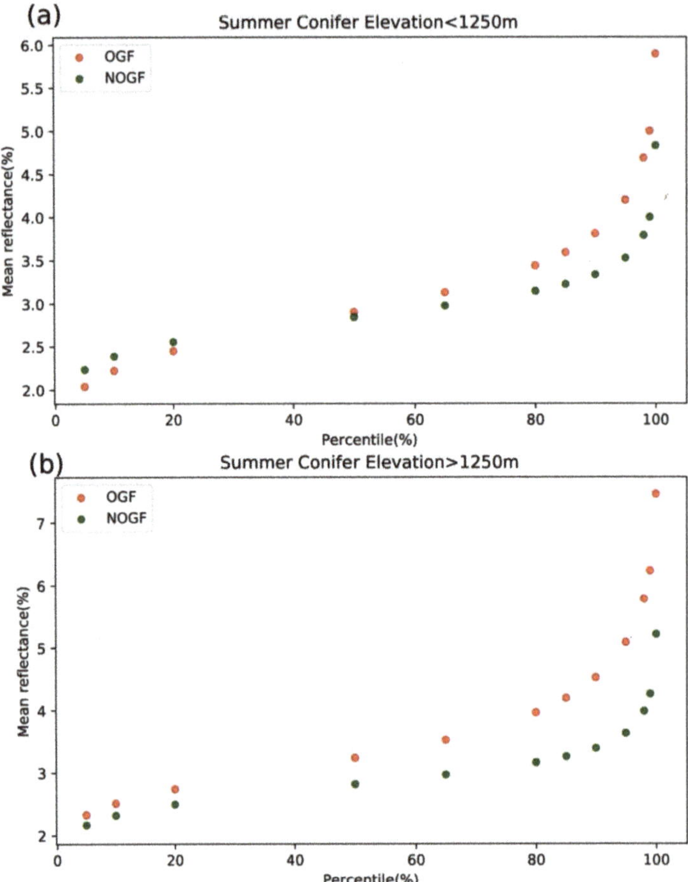

Figure 5. Mean percentile values for the green Band (B3) for conifer Old-Growth Forest and non-Old-Growth Forest polygons with mean elevation (**a**) below 1250 m (**b**) above 1250 m.

The vegetation indices NDVI, AVI, RENDVI and EVI are strongly correlated to chlorophyll content [68,69]. All indices were greater for NOGF than OGF for all forest types (Figure S3b): as forest ages the amount of green vegetation tends to decline from both an increase in dead and dying trees and an increase in the amount of vegetation obscured by shadow from emergent trees. NDII is the difference index between the NIR and SWIR bands and a measure of the canopy water content [70]. It was likewise higher for NOGF than OGF for summer and autumn images, again attributable to greater non-photosynthetic vegetation in the OGF.

The texture of OGF was more heterogeneous, with OGF having higher contrast values for all images in all forest types and bands than NOGF (Figure S3c). The large crowns of OGF cast large shadows, which result in a coarse texture compare to the finer-grained texture of smaller, more densely packed, younger tree stands. In contrast to our results, an investigation [20] of the effects of stand structure on an absolute difference algorithm of tasselled cap vegetation indices found a poor correlation between these textural features and stand characteristics which included age, which the authors attribute to the coarse (30 m) resolution of the Landsat imagery used. A study of forest in Israel [51] found that contrary to our results, mature unmanaged forests had lower GLCM entropy and contrast values than younger forest. The authors explain this as resulting from the very high resolution

(2 m) satellite imagery used, so that the large crown sizes associated with mature forest increased the number of adjacent pixels with similar grey levels.

Figure 6 and Tables S2 and S3 shows the OGF producer's accuracy and overall accuracy for Random Forest models. OGF producer's accuracy is arguably the most important measure—it matters more if existing OGF is misidentified as NOGF and consequently overlooked than if we misclassify NOGF as OGF. Figure S5 shows the ranking of features for importance.

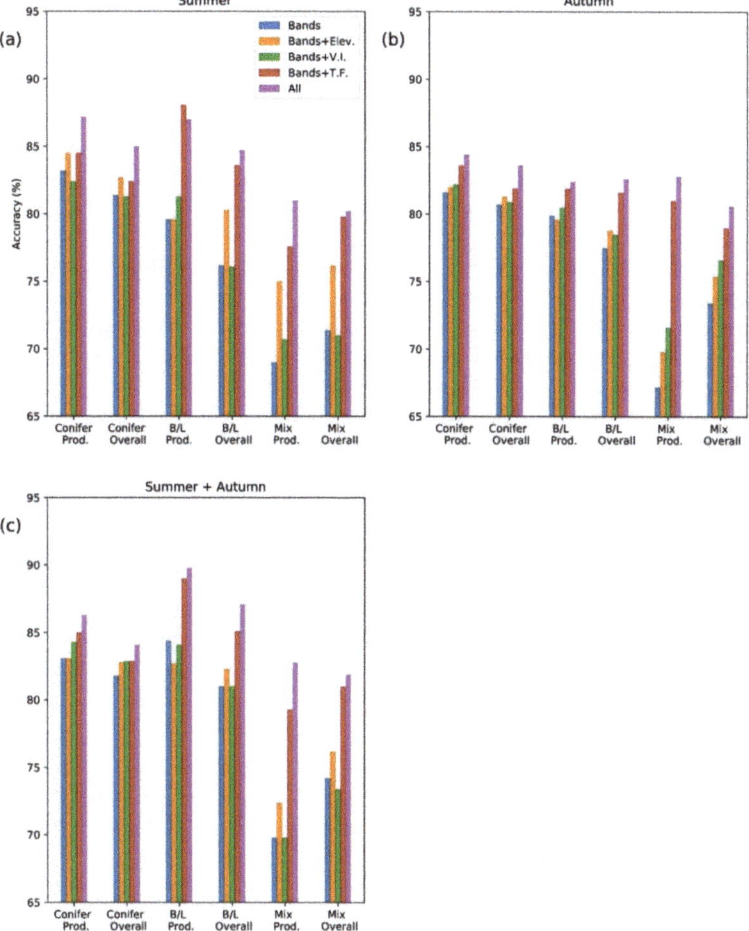

Figure 6. Old-Growth Forest (OGF) Producer's and overall accuracies resulting from Random Forest classification for OGF and Non-OGF polygons for 5 selected models for (**a**) summer (2 August 2017) (**b**) autumn (16 October 2017) and (**c**) summer + autumn.

Classification accuracies for OGF were roughly uniform with both user's and producer's accuracies between 75%–85% for both conifer and broadleaf forest. Mixed forest accuracies were lower with producer's accuracy ranging from 65%–80% and user's accuracy around 70%–85%. Overall, classification accuracy using all features was 84%, on the verge of the 85% success threshold that is often used for machine learning studies. A previous study [19] obtained 75% accuracy in distinguishing OGF from mature forest using regression analysis and Landsat 5 imagery. Landsat 7 imagery and unsupervised classification was used [22] to distinguish old and mature conifer forests with an overall accuracy of 80%–90% depending on the ecoregion. Landsat 5 imagery and unsupervised classification [21] found 78% accuracy for classifying closed canopy conifer OGF. Another study [24] using Random Forest and high resolution LIDAR data to separate old near-natural and old managed conifer forest obtained overall classification accuracies of 85%–90%.

Accuracy was higher for mountain pine OGF stands than Norway spruce (about 92% and 83% respectively). This was consistent regardless of feature selection. Accuracy rose with elevation for both broadleaf and conifers. The model failed significantly with the small number of silver fir polygons—correctly identifying as OGF only about half. Again this result was consistent for all the RF models examined. For mixed forest the classification accuracy for Norway spruce CBMix was high (90%), while the accuracy for beech BCMix was much lower (70%).

A ranking of features (Table S2) indicates that in both summer and autumn, SWIR bands were most important for conifers, with red edge and NIR most important for broadleaves. Overall, for the band spectra accuracy rates were 0.3% higher using the summer than the autumn image. Adding elevation data to the bands usually improved overall accuracy (by an average of 1.8% overall): a large proportion of the surviving OGF ringed mountain summits, so that the adjacent NOGF polygons were generally lower in elevation.

Vegetation indices generally performed worse than Bands, with their use instead of bands reducing accuracy by 0.3%. Textural features performed extremely well and on average were 2.8% more accurate than just using the bands. Adding mean vegetation index, band standard deviations and textural feature data to the band spectra increased overall accuracy by 0.3%, 2.1% and 3.3% respectively (see Table S2 and Figure 6). Using all features improved accuracy by 5% compared to using just the mean band spectra.

We ran the RF classification separately on broadleaf, conifer and mixed forest types to provide insight into how performance and most important features differed for different forest types. However, the improvement in accuracy over running it on all OGF/NOGF forest types lumped together was fairly small (0.15%).

Figure 6 shows the producer's overall accuracies for various season combinations using all features. Combining images improved accuracy in general, with the combined summer and autumn images 0.5% and 1.9% more accurate than summer or autumn on its own respectively. The highest overall accuracy achieved for all OGF polygons was 84.8% using summer and autumn images and all features.

4. Conclusions

The objective of this study was to evaluate the suitability of multitemporal Sentinel-2 data for identifying OGF tree species and distinguishing OGF from other woodland. The Sentinel-2 spectral signatures along with associated vegetation indices, textural features and elevation data were analysed with the Random Forest classifier. The OGF analysed consisted of beech, Norway spruce, mountain pine or mixtures of species in the Carpathian Mountains of Ukraine. An overall accuracy of about 85% was achieved in separating OGF from the surrounding forest, with classification accuracies higher for conifer and broadleaved than mixed forest.

We make a number of recommendations for automated identification of OGF. OGF is more spatially heterogeneous than other forest types. Adding textural features therefore improved classification. The addition of band standard deviations, combining summer and autumn images and adding elevation data also improved overall accuracy. We found limited benefits to using vegetation indices—which if added to the bands gave only a minimal performance improvement. We'd recommend calculating textural features instead as it involves the same amount of effort and since the spatial relationship of the pixels is not strongly correlated to their brightness you are adding useful independent information to the model.

Our method of comparing OGF to adjacent forest is not without weaknesses. It meant that our comparison of OGF to NOGF was not comparing 'like-with-like,' as the control NOGF polygons were usually forests lower in height. However, as remaining OGF in Europe is usually confined to the mountains this will tend to be true of any real-world attempt to classify OGF. Furthermore, ground identification will generally include criteria such as deadwood quantity and quality, presence of non-native tree species and human impact such as livestock grazing that cannot be surveyed remotely. With these caveats we were able to use free publicly available satellite imagery to correctly classify OGF on the ground with an overall accuracy of about 85%. This is at the threshold of what is usually deemed acceptable in machine learning studies [71]. Potential improvements could involve exploring the use of other classification types—for example, Support vector Machines (SVM) has been found to be more accurate than Random Forest in tree species classification studies [13,14,17]. Further studies that cover different OGF types within different biogeographical settings would be useful.

Supplementary Materials: The following are available online at http://www.mdpi.com/1999-4907/10/2/127/s1, Figure S1: Boxplots of (a) spectral signatures, (b) vegetation indices and (c) textural measures for tree species silver fir (AA), beech (FS), Norway spruce (PA), mountain pine (PM) and oak species (Quer) in Old-Growth Forest polygons, Figure S2: Random Forest feature importance for OGF tree species classification, Figure S3: Boxplots of (a) spectral signatures, (b) vegetation indices and (c) textural measures for broadleaf, conifer and mixed Old-Growth Forest and Non Old-Growth Forest polygons, arranged in two pairs of summer (thick line boxplots) and autumn (thin line boxplots) from left to right, Figure S4: Mean percentile values for the green Band (B3) for (a) broadleaf and (b) mixed Old-Growth Forest and Non-Old Growth Forest polygons, Figure S5: Random Forest feature importance for Old-Growth forest and Non-Old Growth forest classification for (a) conifers, (b) broadleaf, (c) mixed forest, Table S1: Producers, users and overall accuracy for Random Forest Old-Growth forest tree species classifications for our 8 models: mean band values (B), Textural features (TF), Vegetation indices(VI), B+TF, B+VI, B+mean elevation (elev), B+band standard deviations (B_sd) and B+B_sd+Elev+TF+VI, Table S2: Producers, users and overall accuracy for Random Forest Old Growth forest classification for our 8 models: mean band values (B), Textural features (TF), Vegetation indices(VI), B+TF, B+VI, B+mean elevation (elev), B+band standard deviations (B_sd) and B+B_sd+Elev+TF+VI, Table S3: Producers, users and overall accuracy for Random Forest Old-Growth Forest classification for summer and autumn images combined and our 8 models: mean band values (B), Textural features (TF), Vegetation indices(VI), B+TF, B+VI, B+mean elevation (elev), B+band standard deviations (B_sd) and B+B_sd+elev+TF+VI.

Author Contributions: Conceptualization, B.D.S. and D.V.S.; Methodology, B.D.S.; Software, B.D.S.; Validation, B.D.S.; Formal Analysis, B.D.S.; Investigation, B.D.S.; Resources, B.D.S. and D.V.S.; Data Curation, B.D.S.; Writing—Original Draft Preparation, B.D.S.; Writing—Review & Editing, D.V.S.; Visualization, B.D.S.; Supervision, D.V.S.; Project Administration, D.V.S.; Funding Acquisition, D.V.S.

Funding: We thank the United Bank of Carbon and the Natural Environment Research Council (NE/L013347/1) for funding. DVS acknowledges a Philip Leverhulme Prize.

Acknowledgments: We wish to thank WWF Ukraine and in particular Valeriia Nemykina, Taras Yamelynets and Roman Volosyanchuk for their assistance and provision of data. We thank three anonymous reviewers whose suggestions greatly improved the paper.

Conflicts of Interest: The authors declare no conflict of interest.

References

1. Merce, O.; Borlea, G.F.; Turcu, D.O. Definitions and structural attributes of the ecosystems from natural forests-short review. *J. Hortic. For. Biotechnol.* **2014**, *18*, 114–120.
2. Wirth, C.; Messier, C.; Bergeron, Y.; Frank, D.; Fankhänel, A. Old-growth forest definitions: A pragmatic view. Wirth, C., Gleixner, G., Heimann, M., Eds.; In *Old-Growth Forests*; Springer: Berlin/Heidelberg, Germany, 2009; Volume 207, pp. 11–33.
3. Burrascano, S.; Keeton, W.S.; Sabatini, F.M.; Blasi, C. Commonality and variability in the structural attributes of moist temperate old-growth forests: A global review. *For. Ecol. Manag.* **2013**, *291*, 458–479. [CrossRef]
4. Paillet, Y.; Bergès, L.; Hjältén, J.; Ódor, P.; Avon, C.; Bernhardt-Römermann, M.; Bijlsma, R.-J.; De Bruyn, L.U.C.; Fuhr, M.; Grandin, U.L.F.; et al. Biodiversity differences between managed and unmanaged forests: Meta-analysis of species richness in Europe. *Conserv. Biol.* **2010**, *24*, 101–112. [CrossRef] [PubMed]
5. Granata, M.U.; Gratani, L.; Bracco, F.; Sartori, F.; Catoni, R. Carbon stock estimation in an unmanaged old-growth forest: A case study from a broad-leaf deciduous forest in the Northwest of Italy. *Int. For. Rev.* **2016**, *18*, 444–451. [CrossRef]
6. Jacob, M.; Bade, C.; Calvete, H.; Dittrich, S.; Leuschner, C.; Hauck, M. Significance of over-mature and decaying trees for carbon stocks in a Central European natural spruce forest. *Ecosystems* **2013**, *16*, 336–346. [CrossRef]

7. Seedre, M.; Kopáček, J.; Janda, P.; Bače, R.; Svoboda, M. Carbon pools in a montane old-growth Norway spruce ecosystem in Bohemian Forest: Effects of stand age and elevation. *For. Ecol. Manag.* **2015**, *346*, 106–113. [CrossRef]

8. Keeton, W.S.; Chernyavskyy, M.; Gratzer, G.; Main-Knorn, M.; Shpylchak, M.; Bihun, Y. Structural characteristics and aboveground biomass of old-growth spruce–fir stands in the eastern Carpathian mountains, Ukraine. *Plant Biosyst.* **2010**, *144*, 148–159. [CrossRef]

9. Luyssaert, S.; Schulze, E.-D.; Börner, A.; Knohl, A.; Hessenmöller, D.; Law, B.E.; Ciais, P.; Grace, J. Old-growth forests as global carbon sinks. *Nature* **2008**, *455*, 213–215. [CrossRef]

10. Frey, S.J.; Hadley, A.S.; Johnson, S.L.; Schulze, M.; Jones, J.A.; Betts, M.G. Spatial models reveal the microclimatic buffering capacity of old-growth forests. *Sci. Adv.* **2016**, *2*, e1501392. [CrossRef]

11. Sabatini, F.M.; Burrascano, S.; Keeton, W.S.; Levers, C.; Lindner, M.; Pötzschner, F.; Verkerk, P.J.; Bauhus, J.; Buchwald, E.; Chaskovsky, O.; et al. Where are Europe's last primary forests? *Divers. Distrib.* **2018**, *24*, 1426–1439. [CrossRef]

12. Knorn, J.A.N.; Kuemmerle, T.; Radeloff, V.C.; Keeton, W.S.; Gancz, V.; BIRIŞ, I.-A.; Svoboda, M.; Griffiths, P.; Hagatis, A.; Hostert, P. Continued loss of temperate old-growth forests in the Romanian Carpathians despite an increasing protected area network. *Environ. Conserv.* **2013**, *40*, 182–193. [CrossRef]

13. Dalponte, M.; Bruzzone, L.; Gianelle, D. Tree species classification in the Southern Alps based on the fusion of very high geometrical resolution multispectral/hyperspectral images and LiDAR data. *Remote Sens. Environ.* **2012**, *123*, 258–270. [CrossRef]

14. Heinzel, J.; Koch, B. Investigating multiple data sources for tree species classification in temperate forest and use for single tree delineation. *Int. J. Appl. Earth Obs. Geoinf.* **2012**, *18*, 101–110. [CrossRef]

15. Immitzer, M.; Atzberger, C.; Koukal, T. Tree species classification with random forest using very high spatial resolution 8-band WorldView-2 satellite data. *Remote Sens.* **2012**, *4*, 2661–2693. [CrossRef]

16. Puletti, N.; Chianucci, F.; Castaldi, C. Use of Sentinel-2 for forest classification in Mediterranean environments. *Ann. Silvic. Res.* **2018**, *42*, 32–38.

17. Sheeren, D.; Fauvel, M.; Josipović, V.; Lopes, M.; Planque, C.; Willm, J.; Dejoux, J.-F. Tree species classification in temperate forests using Formosat-2 satellite image time series. *Remote Sens.* **2016**, *8*, 734. [CrossRef]

18. Leckie, D.G.; Tinis, S.; Nelson, T.; Burnett, C.; Gougeon, F.A.; Cloney, E.; Paradine, D. Issues in species classification of trees in old growth conifer stands. *Can. J. Remote Sens.* **2005**, *31*, 175–190. [CrossRef]

19. Cohen, W.B.; Spies, T.A.; Fiorella, M. Estimating the age and structure of forests in a multi-ownership landscape of western Oregon, USA. *Int. J. Remote Sens.* **1995**, *16*, 721–746. [CrossRef]

20. Cohen, W.B.; Spies, T.A. Estimating structural attributes of Douglas-fir/western hemlock forest stands from Landsat and SPOT imagery. *Remote Sens. Environ.* **1992**, *41*, 1–17. [CrossRef]

21. Fiorella, M.; Ripple, W.J. Determining successional stage of temperate coniferous forests with Landsat satellite data. *Photogramm. Eng. Remote Sens.* **1993**, *59*, 239–246.

22. Jiang, H.; Strittholt, J.R.; Frost, P.A.; Slosser, N.C. The classification of late seral forests in the Pacific Northwest, USA using Landsat ETM+ imagery. *Remote Sens. Environ.* **2004**, *91*, 320–331. [CrossRef]

23. Sabol Jr, D.E.; Gillespie, A.R.; Adams, J.B.; Smith, M.O.; Tucker, C.J. Structural stage in Pacific Northwest forests estimated using simple mixing models of multispectral images. *Remote Sens. Environ.* **2002**, *80*, 1–16. [CrossRef]

24. Sverdrup-Thygeson, A.; Ørka, H.O.; Gobakken, T.; Næsset, E. Can airborne laser scanning assist in mapping and monitoring natural forests? *For. Ecol. Manag.* **2016**, *369*, 116–125. [CrossRef]

25. Kathmann, F.; Ciutea, A.; Biris, I.-A.; Ibisch, P.L.; Salageanu, V. *Potential Primary Forests Map of Romania*; Ibisch, P.L., Ursu, A., Eds.; Greenpeace CEE Romania; Centre for Econics and Ecosystem Management, Eberswalde University for Sustainable Development; Geography Department, A. I. Cuza University of Iaşi: Bucharest, Romania, 2017; Available online: https://www.researchgate.net/publication/321098644_Potential_Primary_Forests_Map_of_Romania_published_by_Greenpeace_CEE_Romania_Centre_for_Econics_and_Ecosystem_Management_Eberswalde_University_for_Sustainable_Development_Geography_Department_A_I_Cuza (accessed on 11 January 2018).

26. Immitzer, M.; Vuolo, F.; Atzberger, C. First experience with Sentinel-2 data for crop and tree species classifications in central Europe. *Remote Sens.* **2016**, *8*, 166. [CrossRef]

27. Forest Code of Ukraine (in Ukrainian). Available online: https://zakon.rada.gov.ua/go/3852-12 (accessed on 10 January 2019).

28. Lavnyy, V.; Lässig, R. Extent of storms in the Ukrainian Carpathians. In Proceedings of the Proceedings of the International Conference on Wind Effects on Trees, University of Karlsruhe, Karlsruhe, Baden-Württemberg, Germany, 16–18 September 2003; pp. 16–18.

29. Simpson, M. Determining the potential distribution of highly invasive plants in the Carpathian Mountains of Ukraine: A species distribution modeling approach under different climate-land-use scenarios and possible implications for natural-resource management. *Environ. Sci. Policy* **2011**, *12*. Available online: http://jhir.library.jhu.edu/handle/1774.2/35707 (accessed on 10 January 2019).

30. Kozak, J.; Estreguil, C.; Troll, M. Forest cover changes in the northern Carpathians in the 20th century: A slow transition. *J. Land Use Sci.* **2007**, *2*, 127–146. [CrossRef]

31. Kuemmerle, T.; Hostert, P.; Radeloff, V.C.; van der Linden, S.; Perzanowski, K.; Kruhlov, I. Cross-border comparison of post-socialist farmland abandonment in the Carpathians. *Ecosystems* **2008**, *11*, 614. [CrossRef]

32. Kuemmerle, T.; Hostert, P.; Radeloff, V.C.; Perzanowski, K.; Kruhlov, I. Post-socialist forest disturbance in the Carpathian border region of Poland, Slovakia and Ukraine. *Ecol. Appl.* **2007**, *17*, 1279–1295. [CrossRef]

33. Kuemmerle, T.; Chaskovskyy, O.; Knorn, J.; Radeloff, V.C.; Kruhlov, I.; Keeton, W.S.; Hostert, P. Forest cover change and illegal logging in the Ukrainian Carpathians in the transition period from 1988 to 2007. *Remote Sens. Environ.* **2009**, *113*, 1194–1207. [CrossRef]

34. Complicit in Corruption. How billion-dollar firms and EU Governments are Failing Ukraine's Forests (2018) Earthsight. Available online: https://docs.wixstatic.com/ugd/624187_673e3aa69ed84129bdfeb91b6aa9ec17.pdf (accessed on 10 January 2019).

35. Irland, L.; Kremenetska, E. Practical economics of forest ecosystem management: The case of the Ukrainian Carpathians. *Ecol. Econ. Sustain. For. Manag. Dev. Trans-Discip. Approach Carpathian Mt. Ukr. Natl. For. Univ. Press* **2009**, *59*, 180–200.

36. Sitko, I.; Troll, M. Timberline changes in relation to summer farming in the Western Chornohora (Ukrainian Carpathians). *Mt. Res. Dev.* **2008**, *28*, 263–271. [CrossRef]

37. Trotsiuk, V.; Hobi, M.L.; Commarmot, B. Age structure and disturbance dynamics of the relic virgin beech forest Uholka (Ukrainian Carpathians). *For. Ecol. Manag.* **2012**, *265*, 181–190. [CrossRef]

38. Trotsiuk, V.; Svoboda, M.; Janda, P.; Mikolas, M.; Bace, R.; Rejzek, J.; Samonil, P.; Chaskovskyy, O.; Korol, M.; Myklush, S. A mixed severity disturbance regime in the primary *Picea abies* (L.) Karst. forests of the Ukrainian Carpathians. *For. Ecol. Manag.* **2014**, *334*, 144–153. [CrossRef]

39. Hobi, M.L.; Commarmot, B.; Bugmann, H. Pattern and process in the largest primeval beech forest of Europe (Ukrainian Carpathians). *J. Veg. Sci.* **2015**, *26*, 323–336. [CrossRef]

40. Volosyanchuk, R.; Prots, B.; Kagalo, A.; Shparyk, Y.; Cherniavskyi, M.; Bondaruk, G. *Criteria and Methology for Virgin and Old-Growth (Quasi-Virgin) Forest Identification*; Volosyanchuk, R., Prots, B., Kagalo, A., Eds.; Liga Press: Lviv, Ukraine, 2017. Available online: http://d2ouvy59p0dg6k.cloudfront.net/downloads/old_growth_forest_identification_methodology.pdf (accessed on 1 December 2018).

41. Roman Volosyanchuk Virgin and Old Growth Forests in Ukraine. Available online: http://www.carpathianconvention.org/tl_files/carpathiancon/Downloads/03%20Meetings%20and%20Events/Working%20Groups/Sustainable%20Forest%20Management/6th%20meeting/presentations/VF_for_CC_Sopron_WWF_UA.pdf (accessed on 11 January 2018).

42. Hansen, M.C.; Potapov, P.V.; Moore, R.; Hancher, M.; Turubanova, S.A.; Tyukavina, A.; Thau, D.; Stehman, S.V.; Goetz, S.J.; Loveland, T.R.; et al. High-resolution global maps of 21st-century forest cover change. *Science* **2013**, *342*, 850–853. [CrossRef] [PubMed]

43. Wulder, M.A.; Skakun, R.S.; Kurz, W.A.; White, J.C. Estimating time since forest harvest using segmented Landsat ETM+ imagery. *Remote Sens. Environ.* **2004**, *93*, 179–187. [CrossRef]

44. Drusch, M.; Del Bello, U.; Carlier, S.; Colin, O.; Fernandez, V.; Gascon, F.; Hoersch, B.; Isola, C.; Laberinti, P.; Martimort, P. Sentinel-2: ESA's optical high-resolution mission for GMES operational services. *Remote Sens. Environ.* **2012**, *120*, 25–36. [CrossRef]

45. Louis, J.; Debaecker, V.; Pflug, B.; Main-Knorn, M.; Bieniarz, J.; Mueller-Wilm, U.; Cadau, E.; Gascon, F. Sentinel-2 Sen2Cor: L2A Processor for Users. In Proceedings of the Proceedings Living Planet Symposium 2016, Prague, Czech Republic, 9–13 May 2016; pp. 1–8.

46. Clark, M.L.; Roberts, D.A.; Clark, D.B. Hyperspectral discrimination of tropical rain forest tree species at leaf to crown scales. *Remote Sens. Environ.* **2005**, *96*, 375–398. [CrossRef]

47. Coburn, C.A.; Roberts, A.C. A multiscale texture analysis procedure for improved forest stand classification. *Int. J. Remote Sens.* **2004**, *25*, 4287–4308. [CrossRef]

48. Zhaoa, P.; Zhaoa, J.; Wub, J.; Yanga, Y.; Xuea, W.; Houa, Y. Integration of multi-classifiers in object-based methods for forest classification in the Loess plateau, China. *SCIENCEASIA* **2016**, *42*, 283–289. [CrossRef]

49. Haralick, R.M. Statistical and structural approaches to texture. *Proc. IEEE* **1979**, *67*, 786–804. [CrossRef]

50. Hall-Beyer, M. *GLCM Texture: A Tutorial*; Technical Report for Department of Geography; University of Calgary: Calgary, AB, Canada, March 2000.

51. Ozdemir, I.; Karnieli, A. Predicting forest structural parameters using the image texture derived from WorldView-2 multispectral imagery in a dryland forest, Israel. *Int. J. Appl. Earth Obs. Geoinf.* **2011**, *13*, 701–710. [CrossRef]

52. Shaban, M.A.; Dikshit, O. Improvement of classification in urban areas by the use of textural features: The case study of Lucknow city, Uttar Pradesh. *Int. J. Remote Sens.* **2001**, *22*, 565–593. [CrossRef]

53. Chen, D.; Stow, D.A.; Gong, P. Examining the effect of spatial resolution and texture window size on classification accuracy: An urban environment case. *Int. J. Remote Sens.* **2004**, *25*, 2177–2192. [CrossRef]

54. Jarvis, A.; Reuter, H.I.; Nelson, A.; Guevara, E. Hole-filled SRTM for the globe Version 4: Data grid. 2008. Available online: https://research.utwente.nl/en/publications/hole-filled-srtm-for-the-globe-version-4-data-grid (accessed on 31 December 2018).

55. Breiman, L. Random forests. *Mach. Learn.* **2001**, *45*, 5–32. [CrossRef]

56. Rodriguez-Galiano, V.F.; Ghimire, B.; Rogan, J.; Chica-Olmo, M.; Rigol-Sanchez, J.P. An assessment of the effectiveness of a random forest classifier for land-cover classification. *ISPRS J. Photogramm. Remote Sens.* **2012**, *67*, 93–104. [CrossRef]

57. Cutler, D.R.; Edwards Jr, T.C.; Beard, K.H.; Cutler, A.; Hess, K.T.; Gibson, J.; Lawler, J.J. Random forests for classification in ecology. *Ecology* **2007**, *88*, 2783–2792. [CrossRef] [PubMed]

58. Pal, M. Random forest classifier for remote sensing classification. *Int. J. Remote Sens.* **2005**, *26*, 217–222. [CrossRef]

59. Sesnie, S.E.; Finegan, B.; Gessler, P.E.; Thessler, S.; Ramos Bendana, Z.; Smith, A.M. The multispectral separability of Costa Rican rainforest types with support vector machines and Random Forest decision trees. *Int. J. Remote Sens.* **2010**, *31*, 2885–2909. [CrossRef]

60. Nelson, M. Evaluating Multitemporal Sentinel-2 data for Forest Mapping using Random Forest. Master's Thesis, Stockholm University, Stockholm, Sweden, December 2017.

61. Pedregosa, F.; Varoquaux, G.; Gramfort, A.; Michel, V.; Thirion, B.; Grisel, O.; Blondel, M.; Prettenhofer, P.; Weiss, R.; Dubourg, V. Scikit-learn: Machine learning in Python. *J. Mach. Learn. Res.* **2011**, *12*, 2825–2830.

62. Lu, D.; Weng, Q. A survey of image classification methods and techniques for improving classification performance. *Int. J. Remote Sens.* **2007**, *28*, 823–870. [CrossRef]

63. Key, T.; Warner, T.A.; McGraw, J.B.; Fajvan, M.A. A comparison of multispectral and multitemporal information in high spatial resolution imagery for classification of individual tree species in a temperate hardwood forest. *Remote Sens. Environ.* **2001**, *75*, 100–112. [CrossRef]

64. Schriever, J.R.; Congalton, R.G. Evaluating seasonal variablity as an aid to cover-type mapping from landsat thematic mapper data in the northwest. *Photogramm. Eng. Remote Sens.* **1995**, *61*, 321–327.

65. Zimmermann, N.E.; Edwards, T.C.; Moisen, G.G.; Frescino, T.S.; Blackard, J.A. Remote sensing-based predictors improve distribution models of rare, early successional and broadleaf tree species in Utah. *J. Appl. Ecol.* **2007**, *44*, 1057–1067. [CrossRef] [PubMed]

66. Crist, E.P.; Laurin, R.; Cicone, R.C. Vegetation and soils information contained in transformed Thematic Mapper data. In *Proceedings of the Proceedings of IGARSS'86 Symposium*; European Space Agency: Paris, France, 1986; pp. 1465–1470.

67. Kimes, D.S.; Newcomb, W.W.; Nelson, R.F.; Schutt, J.B. Directional reflectance distributions of a hardwood and pine forest canopy. *IEEE Trans. Geosci. Remote Sens.* **1986**, *24*, 281–293. [CrossRef]

68. Glenn, E.; Huete, A.; Nagler, P.; Nelson, S. Relationship between remotely-sensed vegetation indices, canopy attributes and plant physiological processes: What vegetation indices can and cannot tell us about the landscape? *Sensors* **2008**, *8*, 2136–2160. [CrossRef] [PubMed]

69. Tuominen, J.; Lipping, T.; Kuosmanen, V.; Haapanen, R. *Remote Sensing of Forest Health. Geoscience and Remote Sensing*; IntechOpen: London, UK, 2009; Available online: https://www.intechopen.com/books/geoscience-and-remote-sensing/remote-sensing-of-forest-health (accessed on 1 January 2019).

70. Yilmaz, M.T.; Hunt, E.R., Jr.; Goins, L.D.; Ustin, S.L.; Vanderbilt, V.C.; Jackson, T.J. Vegetation water content during SMEX04 from ground data and Landsat 5 Thematic Mapper imagery. *Remote Sens. Environ.* **2008**, *112*, 350–362. [CrossRef]
71. Thomlinson, J.R.; Bolstad, P.V.; Cohen, W.B. Coordinating methodologies for scaling landcover classifications from site-specific to global: Steps toward validating global map products. *Remote Sens. Environ.* **1999**, *70*, 16–28. [CrossRef]

 © 2019 by the authors. Licensee MDPI, Basel, Switzerland. This article is an open access article distributed under the terms and conditions of the Creative Commons Attribution (CC BY) license (http://creativecommons.org/licenses/by/4.0/).

 *forests*

Article

Landsat 8 Based Leaf Area Index Estimation in Loblolly Pine Plantations

Christine E. Blinn [1], Matthew N. House [1], Randolph H. Wynne [1,*], Valerie A. Thomas [1], Thomas R. Fox [2] and Matthew Sumnall [1]

[1] Department of Forest Resources and Environmental Conservation, College of Natural Resources and Environment, Virginia Tech, Blacksburg, VA 24061, USA; cblinn@vt.edu (C.E.B.); mhomes84@vt.edu (M.N.H.); thomasv@vt.edu (V.A.T.); msumnall@vt.edu (M.S.)
[2] Rayonier Inc., Yulee, FL 32097, USA; tom.fox@rayonier.com
* Correspondence: wynne@vt.edu; Tel.: +1-540-231-7811

Received: 18 January 2019; Accepted: 28 February 2019; Published: 2 March 2019

Abstract: Leaf area index (LAI) is an important biophysical parameter used to monitor, model, and manage loblolly pine plantations across the southeastern United States. Landsat provides forest scientists and managers the ability to obtain accurate and timely LAI estimates. The objective of this study was to investigate the relationship between loblolly pine LAI measured in situ (at both leaf area minimum and maximum through two growing seasons at two geographically disparate study areas) and vegetation indices calculated using data from Landsat 7 (ETM+) and Landsat 8 (OLI). Sub-objectives included examination of the impact of georegistration accuracy, comparison of top-of-atmosphere and surface reflectance, development of a new empirical model for the species and region, and comparison of the new empirical model with the current operational standard. Permanent plots for the collection of ground LAI measurements were established at two locations near Appomattox, Virginia and Tuscaloosa, Alabama in 2013 and 2014, respectively. Each plot is thirty by thirty meters in size and is located at least thirty meters from a stand boundary. Plot LAI measurements were collected twice a year using the LI-COR LAI-2200 Plant Canopy Analyzer. Ground measurements were used as dependent variables in ordinary least squares regressions with ETM+ and OLI-derived vegetation indices. We conclude that accurately-located ground LAI estimates at minimum and maximum LAI in loblolly pine stands can be combined and modeled with Landsat-derived vegetation indices using surface reflectance, particularly simple ratio (SR) and normalized difference moisture index (NDMI), across sites and sensors. The best resulting model (LAI = −0.00212 + 0.3329SR) appears not to saturate through an LAI of 5 and is an improvement over the current operational standard for loblolly pine monitoring, modeling, and management in this ecologically and economically important region.

Keywords: remote sensing; forestry; phenology; silviculture

---

1. Introduction

Leaf area index (LAI) is widely used in silvicultural decision support [1,2]. Careful silvicultural manipulation of LAI can reduce drought stress [3], and can be used to assess site preparation outcomes, including the proportion of stand LAI post-establishment that is from competing herbaceous and arborescent species [4]. LAI has been shown to be an important predictor of site nutrient status in pine stands and can be used to determine the likelihood of forest management response [5]. Fertilization increases LAI in loblolly pine (*Pinus taeda* L.) stands which leads to faster growth [6–8]. LAI can be used by forest managers to identify stands that would likely benefit from fertilization and/or competition control [9], and, for a given species, a stand's maximum (life cycle) LAI is strongly correlated with other measures of site quality such as site index [10].

Modern forestry thus requires satellite-derived LAI [1,2], but as noted by Gao et al. 2014 [11], the coarse scale of available satellite derived LAI estimates from the Moderate Resolution Imaging Spectroradiometer (MODIS) is not adequate for stand-scale analyses since most pine plantations in the southern U.S. contain very few or no homogeneous MODIS pixels. The promise and potential of LAI estimation in monospecific pine plantations using Landsat data was noted by Vogel et al. [12]. Thus, there is both strong potential and extant need for LAI estimates from medium resolution sensors for use in forest management.

Loblolly pine trees carry two needle cohorts at maximum LAI in late summer and one at minimum LAI in late winter resulting in an almost doubling of pine LAI between minimum and maximum [13]. Since a majority of loblolly pine plantations also have hardwood competition that contributes to LAI in seasons other than winter, empirical modeling has often focused on minimum LAI in the winter when only the pine trees have "leaves" [14]. Previous research conducted by Flores et al. [15] related top-of-atmosphere (TOA) reflectance-based simple ratio vegetation index from Landsat 7's Enhanced Thematic Mapper Plus (ETM+) sensor to loblolly pine LAI in the winter. They indicated that others have also successfully used remote sensing to characterize the LAI of forest stands [15].

Differences between Landsat-derived estimates of LAI in spring and winter have been shown to be useful in the estimation of the relative abundance of deciduous competition in loblolly pine stands [9]. However, models for the estimation of loblolly pine LAI with satellite-based vegetation indices (VIs) have not been as successful at maximum LAI, with the exception of intensively managed slash pine with an ericaceous understory [16–18]. Chen and Cihlar [19] found late spring images to be better than summer images for the estimation of LAI in boreal conifer stands because of the reduced influence of the understory, a result echoed by Tian et al. [20] using the annual leaf area minimum (better model) versus maximum (worse model). For boreal forests, Chen and Cihlar [19] concluded that there is a linear relationship between SR and LAI, while Curran et al. [18] demonstrated that remote sensing could be used to study seasonal dynamics of LAI in southern slash pine plantations.

Vegetation index saturation has been a problem in forest stands with high levels of LAI [21–23], which is especially problematic in the summer across the southern portions of loblolly pine's range in the U.S. Turner et al. [22] evaluated the relationship between in situ LAI estimates and Landsat Thematic Mapper (TM)-derived VIs for several cover types at three different study sites in the summer. Surface reflectance based normalized difference vegetation index (NDVI) resulted in the best model fit, but VI saturation was still noted in the coniferous stands [22]. Sumnall et al. [24] suggest that using LiDAR can reduce or remove the saturation issues in areas of high biomass.

Empirical models for the estimation of LAI in forest stands have often been based on a single point in time measurement of ground LAI and a single Landsat image, but more recent studies have looked at variation in LAI over time [17,23,25–27]. Peduzzi et al. [16] collected ground LAI estimates of both the overstory and understory vegetation in loblolly and slash pine stands (at both maximum and minimum LAI) to investigate the impact of evergreen competition on LAI estimation with Landsat. They found that overstory LAI during growing season maximum was better predicted using VIs from the growing season minimum (winter) than it was using those from the growing season maximum (summer).

The objective of this study was to investigate the relationship between loblolly pine LAI measured in situ (at both leaf area minimum and maximum through two growing seasons at two geographically disparate study areas) and VIs calculated using data from Landsat 7 (ETM+) and Landsat 8 Operational Land Imager (OLI). Sub-objectives included examination of the impact of georegistration accuracy, comparison of TOA and surface reflectance (since the current operational model [15] uses TOA only), development of a new empirical model for the species and region, and comparison of the new empirical model with the current operational standard.

2. Materials and Methods

2.1. Study Sites and Field Data

Twenty-two permanent ground plots (Figure 1) were established in loblolly pine plantations at each of the two study sites (Figure 1C) for the estimation of ground LAI. One site was established at the Appomattox-Buckingham State Forest near Appomattox, Virginia, USA in March of 2013 which is within Landsat Worldwide Reference System 2 (WRS-2) path 16 row 34 (16/34). A second site was established on forest industry land near Tuscaloosa, AL in February of 2014 which is located in WRS-2 path 21 row 37 (21/37). The Virginia site is located in the Piedmont and the Alabama site is located in the Coastal Plain physiographic province. Pine stands were selected to cover the variability in LAI across the sites and to minimize the within stand variability of LAI. Each plot is 30 by 30 m in size (but not installed to align perfectly with a single pixel footprint) and all plot centers were established at least 30 m from stand boundaries. Very little to no evergreen understory existed within any of the ground LAI plots. A Trimble GeoXT was used to collect GPS locations for each plot center with the Trimble TerraSync software version 5.02. All GPS points were differentially corrected in Trimble's GPS Pathfinder Office software version 5.20 to a horizontal accuracy of less than one meter.

Optical ground measurements of LAI were collected using two cross-calibrated LI-COR LAI-2200 or 2200C Plant Canopy Analyzers (and one data logger) with 90 degree view caps [28], cross-checked with litter traps and hemispheric photography in a prior study in this ecosystem [29]. The sensor was held at approximately 3.5 feet off the ground. One LAI-2200 sensor was set up in a large open area in close proximity to the ground plots to continuously collect above canopy readings at a 15 s interval during plot measurements. Both diagonals of each ground plot were sampled with at least 27 below canopy readings per plot. LAI sampling was predominantly conducted during relatively clear sky days or when the sky was uniformly overcast as recommended by LI-COR, Inc. (Lincoln, NE 68504, USA) [30]. LAI estimates for each below canopy reading were obtained using LI-COR's FV2200 software version 2.0 or 2.1 with a horizontal model and the exclusion of ring 5, which is most highly obstructed by trunks and branches [16]. These LAI estimates might be better thought of as plant area index since anything including stems and branches that blocks the sensors view of incoming radiation impacts the estimates. Since the FV2200 software includes a correction for clumping (employed in this study), these estimates are closer to true LAI than effective LAI [30]. All LAI estimates for a given plot and measurement period were averaged and used as the dependent variables in subsequent analyses.

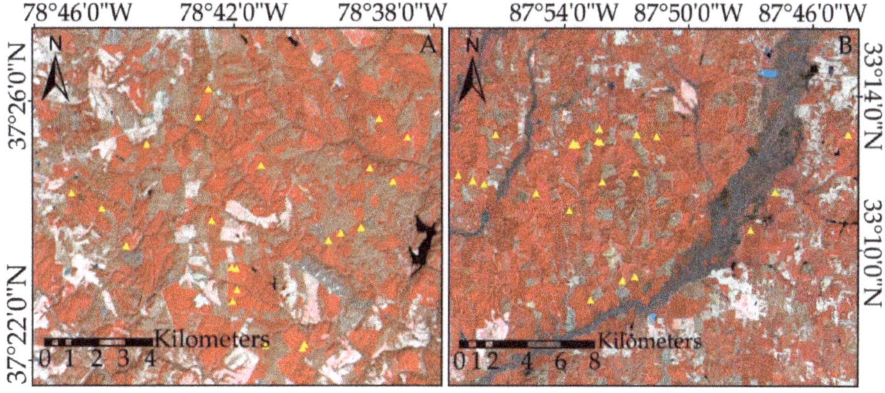

Figure 1. *Cont.*

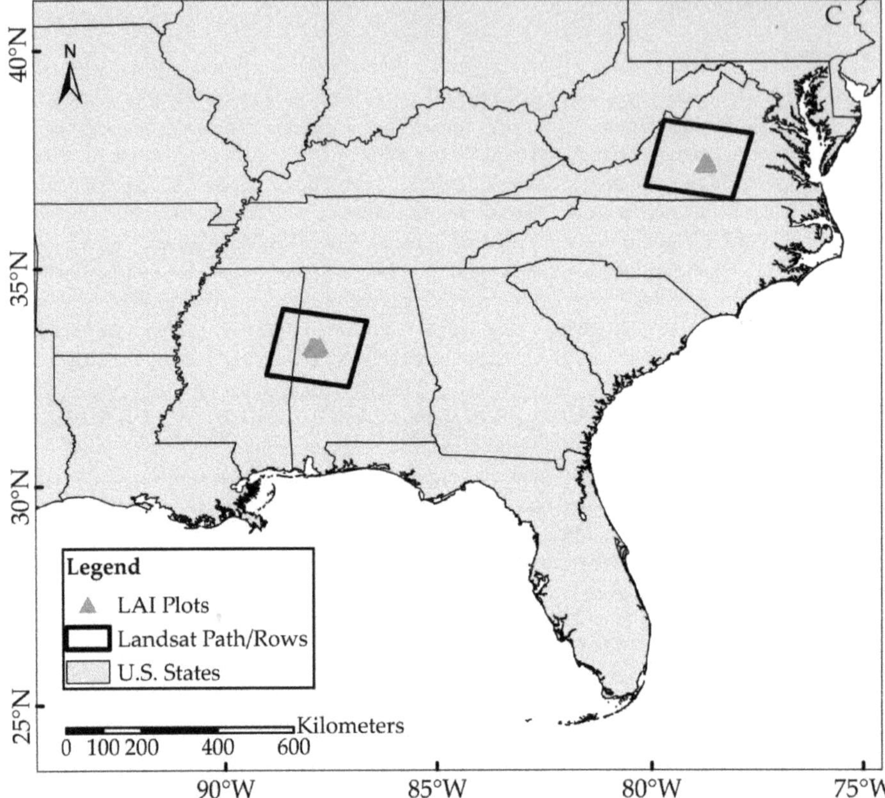

Figure 1. Ground LAI plot locations shown as yellow triangles over Landsat 8 OLI imagery displayed with the color-infrared band combination NIR, Red, Green as RGB. The Virginia site is on the top left (**A**) over a 16/34 OLI image from 14 March 2014 and the Alabama site is on the top right (**B**) over a 21/37 OLI image from 13 February 2014. (**C**) shows the study site locations within Virginia and Alabama, U.S. and Landsat path/rows used.

Plots were measured two times a year since installation in close proximity to maximum and minimum LAI (Table 1) resulting in six separate field data collections. Since LAI variations occur more rapidly in the growing season than in the winter, and there is often reduced image availability in the summer due to cloud cover, we attempted to match ground data collection as closely as logistically possible to image acquisition dates. LAI estimates for each ground plot at a given site from maximum and minimum LAI within a given year were plotted against one another to identify erroneous ground data points. For the VA site, interannual plots of minimum and maximum LAI were also created. Any plot from a given measurement period that was clearly outside realistic bounds (i.e., less than zero or greater than ten) was removed from the data set. For the thirteen plots with repeat measurements across three seasons, inter-season residual correlation never exceeded 0.31, well within the 95% threshold of 0.53. As such, error independence was assumed.

Table 1. Ground LAI measurement details by study location (Virginia: 16/34, Alabama: 21/37) and LAI status (maximum or minimum). The maximum number of plots that were used in subsequent analyses is included in parentheses in column 4. Some plots were not used due to ground data errors and atmospheric conditions (clouds and shadows). Updated values are provided in parentheses in column 3 when the LAI range changed due to plot removal.

Location of LAI & Status	Ground Measurement Dates	Ground LAI Range (Min–Max)	Plots Measured (Max Used)	OLI Image Date	ETM+ Image Date
VA Min	1 April 2013 & 3 April 2013	1.14–3.07	22 (20)	28 March 2013	19 March 2013
VA Max	13 September 2013 & 14 September 2013	2.04–5.39 (5.33)	10 (8)	5 October 2013	11 September 2013
AL Min	21 February 2014, 24 February 2014 & 27 February 2014	0.68–4.42 (3.93)	22 (21)	13 February 2014	21 February 2014
VA Min	26 March 2014	1.13–4.39 (3.43)	19 (18)	14 March 2014	23 April 2014
AL Max	14 September 2014 & 25 September 2014	0.63–7.34 (5.14)	9 (7)	25 September 2014	1 September 2014
VA Max	20 September 2014	2.04–5.94 (5.60)	20 (19)	22 September 2014	29 August 2014

2.2. Landsat Data and LAI Regression Models

2.2.1. Landsat Sensors

The two different sensors used, Landsat 7 ETM+ and Landsat 8 OLI, have different band orders, as illustrated in Table 2. The imagery from Landsat 7 ETM+ is subject to the Scan Line Corrector (SLC) failure and one plot from the Virginia study site location fell within the resulting data gap, resulting in 21 available plots for ETM+ scenes.

Table 2. Landsat sensors and specific bands used in VIs. All bands listed have a resolution of 30 m. This highlights the differences in wavelength of the bands between Landsat 7 and Landsat 8 [31].

Band Name	Landsat 7 ETM+ Band Number	Landsat 7 ETM+ Wavelength (Micrometers)	Landsat 8 OLI Band Number	Landsat 8 OLI Wavelength (Micrometers)
Blue	1	0.441–0.514	2	0.452–0.512
Red	3	0.631–0.692	4	0.636–0.673
Near Infrared (NIR)	4	0.772–0.898	5	0.851–0.879
Shortwave Infrared (SWIR) 1	5	1.547–1.749	6	1.566–1.651

From Landsat 7 ETM+ to Landsat 8 OLI, the wavelengths of the bands have changed. The OLI sensor's bands are much narrower than the ETM+ bands as shown in Table 2. This improvement allows the OLI sensor to potentially experience less saturation within the bands and to avoid atmospheric absorption features to which ETM+ bands were subjected [32].

2.2.2. Vegetation Indices and Reflectance

Landsat ETM+ and OLI imagery from 16/34 and 21/37 with the least cloud cover (of the available dates near ground measurement dates) over the LAI plots and closest in time and phenological stage to the ground LAI measurement dates (Table 1) were ordered from the USGS ESPA ordering interface (https://espa.cr.usgs.gov/). The ETM+ Level-1 scenes were corrected to surface reflectance using the Landsat Ecosystem Disturbance Adaptive Processing System (LEDAPS) and all OLI scenes were corrected to surface reflectance using the Landsat 8 Surface Reflectance Code (LaSRC) algorithm [31]. Both top-of-atmosphere (TOA) and bottom-of-atmosphere (surface) reflectance data were obtained. VIs available through the ESPA Ordering Interface, including normalized difference vegetation index (NDVI), enhanced vegetation index (EVI), soil adjusted vegetation index (SAVI), modified soil adjusted vegetation index (MSAVI), and normalized difference moisture index (NDMI), were obtained or calculated for surface reflectance for each image date [31] (see equations below). The simple ratio (SR) vegetation index was also calculated using ESRI's ArcGIS 10.2 for Desktop's raster calculator for both TOA and surface reflectance for each image date. One OLI image was collected on 28 March 2013 over the VA study area for 16/34 before Landsat 8 reached final WRS-2 orbit and was not available through the ESPA site. This image was processed to TOA reflectance and used in subsequent analyses. Each index below explains which bands of each sensor are used in the equation.

Normalized Difference Vegetation Index [33]:

$$\text{NDVI} = \frac{\rho_{NIR} - \rho_{Red}}{\rho_{NIR} + \rho_{Red}} \tag{1}$$

where ρ_{NIR} is the reflectance in the near infrared band (band 4 with ETM+ and band 5 with OLI) and ρ_{Red} is the red band reflectance (band 3 with ETM+ and band 4 with OLI).

Enhanced Vegetation Index [34]:

$$\text{EVI} = \frac{\rho_{NIR} - \rho_{Red}}{(\rho_{NIR} + C_1 * \rho_{Red} - C_2 * \rho_{Blue} + L)} \tag{2}$$

where ρ_{Blue} is the reflectance in the blue band (band 1 with ETM+ and band 2 with OLI), L is an adjustment for canopy background and set to 1 for both sensors, and C_1 and C_2 are coefficients for atmospheric resistance. $C_1 = 6$ and $C_2 = 7.5$ for both ETM+ and OLI.

Soil Adjusted Vegetation Index [35]:

$$\text{SAVI} = \frac{(\rho_{NIR} - \rho_{Red})(1 + L)}{\rho_{NIR} + \rho_{Red} + L} \tag{3}$$

where L is a soil brightness correction factor that is set to 0.5 for both ETM+ and OLI.

Modified Soil Adjusted Vegetation Index [36]:

$$\text{MSAVI} = \frac{2\,(\rho_{NIR} + 1) - \sqrt{(2\,\rho_{NIR} + 1)^2 - 8(\rho_{NIR} - \rho_{Red})}}{2} \tag{4}$$

Normalized Difference Moisture Index [37]:

$$\text{NDMI} = \frac{\rho_{NIR} - \rho_{SWIR1}}{\rho_{NIR} + \rho_{SWIR1}} \tag{5}$$

where ρ_{SWIR1} is the reflectance in the shortwave infrared band (band 5 with ETM+ and band 6 with OLI).

Simple Ratio:

$$\text{SR} = \frac{\rho_{NIR}}{\rho_{Red}} \tag{6}$$

2.2.3. LAI Regression Models

Simple linear regression models were calculated using the ground LAI estimates from a given site, year, and time period (minimum or maximum LAI) as the dependent variable and SR calculated with either TOA reflectance or surface reflectance as the independent variable. (Only SR was used for the comparison of TOA and surface reflectance, since SR is the only VI used in the current operational standard [15]). For the TOA and surface reflectance comparison, separate models were created for a single measurement period/site. Combinations of the ground measurements, including all maximum LAI, all minimum LAI, and both maximum and minimum LAI combined (the *combined model*), were compared across all VIs and for TOA- and surface reflectance-based SR models. Plots that either fell inside ETM+ data gaps or that were impacted by clouds or cloud shadows in a given image were excluded from the data set, as indicated in Table 1. Bootstrap statistics for the best overall model were computed with 100,000 replicates using the boot package (version 1.3–20) in R (version 3.5.1 "Feather Spray"). Bootstrap confidence intervals were calculated using boot.ci with 100,000 replicates and the adjusted bootstrap percentile method (type = "bca").

2.2.4. Georegistration Accuracy

To determine if the georegistration accuracy of OLI images had any impact on the LAI regression results, each set of ground plots at the two study sites were shifted to the location of one of their 8 neighbors. For each of the 8 sets of locations, the SR vegetation index based on TOA reflectance were extracted and used in regression models. For example, all of the plots at the VA study site were shifted up one pixel, the SR pixel values at these new locations were then extracted for each plot and used as the independent variable in a regression model with the ground plot LAI estimates. This was repeated for each of the remaining 7 pixel locations in a 3 by 3 window around the plot center pixel. The regression results from the 8 neighboring pixels were then compared with those of the plot center pixel. These calculations were made using one OLI image date (Table 1) for each set of ground LAI data.

2.2.5. Comparison to Current Operational Standard

To assess whether our updated model has potential to replace the one in common operational use, we mapped LAI with our updated combined SR model and with the model presented by Flores et al., commonly used for pine silviculture decision support in the U.S. South (Equation (7)) [15].

$$LAI = 0.56\,SR - 0.83 \tag{7}$$

Two scenes were acquired covering the Appomattox Buckingham State Forest in Virginia, one for maximum and one for minimum LAI. The scenes were from the analysis ready data Landsat 8 data products and the bottom-of-atmosphere (surface) reflectance was retained for subsequent analysis. SR was derived for each image and each image was clipped to the shape of the Appomattox Buckingham State Forest. The two empirical models were then calculated for each pixel within the state forest.

3. Results

3.1. Georegistration Impacts

The 16/34 image acquired on 14 March 2014 (Figure 2c) and 21/37 image acquired on 13 February 2014 (Figure 2e) resulted in the best LAI regression model with the plot center pixel (Figure 2). There was not a consistent trend in terms of which adjacent pixel resulted in the most accurate LAI model with the four other images (Figure 2a,b,d,f). In every instance, at least one of the adjacent pixels to the plot centers resulted in models that explained significantly less of the variability within the ground measurements (Figure 2). Accurate plot and pixel locations are clearly a critical component of model development with any combination of ground and remotely-sensed measurements.

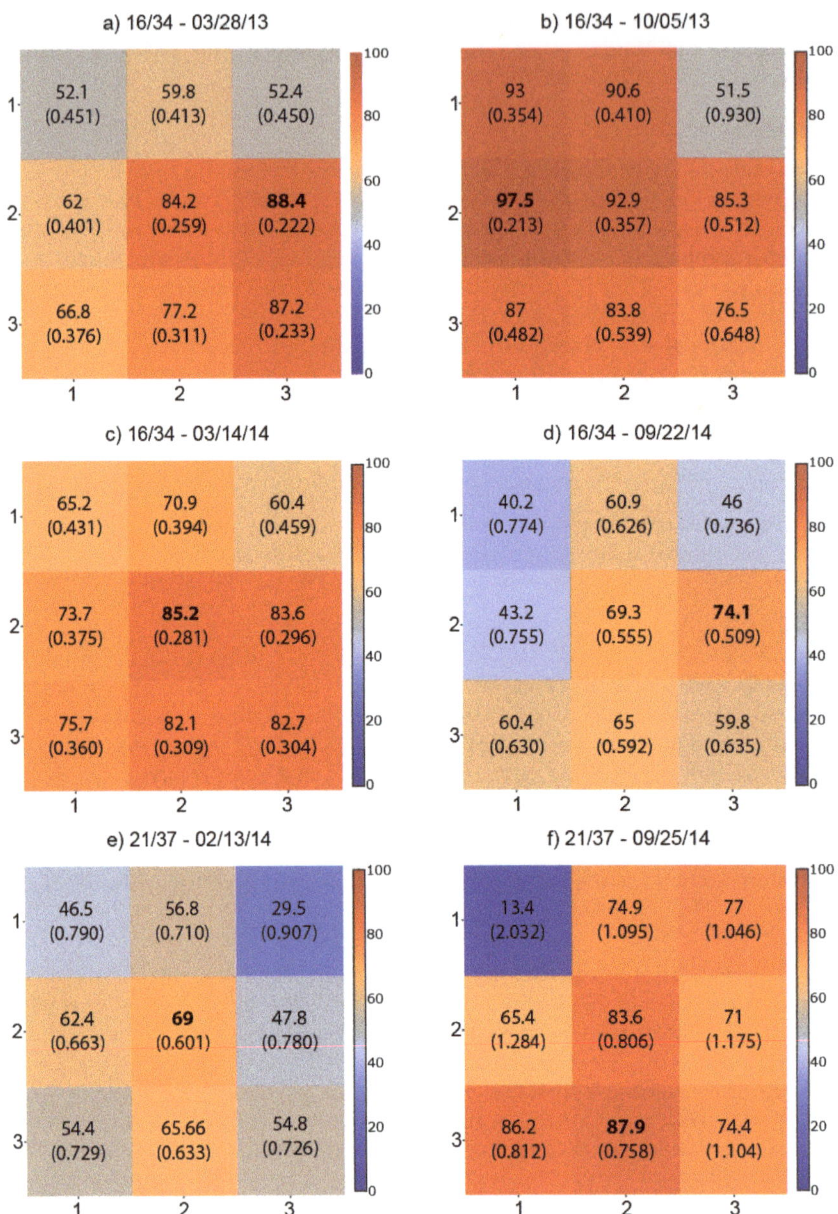

Figure 2. Regression results for Landsat OLI TOA reflectance based simple ratio vegetation index (SR) for each of the nine pixels surrounding or containing the LAI plot center and ground LAI estimates (bold R^2s denote the regression with the best fit for the given ground data collection and image). Each cell contains the R^2 and (RMSE) for the given pixel (row, column) surrounding the plot center in pixel (2, 2).

3.2. TOA versus Surface Reflectance

Regression results for the SR models comparing surface to TOA reflectance are shown in Table 3. Surface reflectance resulted in models that were more accurate for the individual measurement periods with the Alabama study area (21/37) while TOA reflectance outperformed surface reflectance at the Virginia study area. While TOA reflectance showed better results than surface reflectance three out of five times with the ground data from individual measurement periods, surface reflectance showed better results for each of the combined ground data sets, minimum LAI, maximum LAI and combined LAI (Table 4). If the VA minimum LAI 2013 plots are excluded from the TOA analysis, the R^2 s and RMSEs change to 72.4 (0.480) for TOA minimum LAI and 73.5 (0.682) for TOA combined LAI, which are still not as good as the surface reflectance results (Table 4).

Table 3. Regression results between ground measured LAI and Landsat OLI TOA- and surface reflectance-based SR (bold R^2s denote the regression with the best fit for the given ground data collection). * These plot numbers are just for TOA reflectance since the surface reflectance plot numbers were reduced to 39 and 73 for minimum and combined, respectively, due to the surface reflectance image of one 16/34 image date being unavailable.

Path/Row Time Period	Closest OLI Date	OLI R-sq (RMSE) TOA	OLI R-sq (RMSE) Surface	Min/Max R-sq (RMSE) TOA	Combined R-sq (RMSE) TOA	Min/Max R-sq (RMSE) Surface	Combined R-sq (RMSE) Surface	Max Number of Plots Used Separate/Min or Max/Combined
16/34 2013 Min	28 March 2013	84.2 (0.259)	N/A					19/58 */92 *
21/37 2014 Min	13 February 2014	69.0 (0.601)	**76.1** (0.528)	67.8 (0.468)		**73.3** (0.472)		21/58 */92 *
16/34 2014 Min	14 March 2014	**85.2** (0.281)	84.7 (0.285)					18/58 */92 *
16/34 2013 Max	5 October 2013	**92.9** (0.357)	92.1 (0.376)		73.2 (0.641)		**78.7** (0.613)	8/34/92 *
16/34 2014 Max	22 September 2014	**69.3** (0.555)	66.0 (0.584)	58.6 (0.840)		68.7 (0.730)		19/34/92 *
21/37 2014 Max	25 September 2014	86.4 (0.806)	**90.8** (0.663)					7/34/92 *

Table 4. Bootstrap statistics (100,000 replicates) for best overall regression model. CI = 95% confidence interval.

	Original	Bias	Standard Error	Lower CI	Upper CI
Intercept	−0.002119728	0.0004935929	0.13349108	−0.2755	0.2498
Slope	0.332915393	−0.0001120626	0.01585535	0.3021	0.3646

3.3. Vegetation Indices Comparison

NDMI resulted in the most accurate LAI model for six of the nine combinations of LAI time period and sensor considered with the SR vegetation index being most accurate three out of nine times (Table 4). NDVI, EVI, SAVI and MSAVI never resulted in the most accurate LAI model out of the six VIs considered. Unlike EVI, SAVI and MSAVI, NDVI produced the second most accurate model seven out of nine times. NDMI resulted in the best model for all time periods with the ETM+ sensor and for all of the maximum LAI models regardless of sensor (Table 4). Landsat OLI produced the most accurate LAI model for a given time period two out of three times with the SR vegetation index. SR also resulted in the best all time periods model two out of three times. For five of the nine time-period and sensor combinations, NDVI yielded a more accurate model than SR. NDMI, NDVI and SR always produced more accurate LAI models than EVI, SAVI and MSAVI for a given ground data set (time period) across sensors when calculated with surface reflectance. Using the closest in time image to the ground data collection regardless of sensor (the "Both" sensor option in Table 4) resulted in more accurate LAI models than using data from either of the sensors separately.

3.4. LAI Model

Equation (8) is the best regression model for ground data from all time periods and VIs from both sensors (see also Figure 3; Tables 4 and 5).

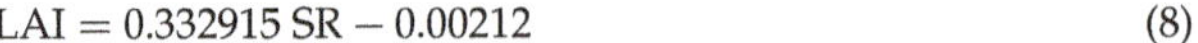

$$\text{LAI} = 0.332915\,\text{SR} - 0.00212 \tag{8}$$

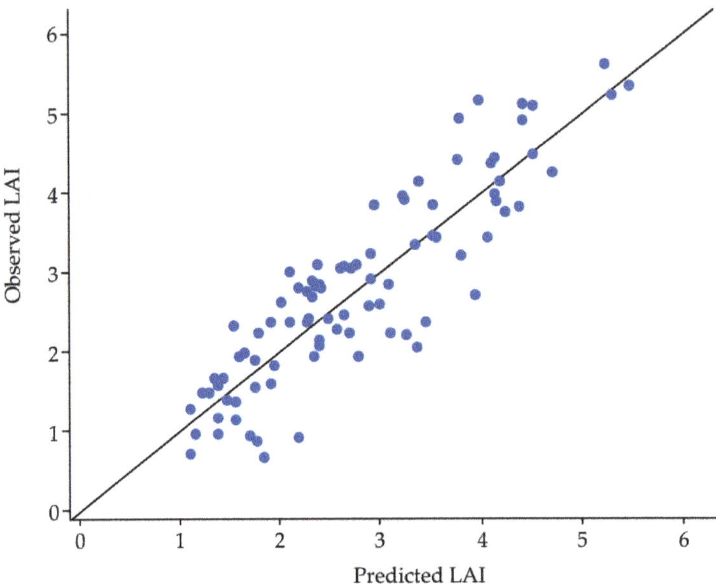

Figure 3. One to one graph of predicted LAI vs observed LAI.

Table 5. Regression results between ground measured LAI and Landsat surface reflectance-based VIs (bold R^2s denote the regression with the best fit for the given time period and Landsat sensor). * These results do not include an OLI image for the VA study site at minimum LAI in 2013.

Time Period	Landsat Sensor	NDMI R^2 (RMSE)	NDVI R^2 (RMSE)	SR R^2 (RMSE)	EVI R^2 (RMSE)	MSAVI R^2 (RMSE)	SAVI R^2 (RMSE)	Number of Plots
Minimum *	OLI	66.6 (0.528)	70.8 (0.494)	**73.3** (0.472)	62.6 (0.558)	59.8 (0.579)	62.8 (0.557)	39
Maximum	OLI	**80.7** (0.573)	69.5 (0.721)	68.7 (0.730)	48.3 (0.939)	47.0 (0.951)	49.0 (0.933)	34
All	OLI	65.9 (0.774)	69.2 (0.736)	**78.7** (0.613)	58.2 (0.858)	56.6 (0.874)	58.5 (0.854)	73
Minimum	ETM+	**72.4** (0.432)	65.3 (0.485)	63.6 (0.496)	55.6 (0.548)	56.3 (0.544)	59.4 (0.524)	56
Maximum	ETM+	**81.1** (0.582)	60.0 (0.847)	49.3 (0.954)	65.7 (0.784)	65.5 (0.786)	66.7 (0.772)	26
All	ETM+	**74.2** (0.605)	65.6 (0.698)	67.1 (0.684)	64.2 (0.713)	65.0 (0.705)	65.0 (0.704)	82
Minimum	Both	**74.5** (0.415)	69.8 (0.452)	69.4 (0.455)	63.8 (0.494)	58.4 (0.530)	60.7 (0.515)	57
Maximum	Both	**82.8** (0.554)	77.8 (0.630)	77.2 (0.639)	32.0 (1.103)	38.8 (1.046)	41.3 (1.025)	32
All	Both	73.3 (0.635)	68.9 (0.686)	**79.2** (0.561)	57.2 (0.805)	57.9 (0.799)	59.0 (0.788)	89

The regression differences between using only maximum versus only minimum and of OLI (L8) versus ETM+ (L7) are shown as Figure 4.

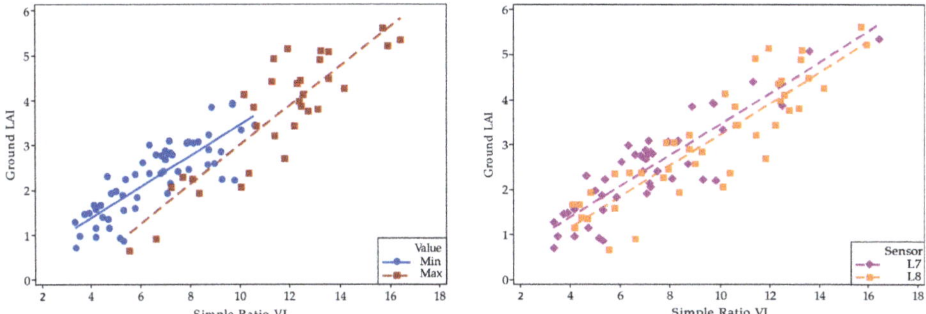

Figure 4. Regression models with closest in time imagery regardless of sensor and all ground data from both minimum and maximum LAI. The blue circles are the minimum LAI and the red triangles are the maximum LAI data points. The different sensor data points are differentiated by purple diamonds (ETM+) and orange squares (OLI).

3.5. Comparison to Current Operational Standard

The sample data used to derive the Flores et al. model do not have as wide a range as the sample data used to derive our combined model using SR as shown in Figure 5 [15].

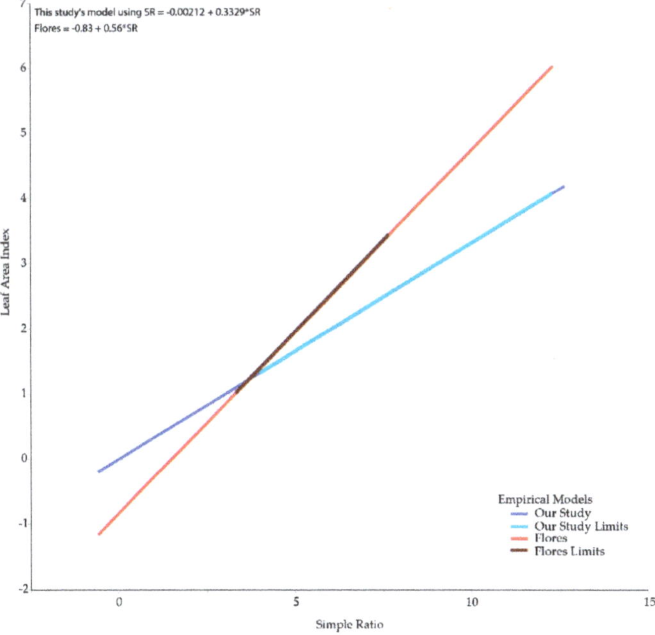

Figure 5. Final models and effective ranges for the two compared models.

Flores et al. [15] found when using winter scenes, when phenological differences are minimized, LAI of loblolly pine plantations could be accurately estimated using Landsat-derived SR [15]. Using summer scenes exclusively has not resulted in robust LAI models. However, when both maximum and minimum LAI data are combined the results are better than using the data from winter alone, as shown in Table 4. By using winter data exclusively in model development, LAI may be overestimated when estimating maximum LAI as shown in Figures 6 and 7. It should also be noted that the model from Flores et al. [12] was developed using TOA rather than surface reflectance and measured many fewer plots in situ. The Flores et al. [15] model has a large number of values that are considered to be overly high (above seven) for the species at this site (Figure 7). In our combined model, there are very few LAI values above seven, another indication of its potential utility for operational use.

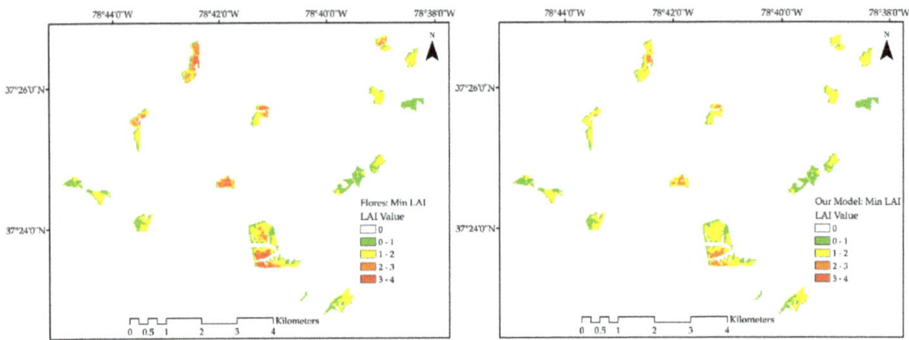

Figure 6. Minimum LAI shown for both the Flores et al. [15] model (**left**) and our combined model using SR (**right**).

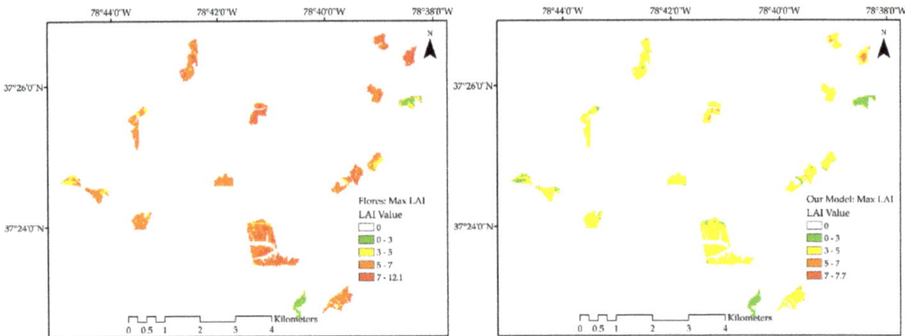

Figure 7. Maximum LAI shown for both the Flores et al. model (**left**) and our combined model using SR (**right**).

4. Discussion

4.1. Georegistration Impacts

Landsat misregistration is a source of error for studies assessing change through time [38] or relating plot-level to satellite-derived measurements [39]. McRoberts et al. [39] note that "for common magnitudes of rectification and GPS errors, as many as half the ground plots may be assigned to incorrect pixels." With these and related studies in mind, the size of the LAI ground plots in this study was selected to match the size of a Landsat pixel with the goal of improving the relationship between the ground and satellite data. However, the ground plots were not installed to align perfectly with a single pixel footprint. This is why stands that were as homogeneous as possible with respect to LAI were used and plot locations were selected in areas with similar LAI estimates in the surrounding 3 × 3 pixel area as illustrated in Figure 2. Plots were also located at least 30 m from a stand boundary to avoid having the plot center fall inside a mixed pixel. Additionally, the LAI-2200 measured beyond the plot boundaries when readings were collected in close proximity to the plot boundaries, as they were at the start and end of each transect. A relatively large number of LAI-2200 readings were collected per plot transect to improve the estimated ground LAI values for a given plot and to minimize the impact of any single erroneous readings. The geometric accuracy requirement for OLI is 12 m (less than half the width of the ground resolution cell size), met or exceeded in operations [40]. Furthermore, the Thematic Mapper, ETM+, and OLI *all* exceed the 12 m geometric accuracy requirement at least 92% of the time [41]. This realized geometric accuracy requirement, coupled with careful plot placement and submeter GPS accuracy, was sufficient to relate ground- to satellite-derived measurements on a pixel-specific basis (Figure 2).

4.2. TOA versus Surface Reflectance

Both TOA and surface reflectance performed well across models. However, with the combined data sets, surface reflectance (R^2 = 78.7%) outperformed TOA reflectance (R^2 = 73.2%), as shown in Table 3. Surface reflectance also performed better than TOA reflectance across three sensors (IKONOS, SPOT, and ETM+) in the Soudani et al. [42] study. A recent study [43] comparing Landsat TOA and surface reflectances to estimate the fraction of absorbed photosynthetically active radiation (FPAR; physiologically related to LAI) found that surface reflectance also slightly outperformed TOA reflectance (surface reflectance error, 0.03; TOA reflectance error, 0.06). Our results, in conjunction with the prior literature, strongly suggest that Landsat surface reflectance data be used for empirical LAI estimation.

4.3. Vegetation Indices Comparison

The relationships between VIs and LAI remained linear (and thus did not saturate) at any field-measured LAI value in this study, unlike many that preceded it. Some field LAI estimates exceeded 5 (Table 1). Chen et al. [44] did not find a saturation point at high LAI in boreal conifer stands and also found the relationship between optical ground LAI estimates and the Landsat SR to be linear in the summer. Data measured across multiple Landsat path/rows were also combined were found necessary to develop reliable models [44]. Similarly, Potithep et al. [27] found a two-period relationship better than a single period between in situ LAI estimates and in situ VIs derived from a hemispherical spectroradiometer in a deciduous forest. Although our combined data did not result in a better fit than the single image models, they did have strong relationships (Table 3) and the combined model across sites and times has greater utility for forest managers.

Cohen et al. [45] also advocated for the use of VIs from multiple dates of imagery in regression analyses for the estimation of LAI. Exploration of more than one spectral variable or VI in the LAI regression models may be warranted in future work based on the findings of [45] and Fassnacht et al. [46]. Evaluation of LAI estimates for a given site at two points in time close to minimum and maximum LAI was also explored by [25] in mixed natural forests in the southeastern U.S.

Dube and Mutanga [47] showed that texture parameters could enhance aboveground biomass estimation in deciduous forests. Since Madugundu et al. [48] previously noted a strong relationship between aboveground biomass and LAI in both pine and eucalyptus plantations, it follows that texture might also improve LAI models. However, Gray and Song [49] mapped minimum and maximum LAI by combining the spectral, spatial and temporal information of Landsat, IKONOS and MODIS, respectively, but did not find a significant relationship between texture variables and effective LAI in evergreen stands. Since (1) texture variables are neighborhood-derived variables and are thus less useful for sub-stand, pixel-specific estimates; and (2) they have not yet been directly shown to be useful for LAI estimation in pines, we did not employ them in this study. However, it seems prudent that future studies focused on stand-scale LAI estimation consider texture parameters as predictor variables.

Chen et al. [50] found that when using multiple sensors, the ones that had a larger discrepancy between plot size and sensor resolution did not perform as well as the sensors that were closer to the plot size. Middinti et al. [51] found that when combining MODIS and OLI data there were a significant number of high LAI values when compared to a solely OLI-derived map. This supports the notion that the larger the sensor resolution difference when combining multiple sensor types, the less apt the combination is at estimating LAI. Korhonen et al. [52] compared Sentinel 2 to Landsat 8 and found that when using conventional indices, the sensors performed similarly (for scale-appropriate analyses).

Similar to our results, SAVI and EVI (both of which contain the blue band, which is often more difficult to convert to surface reflectance) did not perform well in comparison to NDVI and SR for the estimation of LAI in Scots pine stands [42]. Lee et al. [53] compared hyperspectral and multispectral data for LAI estimation and found bands in the red-edge and shortwave-infrared regions to be more important than near-infrared bands. Similarly, Eklundh et al. [54] found both SWIR1 and SWIR2 more correlated with boreal pine LAI than any other ETM+ spectral channel. This might partially explain why NDMI, which uses a shortwave infrared band, outperformed both NDVI and simple ratio a majority of the time in this study (Table 5). This echoes the results of Eklundh [55] who found the moisture stress index (SWIR1 − NIR) the best index, out of many tested, for coniferous stands. Further, Kodar et al. [56] found NDVI to be the worst performing of all tested VIs for boreal LAI estimation.

Tillack et al. [26] looked at seasonal relationships between ground LAI estimates and high-resolution VIs from RapidEye. The strength of the relationships and the VI that produced the strongest relationships varied with the seasons, but NDVI worked best when all seasons were combined. In this study, simple ratio and NDMI worked best when minimum and maximum LAI were combined. Like [26], our field data only measured the LAI stand component above the sensor which was held at approximately 3.5 feet off the ground and above any low understory vegetation. The exclusion of portions of the understory LAI in our maximum LAI ground estimates likely causes some of the unexplained variation in our models, especially in areas with incomplete canopy closure.

4.4. LAI Model

This study's combined model used data from both Landsat 7 ETM+ and Landsat 8 OLI. Flores et al. [15] used Landsat 7 ETM+ exclusively. Masemola et al. [57] found, when comparing Landsat 7 ETM+ to Landsat 8 OLI data, that Landsat 8 OLI estimated grass LAI with better accuracy than Landsat 7 ETM+. Through the use of a combined radiative transfer model (one-dimensional leaf reflectance, PROSPECT and a canopy reflectance model, SAIL) called PROSAILH, they found Landsat 8 OLI data showed significant improvement over Landsat 7 ETM+ data and concluded that OLI data could be used to accurately estimate LAI when compared to ETM+. When estimating LAI, Schott et al. [58] found that regression model fits improved with the increase in signal-to-noise ratio from ETM+ imagery to OLI imagery and that the increases in signal-to-noise ratio were significant.

Time between the ground data collection and image acquisition is clearly an important factor in the estimation of LAI. Although the OLI sensor has a number of improvements [59] over the ETM+ sensor, timing mattered more than the sensor in this study. Similarly, Zhang et al. [60] found that an estimated Landsat VI based on MODIS fusion that matched the date of the summer ground LAI estimates from a study site in Canada resulted in better LAI models than the three available cloud free Landsat images that were collected 17 or more days before or after the ground data. Since LAI variations occur more rapidly in the growing season than in the winter, and there is often reduced image availability in the summer due to cloud cover, these findings are not surprising. For this reason, we attempted to match ground data collection as closely as logistically possible to image acquisition dates. Since both ETM+ and OLI data can be used effectively for LAI estimation, a given site has the potential of a good Landsat image every 8 days instead of 16 days. Based on this, future studies should focus on acquiring ground LAI estimates on or within a few days of image acquisition.

4.5. Comparison to Current Operational Standard

The model from Flores et al. [15] (Equation (7)) used as the current operational standard was derived using TOA reflectance data in concert with in situ data from a fertilization study at a unique site (SETRES-2) on the North Carolina coastal plain. By using both fertilized and control plots across differing soil conditions they compiled 12 observations, which resulted in no observed LAI in the winter between 1.75 and 2.5 [15]. Our study used two different regions in the southeastern United States (industry land in Alabama and managed state forest in Virginia) across two years and included two winter and two summer seasons. By expanding to two regions, multiple years, and multiple sensors, we were able to include 89 observations to derive our model, which used surface rather than TOA reflectance. More than a third of our winter LAI data fell between 1.75 and 2.5. The fact that this study had more samples over a wider range of LAI than that of Flores et al. [15] likely improved the robustness of the resulting model. Qualitatively, as seen in Figures 6 and 7, the resulting mapped estimates are more realistic with the new model than with the current operational standard.

5. Conclusions

This study has shown that ground LAI estimates at minimum and maximum LAI in loblolly pine stands can be combined and modeled with Landsat VIs across sites and sensors. SR and NDMI both produce strong relationships with LAI data combined across seasons. Collecting ground data in close proximity to image acquisition has a greater influence on LAI-VI modeling than Landsat sensor does. Data from both current Landsat sensors, ETM+ and OLI, can be combined and used in the same model. Accurate plot and pixel locations are important for minimizing unexplained variability within LAI-VI models. The resulting combined model using SR is recommended for operational use in the southeastern USA.

Author Contributions: Conceptualization, C.E.B., R.H.W., V.A.T., and T.R.F.; Analysis, C.E.B. and R.H.W.; Funding acquisition, R.H.W., V.A.T., and T.R.F.; Investigation, C.E.B.; Methodology, C.E.B.; Supervision, R.H.W., V.A.T., T.R.F. and M.S.; Writing—original draft, C.E.B.; Writing—review & editing, M.N.H., M.S., V.A.T., and R.H.W.

Funding: This work was supported in part by the US Geological Survey Grant G12PC00073, "Making Multitemporal Work", NASA Synthesis project (NASA Grant NNX17AI09G, "Regionally Specific Drivers of Land-Use Transitions and Future Scenarios: A Synthesis Considering the Land Management Influence in the Southeastern US"), McIntire Stennis Cooperative Forestry Program (USDA-NIFA McIntire-Stennis Grant Award 136633), PINEMAP (USDA NIFA award number 2011-68002-30185), and the Forest Productivity Cooperative.

Conflicts of Interest: The authors declare no conflicts of interest.

References

1. Rubilar, R.A.; Lee Allen, H.; Fox, T.R.; Cook, R.L.; Albaugh, T.J.; Campoe, O.C. Advances in silviculture of intensively managed plantations. *Curr. For. Rep.* **2018**, *4*, 23–34. [CrossRef]

2. Allen, H.L.; Fox, T.R.; Campbell, R.G. What is ahead for intensive pine plantation silviculture in the South? *South. J. Appl. For.* **2005**, *29*, 62–69.

3. Osem, Y.; O'Hara, K. An ecohydrological approach to managing dryland forests: Integration of leaf area metrics into assessment and management. *Forestry* **2016**, *89*, 338–349. [CrossRef]

4. Sampson, D.A.; Amatya, D.M.; Lawson, C.D.B.; Skaggs, R.W. Leaf area index (LAI) of loblolly pine and emergent vegetation following a harvest. *Trans. ASABE* **2011**, *54*, 2057–2066. [CrossRef]

5. Fox, T.R.; Allen, L.H.; Albaugh, T.J.; Rubilar, R.; Carlson, C.A. Tree nutrition and forest fertilization of pine plantations in the southern United States. *South. J. Appl. For.* **2007**, *31*, 5–11.

6. Albaugh, T.J.; Allen, H.L.; Dougherty, P.M.; Kress, L.W.; King, J.S. Leaf area and above- and belowground growth responses of loblolly pine to nutrient and water additions. *For. Sci.* **1998**, *44*, 317–328.

7. Campoe, O.C.; Stape, J.L.; Albaugh, T.J.; Allen, H.L.; Fox, T.R.; Rubilar, R.; Binkley, D. Fertilization and irrigation effects on tree level aboveground net primary production, light-interception and light use efficiency in loblolly pine plantation. *For. Ecol. Manag.* **2013**, *288*, 43–48. [CrossRef]

8. Samuelson, L.J.; Pell, C.J.; Stokes, T.A.; Bartkowiak, S.M.; Akers, M.K.; Kane, M.; Markewitz, D.; McGuire, M.A.; Teskey, R.O. Two-year throughfall and fertilization effects on leaf physiology and growth of loblolly pine in the Georgia Piedmont. *For. Ecol. Manag.* **2014**, *330*, 29–37. [CrossRef]

9. Blinn, C.E.; Albaugh, T.J.; Fox, T.R.; Wynne, R.H.; Stape, J.L.; Rubilar, R.A.; Allen, H.L. A method for estimating deciduous competition in pine stands using Landsat. *South. J. Appl. For.* **2012**, *36*, 71–78. [CrossRef]

10. Gonzalez-Benecke, C.A.; Jokela, E.J.; Martin, T.A. Modeling the effects of stand development, site quality, and silviculture on leaf area index, litterfall, and forest floor accumulation in loblolly and slash pine plantations. *For. Sci.* **2012**, *58*, 457–471. [CrossRef]

11. Gao, F.; Anderson, M.C.; Kustas, W.P.; Houborg, R. Retrieving leaf area index from Landsat using MODIS LAI products and field measurements. *IEEE Geosci. Remote Sens.* **2014**, *11*, 773–777.

12. Vogel, S.A.; McKelvey, K.; Gholz, H.L.; Curran, P.J.; Ewel, K.C.; Cropper, W.P.; Teskey, R.O. Dynamics of canopy structure and light interception in *Pinus elliottii* stands, north Florida. *Ecol. Monogr.* **2006**, *61*, 33–51.

13. Sampson, D.A.; Albaugh, T.J.; Johnsen, K.H.; Allen, H.L.; Zarnoch, S.J. Monthly leaf area index estimates from point-in-time measurements and needle phenology for *Pinus taeda*. *Can. J. For. Res.* **2003**, *33*, 2477–2490. [CrossRef]

14. Eriksson, H.M.; Eklundh, L.; Kuusk, A.; Nilson, T. Impact of understory vegetation on forest canopy reflectance and remotely sensed LAI estimates. *Remote Sens. Environ.* **2006**, *103*, 408–418. [CrossRef]

15. Flores, F.J.; Allen, H.L.; Cheshire, H.M.; Davis, J.M.; Fuentes, M.; Kelting, D. Using multispectral satellite imagery to estimate leaf area and response to silvicultural treatments in loblolly pine stands. *Can. J. For. Res.* **2006**, *36*, 1587–1596. [CrossRef]

16. Peduzzi, A.; Allen, H.L.; Wynne, R.H. Leaf area of overstory and understory in pine plantations in the Flatwoods. *South. J. Appl. For.* **2010**, *34*, 154–160.

17. Iiames, J.S.; Congalton, R.G.; Pilant, A.N.; Lewis, T.E. Leaf area index (LAI) change detection analysis on loblolly pine (*Pinus taeda*) following complete understory removal. *Photogramm. Eng. Remote Sens.* **2008**, *74*, 1389–1400. [CrossRef]

18. Curran, P.J.; Dungan, J.L.; Gholz, H.L. Seasonal LAI in slash pine estimated with Landsat TM. *Remote Sens. Environ.* **1992**, *39*, 3–13. [CrossRef]

19. Chen, J.M.; Cihlar, J. Retrieving leaf area index of boreal conifer forests using Landsat TM images. *Remote Sens. Environ.* **1996**, *55*, 153–162. [CrossRef]

20. Tian, Q.; Luo, Z.; Chen, J.M.; Chen, M.; Hui, F. Retrieving leaf area index for coniferous forest in Xingguo County, China with Landsat ETM+ images. *J. Environ. Manag.* **2007**, *85*, 624–627. [CrossRef] [PubMed]

21. Franklin, S.E.; Lavigne, M.B.; Deuling, M.J.; Wulder, M.A.; Hunt, E.R. Estimation of forest leaf area index using remote sensing and GIS data for modelling net primary production. *Int. J. Remote Sens.* **1997**, *18*, 3459–3471. [CrossRef]

22. Turner, D.P.; Cohen, W.B.; Kennedy, R.E.; Fassnacht, K.S.; Briggs, J.M. Relationships between leaf area index and Landsat TM spectral vegetation indices across three temperate zone sites. *Remote Sens. Environ.* **1999**, *70*, 52–68. [CrossRef]

23. Song, C.H. Optical remote sensing of forest leaf area index and biomass. *Prog. Phys. Geogr.* **2013**, *37*, 98–113. [CrossRef]

24. Sumnall, M.; Peduzzi, A.; Fox, T.R.; Wynne, R.H.; Thomas, V.A.; Cook, B. Assessing the transferability of statistical predictive models for leaf area index between two airborne discrete return LiDAR sensor designs within multiple intensely managed loblolly pine forest locations in the south-eastern USA. *Remote Sens. Environ.* **2016**, *176*, 308–319. [CrossRef]

25. Pu, R.L. Mapping leaf area index over a mixed natural forest area in the flooding season using ground-based measurements and Landsat TM imagery. *Int. J. Remote Sens.* **2012**, *33*, 6600–6622. [CrossRef]

26. Tillack, A.; Clasen, A.; Kleinschmit, B.; Förster, M. Estimation of the seasonal leaf area index in an alluvial forest using high-resolution satellite-based vegetation indices. *Remote Sens. Environ.* **2014**, *141*, 52–63. [CrossRef]

27. Potithep, S.; Nagai, S.; Nasahara, K.N.; Muraoka, H.; Suzuki, R. Two separate periods of the LAI-VIs relationships using in situ measurements in a deciduous broadleaf forest. *Agric. For. Meteorol.* **2013**, *169*, 148–155. [CrossRef]

28. Welles, J.M.; Norman, J.M. Instrument for indirect measurement of canopy architecture. *Agron. J.* **1991**, *83*, 818–825. [CrossRef]

29. Peduzzi, A.; Wynne, R.H.; Fox, T.R.; Nelson, R.F.; Thomas, V.A. Estimating leaf area index in intensively managed pine plantations using air- borne laser scanner data. *Forest Ecol. And Mgmt.* **2012**, *270*, 54–65. [CrossRef]

30. LI-COR, Inc. *LAI-2200C Plant Canopy Analyzer Instructions Manual*; Publication Number 984-14112; LI-COR, Inc.: Lincoln, NE, USA, 2013.

31. USGS. Product Guide: Landsat Surface Reflectance-Derived Spectral Indices Version 2.6. Available online: landsat.usgs.gov/documents/si_product_guide.pdf (accessed on 12 April 2015).

32. Irons, J.R.; Dwyer, J.L.; Barsi, J.A. The Next Landsat Satellite: The Landsat Data Continuity Mission. *Remote Sens. Environ.* **2012**, *122*, 11–21. [CrossRef]

33. Rouse, J.W., Jr.; Haas, R.H.; Deering, D.W.; Schell, J.A.; Harlan, J.C. *Monitoring the Vernal Advancement and Retrogradation (Green Wave Effect) of Natural Vegetation, NASA/GSFC Type III Final Report*; Texas A&M University: College Station, TX, USA, 1974; 371p.

34. Huete, A.; Didan, K.; Miura, T.; Rodriguez, E.P.; Gao, X.; Ferreira, L.G. Overview of the radiometric and biophysical performance of the MODIS vegetation indices. *Remote Sens Environ.* **2002**, *83*, 195–213. [CrossRef]

35. Huete, A.R. A soil-adjusted vegetation index (SAVI). *Remote Sens Environ.* **1988**, *25*, 295–309. [CrossRef]

36. Qi, J.; Chehbouni, A.; Huete, A.R.; Kerr, Y.H.; Sorooshian, S. A modified soil adjusted vegetation index. *Remote Sens. Environ.* **1994**, *48*, 119–126. [CrossRef]

37. Hardisky, M.A.; Klemas, V.; Smart, R.M. The influence of soil salinity, growth form, and leaf moisture on the spectral radiance of *Spartina alterniflora* canopies. *Photogramm. Eng. Remote Sens.* **1983**, *49*, 77–83.

38. Rishmawi, K.; Goward, S.N.; Schleeweis, K.; Huang, C.; Dwyer, J.L.; Masek, J.G.; Dungan, J.L.; Michaelis, A.; Lindsey, M.A. Selection and quality assessment of Landsat data for the North American forest dynamics forest history maps of the US. *Int. J. Digit. Earth* **2016**, *9*, 963–980.

39. McRoberts, R.E. The effects of rectification and Global Positioning System errors on satellite image-based estimates of forest area. *Remote Sens. Environ.* **2010**, *114*, 1710–1717. [CrossRef]

40. Storey, J.; Choate, M.; Lee, K. Landsat 8 operational land imager on-orbit geometric calibration and performance. *Remote Sens.* **2014**, *6*, 11127–11152. [CrossRef]

41. USGS. Geometry | Landsat Missions. Available online: https://landsat.usgs.gov/geometry (accessed on 25 February 2019).

42. Soudani, K.; Francois, C.; Le Maire, G.; Le Dantec, V.; Dufrene, E. Comparative analysis of IKONOS, SPOT, and ETM+ data for leaf area index estimation in temperate coniferous and deciduous forest stands. *Remote Sens. Environ.* **2006**, *102*, 161–175. [CrossRef]

43. Liu, R.; Ren, H.; Liu, S.; Liu, Q.; Yan, B.; Gan, F. Generalized FPAR estimation methods from various satellite sensors and validation. *Agric. For. Meteorol.* **2018**, *260–261*, 55–72. [CrossRef]

44. Chen, J.M.; Pavlic, G.; Brown, L.; Cihlar, J.; Leblanc, S.G.; White, H.P.; Hall, R.J.; Peddle, D.R.; King, D.J.; Trofymow, J.A.; et al. Derivation and validation of Canada-wide coarse-resolution leaf area index maps using high-resolution satellite imagery and ground measurements. *Remote Sens. Environ.* **2002**, *80*, 165–184. [CrossRef]

45. Cohen, W.B.; Maiersperger, T.K.; Gower, S.T.; Turner, D.P. An improved strategy for regression of biophysical variables and Landsat ETM+ data. *Remote Sens. Environ.* **2003**, *84*, 561–571. [CrossRef]

46. Fassnacht, K.S.; Gower, S.T.; MacKenzie, M.D.; Nordheim, E.V.; Lillesand, T.M. Estimating the leaf area index of north central Wisconsin forests using the Landsat Thematic Mapper. *Remote Sens. Environ.* **1997**, *61*, 229–245. [CrossRef]

47. Dube, T.; Mutanga, O. Investigating the robustness of the new Landsat-8 Operational Land Imager derived texture metrics in estimating plantation forest aboveground biomass in resource constrained areas. *ISPRS J. Photogramm. Remote Sens.* **2015**, *108*, 12–32. [CrossRef]

48. Madugundu, R.; Nizalapur, V.; Jha, C.S. Estimation of LAI and above-ground biomass in deciduous forests: Western Ghats of Karnataka, India. *Int. J. Appl. Earth Obs. Geoinf.* **2008**, *10*, 211–219. [CrossRef]

49. Gray, J.; Song, C. Mapping leaf area index using spatial, spectral, and temporal information from multiple sensors. *Remote Sens. Environ.* **2012**, *119*, 173–183. [CrossRef]

50. Chen, W.; Yin, H.; Moriya, K.; Sakai, T.; Cao, C. Retrieval and comparison of forest leaf area index based on remote sensing data from AVNIR-2, Landsat-5 TM, MODIS, and PALSAR sensors. *ISPRS Int. J. Geoinf.* **2017**, *6*, 179. [CrossRef]

51. Middinti, S.; Thumaty, K.C.; Gopalakrishnan, R.; Jha, C.S.; Thatiparthi, B.R. Estimating the leaf area index in Indian tropical forests using Landsat-8 OLI data. *Int. J. Remote Sens.* **2017**, *38*, 6769–6789. [CrossRef]

52. Korhonen, L.; Packalen, P.; Rautiainen, M. Comparison of Sentinel-2 and Landsat 8 in the estimation of boreal forest canopy cover and leaf area index. *Remote Sens. Environ.* **2017**, *195*, 259–274. [CrossRef]

53. Lee, K.-S.; Cohen, W.B.; Kennedy, R.E.; Maiersperger, T.K.; Gower, S.T. Hyperspectral versus multispectral data for estimating leaf area index in four different biomes. *Remote Sens. Environ.* **2004**, *91*, 508–520. [CrossRef]

54. Eklundh, L.; Harrie, L.; Kuusk, A. Investigating relationships between Landsat ETM plus sensor data and leaf area index in a boreal conifer forest. *Remote Sens. Environ.* **2001**, *78*, 239–251. [CrossRef]

55. Eklundh, L. Estimating leaf area index in coniferous and deciduous forests in Sweden using Landsat optical sensor data. *Proc. SPIE* **2003**, *4879*, 379–390.

56. Kodar, A.; Kutsar, R.; Lang, M.; Lukk, T.; Nilson, T. Leaf area indices of forest canopies from optical measurements. *Balt. For.* **2008**, *14*, 185–194.

57. Masemola, C.; Cho, M.A.; Ramoelo, A. Comparison of Landsat 8 OLI and Landsat 7 ETM+ for estimating grassland LAI using model inversion and spectral indices: Case study of Mpumalanga, South Africa. *Int. J. Remote Sens.* **2016**, *37*, 4401–4419. [CrossRef]

58. Schott, J.R.; Gerace, A.; Woodcock, C.E.; Wang, S.; Zhu, Z.; Wynne, R.H.; Blinn, C.E. The impact of improved signal-to-noise ratios on algorithm performance: Case studies for Landsat class instruments. *Remote Sens. Environ.* **2016**, *185*, 37–45. [CrossRef]

59. Roy, D.P.; Wulder, M.A.; Loveland, T.R.; Woodcock, C.E.; Allen, R.G.; Anderson, M.C.; Helder, D.; Irons, J.R.; Johnson, D.M.; Kennedy, R.; et al. Landsat-8: Science and product vision for terrestrial global change research. *Remote Sens. Environ.* **2014**, *145*, 154–172. [CrossRef]

60. Zhang, H.; Chen, J.M.; Huang, B.; Song, H.; Li, Y. Reconstructing seasonal variation of Landsat vegetation index related leaf area index by fusing with MODIS data. *IEEE J. Sel. Top. Appl. Earth Obs. Remote Sens.* **2014**, *7*, 950–960. [CrossRef]

© 2019 by the authors. Licensee MDPI, Basel, Switzerland. This article is an open access article distributed under the terms and conditions of the Creative Commons Attribution (CC BY) license (http://creativecommons.org/licenses/by/4.0/).

Article

Forest Growing Stock Volume Estimation in Subtropical Mountain Areas Using PALSAR-2 L-Band PolSAR Data

Haibo Zhang [1], Jianjun Zhu [1,*], Changcheng Wang [1], Hui Lin [2], Jiangping Long [2], Lei Zhao [3], Haiqiang Fu [1] and Zhiwei Liu [1]

[1] School of Geosciences and Info-Physics, Central South University, Changsha 410083, China; haibozhang@csu.edu.cn (H.Z.); wangchangcheng@csu.edu.cn (C.W.); haiqiangfu@csu.edu.cn (H.F.); liuzhiwei@csu.edu.cn (Z.L.)

[2] Research Center of Forestry Remote Sensing & Information Engineering, Central South University & Technology, Changsha 410004, China; linhui@csuft.edu.cn (H.L.); longjiangping11@163.com (J.L.)

[3] Institute of Forest Resources Information Technique, Chinese Academy of Forestry, Beijing 100091, China; zhaoleiiam@gmail.com

* Correspondence: zjj@csu.edu.cn; Tel.: +86-731-8883-6931

Received: 26 February 2019; Accepted: 16 March 2019; Published: 20 March 2019

Abstract: Forest growing stock volume (GSV) extraction using synthetic aperture radar (SAR) images has been widely used in climate change research. However, the relationships between forest GSV and polarimetric SAR (PolSAR) data in the mountain region of central China remain unknown. Moreover, it is challenging to estimate GSV due to the complex topography of the region. In this paper, we estimated the forest GSV from advanced land observing satellite-2 (ALOS-2) phased array-type L-band synthetic aperture radar (PALSAR-2) full polarimetric SAR data based on ground truth data collected in Youxian County, Central China in 2016. An integrated three-stage (polarization orientation angle, POA; effective scattering area, ESA; and angular variation effect, AVE) correction method was used to reduce the negative impact of topography on the backscatter coefficient. In the AVE correction stage, a strategy for fine terrain correction was attempted to obtain the optimum correction parameters for different polarization channels. The elements on the diagonal of covariance matrix were used to develop forest GSV prediction models through five single-variable models and a multi-variable model. The results showed that the integrated three-stage terrain correction reduced the negative influence of topography and improved the sensitivity between the forest GSV and backscatter coefficients. In the three stages, the POA compensation was limited in its ability to reduce the impact of complex terrain, the ESA correction was more effective in low-local incidence angles area than high-local incidence angles, and the effect of the AVE correction was opposite to the ESA correction. The data acquired on 14 July 2016 was most suitable for GSV estimation in this study area due to its correlation with GSV, which was the strongest at HH, HV, and VV polarizations. The correlation coefficient values were 0.489, 0.643, and 0.473, respectively, which were improved by 0.363, 0.373, and 0.366 in comparison to before terrain correction. In the five single-variable models, the fitting performance of the Water-Cloud analysis model was the best, and the correlation coefficient R^2 value was 0.612. The constructed multi-variable model produced a better inversion result, with a root mean square error (RMSE) of 70.965 m^3/ha, which was improved by 22.08% in comparison to the single-variable models. Finally, the space distribution map of forest GSV was established using the multi-variable model. The range of estimated forest GSV was 0 to 450 m^3/ha, and the mean value was 135.759 m^3/ha. The study expands the application potential of PolSAR data in complex topographic areas; thus, it is helpful and valuable for the estimation of large-scale forest parameters.

Keywords: forest growing stock volume (GSV); full polarimetric SAR; subtropical forest; topographic effects; environment effects

1. Introduction

Forest carbon stocks are essential to our understanding of global climate change, and can be represented through extracting forest parameters [1]. The forest growing stock volume (GSV) is a key forest variable in the context of forest management and monitoring. Also, the forest GSV is referred to as the total volume (m^3/ha) of the boles or stems of all living trees, and can be converted into above-ground biomass (AGB) by its density factor [2]. Therefore, the accurate quantification of forest biomass or GSV is essential for understanding the spatial distribution of carbon in vegetation areas, which can also provide effective predictions for the change trend of carbon stock [3]. In particular, large-scale forest GSV retrieval has become a research hotspot in recent years.

At present, many methods have been reported for estimating forest GSV. Traditional forest inventory approaches rely upon ground surveys by manually collecting the forest parameters of a single tree at sample plots [4]. However, the large amount of time required, labor intensity, and cost limit its application on a larger scale [5]. Remote sensing technology provides a possible solution to overcome such limitations, in particular the spaceborne remote sensing technique, which plays an important role in forest monitoring and management [6]. Optical remote sensing datasets (e.g., Landsat Thematic Mapper (TM) and Moderate Resolution Imaging Spectroradiometer, MODIS) can be used to estimate forest GSV [7], mainly by analyzing the relationship between forest parameters and vegetation indices (e.g., enhanced vegetation index (EVI), normalized difference vegetation index (NDVI) and perpendicular vegetation index (PVI)) [8–11]. However, the retrieved GSV values using optical remote sensing data are usually troubled with saturation effects, especially in the high carbon stock forests [12]. Another problem is the impact of cloud cover on image collection, constraining its application to moist regions (e.g., the tropical region) [13]. Light detection and ranging (LiDAR) data provides high accurate forest parameters for GSV estimation [14–16]. However, due to space discontinuous problems and complex data processing, LiDAR-derived GSV estimates usually can only be obtained over limited areas [17]. HHHSynthetic aperture radar (SAR) enables imaging in all-weather conditions and with continuous temporal coverage [18]. Now, it has been successfully applied in various fields [19–22]. Especially, the long-wavelength SAR has a wide potential in forestry applications [23–25]. Currently, the SAR techniques that have been utilized for the retrieval of forest parameters mainly are polarimetric SAR (PolSAR) [23,26], interferometric SAR (InSAR) [27,28], polarimetric interferometric SAR (PolInSAR) [29], polarization coherence tomography (PCT) [30–32], and tomography SAR (TomoSAR) [33–35]. In this paper, the full polarimetric SAR technique will be further used for retrieving forest GSV in subtropical mountain areas.

Radar polarimetry is the technique of acquiring, processing, and analyzing the polarization state of an electromagnetic field [36]. Forest characteristic information about the geometrical structure and geophysical properties can be obtained by analyzing polarimetric SAR signatures [37]. In an earlier study, the relationship between polarimetric signatures and forest GSV or AGB was studied by using high-frequency SAR data (e.g., X and C-band). Due to the low penetration, the short wavelengths interacted primarily with the forest canopy, and were suitable for low-carbon stock areas [38]. Lower frequency SAR data have stronger penetrating capability and can interact with various components of vegetation, and have been discovered to be more preferable than higher frequencies in high-carbon stock forests [39,40]. The phased array-type L-band synthetic aperture radar (PALSAR-2) can capture images in quad polarization modes, which provides an opportunity to study forest parameters using multi-polarization and multi-temporal data [41–43].

The L-band backscatter coefficient is sensitive to the biophysical parameters of forest [44–46]. However, the sensitivity is affected by many factors (e.g., radar polarization, forest structure, environment conditions, and topography) [39]. Among these factors, the complex terrain conditions can affect full polarimetric SAR data regarding both azimuth and distance, which is mainly reflected in the following three aspects. (1) The azimuthal slope causes a change in the polarization state, which leads to polarization orientation angles (POA) offsetting [47]. (2) Local terrain undulations cause a change in the effective scattering area (ESA), which leads to the change of actual backscatter [48]. (3) In vegetation-covered areas, the local terrain causes variation in penetration depth and scattering mechanisms, which are reflected in the angular variation effect (AVE) [49]. Zhao et al. [50] showed that the correction of these three aspects (POA compensation, ESA correction, and AVE correction) could reduce the topographic effect. However, in AVE correction, the critical correction factor n is obtained only according to the impact of the entire forest area, without considering the impact of different forest cover types. Meanwhile, the range of n values in different polarization channels in subtropical mountain areas need to be further explored.

The main purposes of this study are to (1) understand the role of terrain correction in forest GSV estimation; and (2) investigate the potential of PALSAR-2 L-band full polarimetric data for the retrieval of forest GSV in the subtropical mountain regions.

2. Materials

2.1. Study Area

The work was carried out in the Youxian county in Hunan of central China (27°05′ to 27°24′ N, 113°35′ to 113°55′ E, see Figure 1). It is a field site ground for forest research at Central South University of Forestry and Technology. The topography varies between 60–1386 m. The slope ranges from 0° to 84°. The climate type is a subtropical monsoon humid climate. The annual mean temperature is 17.8 °C. The average annual rainfall is 1410.8 mm, and most of rainfall occurs in the summer. The dominating forest type is coniferous forest, including fir and pine. In addition, there are some other vegetation types, such as bamboo and camphorwood.

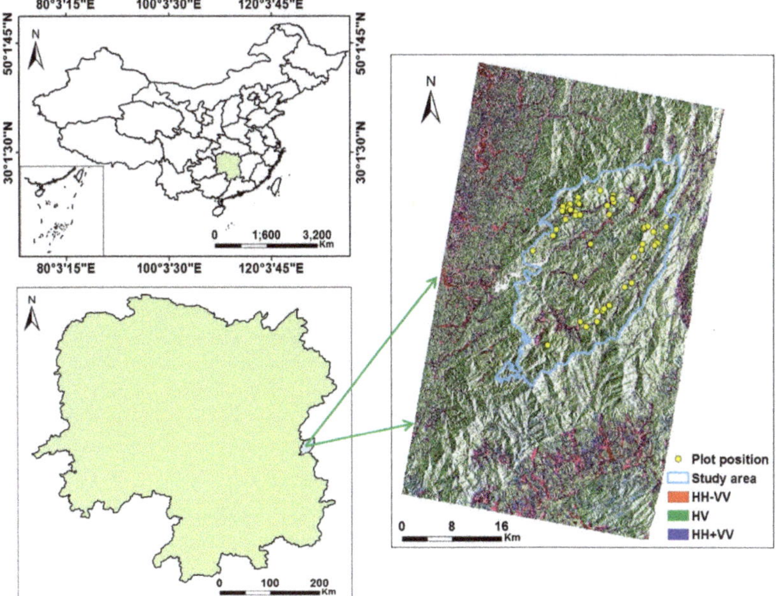

Figure 1. The test site: phased array-type L-band synthetic aperture radar (PALSAR-2) image in the Pauli basis. The yellow circles are field sampling plots.

2.2. Field Inventory Data

Field data collection was conducted from June to July 2016, with the help of the Central South University of Forestry and Technology, and the Chinese Academy of Forestry. A total of 60 forest plots with the size of 30 m × 30 m were surveyed for the experiment (Figure 1). The center of each plot was located by using a global positioning satellite (GPS) receiver, and the location (latitude and longitude) of the central point was recorded. These plots were independent from one another to avoid the spatial autocorrelation. Within each plot, the diameter at 1.3 m above the ground of each individual tree was measured by the diameter at breast height (DBH) ruler, and the tree height measurement was performed by the laser altimeter. The GSV was calculated using the method presented by Fang et al. [51,52], and the GSV of 60 plots ranged from 6.88 m^3/ha to 434.42 m^3/ha, with an average value of 194.75 m^3/ha (Table 1). By random sampling, the plots were divided for the training (n = 44) and validation (n = 16) of models into two groups.

Table 1. Main biophysical properties of 60 plots in the study area. DBH: diameter at breast height.

	Range	Mean
DBH	4.06 to 30.10 cm	17.84 cm
Height	4.60 to 20.20 m	13.24 m
Number of Stems	30 to 350	96
Growing Stock Volume	6.88 to 434.42 m^3/ha	194.75 m^3/ha

2.3. Polarimetric SAR Data and Pre-Processing

Full polarimetric (HH, HV, VH, and VV polarizations) L-band SAR data over this experiment site were acquired by the Japanese Aerospace Exploration Agency (JAXA) using the PALSAR-2. A total of seven scenes data were ordered as L1.1 level with the single-look complex (SLC) format in slant range geometry, and were acquired from June to October 2016 at approximately 4:22 local time. The central location of these datasets is approximately 27.18° N–113.68° E, and the dimensions are 69 km in azimuth and 25.8 km in range. The incidence angle ranges from 37.8° to 40.1°. The azimuth resolution is 2.97 m, and the range resolution is 2.86 m.

The basic data pre-processing steps were applied to reduce the geometric and radiometric distortions and speckle effects. The radiometric calibration of these data was first performed [53]. The coherency matrix $[T_3]$ was generated and converted into the covariance matrix $[C_3]$, which could represent the full polarimetric data. Then, these data were multi-looked with 7 × 10 in the azimuth and range directions. A Lee filter with a 3 × 3 window was applied to reduce the speckle effects. Geocoding was performed using shuttle radar topography mission (SRTM) elevation data (30-m spatial resolution). Finally, the SAR images were re-sampled to 30-m spatial resolution. PolSARpro software (Version 5.1.3, European Space Agency, Paris, France) was used to pre-process the SAR data. Gamma software was used to perform geocoding and resampling.

2.4. Ancillary Data

The ancillary data used in this study mainly include the SRTM digital elevation model (DEM) (https://earthexplorer.usgs.gov/) and land-use data product (http://www.dsac.cn/). The SRTM DEM (Figure 2a) has a 30-m resolution and was created by the National Geospatial Intelligence Agency and Jet Propulsion Laboratory. We used it to assist the SAR dataset geocoding. Besides, based on the SAR imaging geometry, terrain correction factors (i.e., projection angle and local incidence angle) could also be obtained from the DEM data. The land-use classification data (Figure 2b) with a spatial resolution of 30 m was provided by the Geographical Information Monitoring Cloud Platform at the same time as the PALSAR-2 dataset acquisition. According to the secondary classification of land use, the forest was divided into four types: woodland, shrubbery, sparse woodland (S-Woodland), and other forest (O-Forest), which would be used to assist the terrain correction factors.

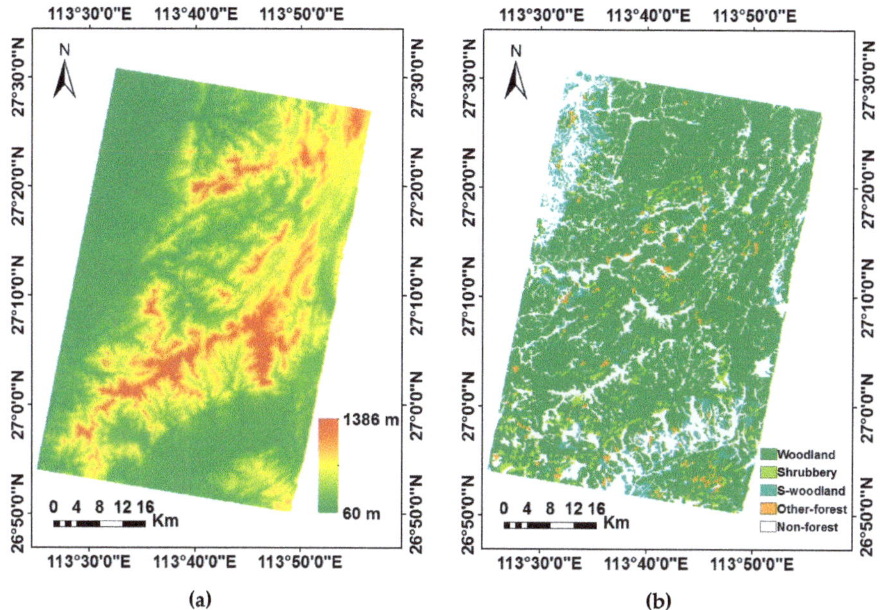

Figure 2. Ancillary data. (**a**) Digital elevation model (DEM); (**b**) Land-use data product.

3. Methodology

The processing flow chart presented in Figure 3 illustrates the framework of analysis steps. We firstly carried out basic pre-processing for the SLC level 1.1 datasets, including radiometric calibration, multi-looking, filtering, and geocoding. Secondly, POA and ESA correction were performed. Then, we calculated the values of n for different polarization channels by analyzing the correlation coefficients between the local incidence angles and backscatter coefficients. Thirdly, AVE correction was performed for different forest cover types, and then the results were spliced for later analysis. Fourthly, the correlations were analyzed between multi-temporal PolSAR data backscatter and forest GSV of all the sample plots, and the optimal data was selected for GSV estimation. Finally, the estimation models were constructed and compared by using the primary diagonal elements of the covariance matrix. Then, we estimated and mapped the GSV for the whole experiment area.

3.1. Terrain Correction

3.1.1. Polarization Orientation Angle Correction

The azimuth slope was the main factor that caused the polarization ellipse to rotate, and then affected the polarization state of the electromagnetic wave. To compensate for the impacts of the azimuth slope, the polarization orientation angles could be obtained by the circular polarization algorithm [54], as shown in Equation (1):

$$\eta = \frac{1}{4}\left[arctan\left(\frac{2Re[T_{23}]}{T_{22} - T_{33}} \right) + \pi \right] \tag{1}$$

where η is the shift angle, and T_{22}, T_{23}, T_{33} are the corresponding elements of matrix $[T]$. After acquiring the shift angle, a new rotated polarimetric covariance matrix (C_{POA}) can be formed by Equation (2):

$$C_{POA} = \left[U_{3(\eta)} \right] [C_3] \left[U_{3(\eta)} \right]^T \tag{2}$$

where C_3 denotes a polarimetric covariance matrix that represents multi-looked PolSAR data, and $U_{3(\eta)}$ is a rotation matrix.

$$\left[U_{3(\eta)}\right] = \frac{1}{2}\begin{bmatrix} 1+cos2\eta & \sqrt{2}sin2\eta & 1-cos2\eta \\ -\sqrt{2}sin2\eta & 2cos2\eta & \sqrt{2}sin2\eta \\ 1-cos2\eta & -\sqrt{2}sin2\eta & 1+cos2\eta \end{bmatrix} \tag{3}$$

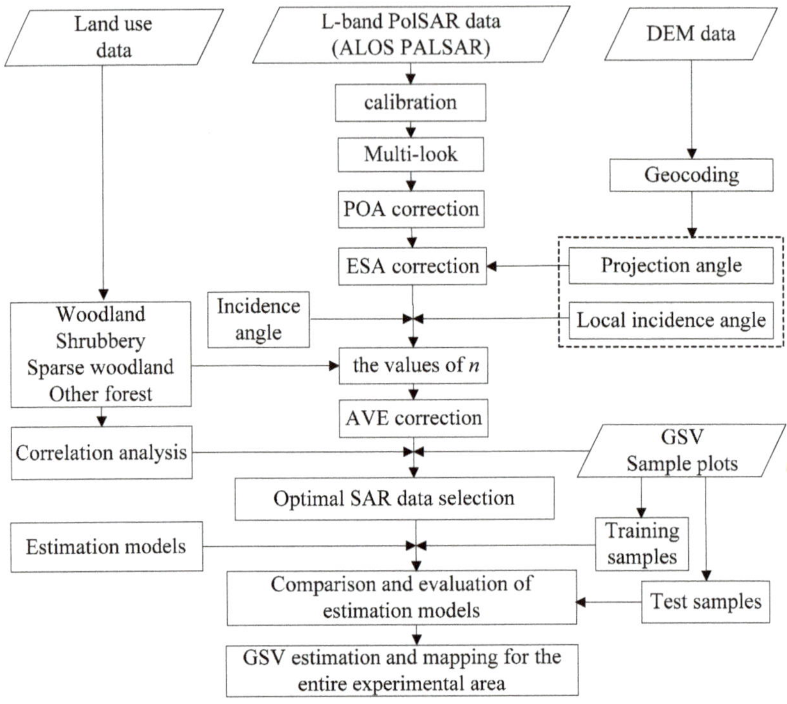

Figure 3. Flowchart of analysis steps.

3.1.2. Effective Scattering Area Correction

The ratio between the radar cross-section and reference area was usually used to express the backscatter coefficient [48]. The theoretical reference area was the pixel area, which did not change with topographic fluctuation. However, in most practical applications, the reference area was defined to be ground area (i.e., the effective area) that was affected by topographic fluctuation. The relation between the effective area and the theoretical reference area is shown as Equation (4) [55]:

$$A_\sigma = A_\beta / cos\varphi \tag{4}$$

where A_β and A_σ represent the theoretical reference area and the effective area, respectively. φ is the projection angle, and $cos\varphi$ is the correction factor for this step.

Then, a general correction equation for σ_0 could be obtained, which was the product of β_0 and $cos\varphi$. For full PolSAR data, we corrected each element in the polarimetric covariance matrix using the same correction factor, and the equation could be written as follows:

$$C_{ESA} = [C_3] \times cos\varphi \tag{5}$$

Here, C_3 is the polarimetric covariance matrix that has been POA corrected and geocoded. The projection angle φ is complementary to the smallest angle between the surface normal and the image plane, and can be obtained by DEM and orbit information.

3.1.3. Angular Variation Effect Correction

Since the local scattering mechanisms within the forest structure vary with the local incidence angles, further AVE correction was needed after the ESA correction. A simple cosine model was derived to reduce the angular effect [56]:

$$\sigma_{corr}^0(\theta_{loc}) = \sigma_0 \times \left(\frac{cos\theta_{ref}}{cos\theta_{loc}} \right)^n \tag{6}$$

where σ_{corr}^0 represents the terrain corrected backscatter, θ_{loc} denotes the local incidence angle, σ_0 denotes the uncorrected radar backscatter coefficient, θ_{ref} is the radar incidence angle, and n is a parameter that needs further discussion.

In a similar way, the polarimetric covariance matrix can be corrected by a 3×3 correction coefficient matrix $[K_3]$ [50], and the expression is given as:

$$C_{AVE} = [C_3] \times [K_3] \tag{7}$$

where C_3 is the polarimetric covariance matrix that has been corrected by the previous two steps, and K_3 is the correction coefficient matrix.

According to Equation (6), the local scattering mechanisms are mainly affected by the local incidence angles. Therefore, the optimal correction factor n of different polarizations can be obtained through calculating and evaluating the correlation results between the local incidence angle and the corrected backscatter coefficient, which is:

$$n_{p,q}(z) = argmin\{|\rho(\theta_{loc}, C_{i,j})|\} \tag{8}$$

where ρ denotes the correlation between two parameters, $C_{i,j}$ represents the elements of the corrected covariance matrix, z represents the different types of forest cover, and p and q represent different polarization channels. According to Equation (7), only the values of n corresponding to the primary diagonal elements of the covariance matrix need to be obtained. Considering the impact of forest characteristics on terrain correction, we try to calculate n corresponding to different types of forests in this paper in order to effectively reduce the topographic effect. The initial ranges of the n values are from zero to two. In addition, considering the computational complexity and accuracy of the n value, we set the interval of n to 0.01. The optimal n is determined by the absolute value of the correlation.

3.2. Retrieval of GSV

We performed terrain correction on all the full PolSAR data, and then selected the most relevant data for forest GSV estimation through time series analysis. The elements on the diagonal of the covariance matrix (corresponding to backscattering intensity of different polarization channels) were used as variables of estimation models. A few studies reported that individual backscattering measurements could be used to extract forest biological parameters [57,58]. Therefore, five single-variable models (Equations (9) to (13)) were first fitted to analyze the relationship between the single variables and the GSV. Among the five models, model (e) was derived from the parameterization of the improved Water-Cloud model, which we named the Water-Cloud analysis model.

(a) Linear function:

$$\sigma^0 = \beta_1 + \beta_2 GSV \tag{9}$$

(b) Logarithmic function:

$$\sigma^0 = \beta_1 + \beta_2 ln(GSV) \tag{10}$$

(c) Quadratic function:

$$\sigma^0 = \beta_1 + \beta_2 ln(GSV) + \beta_3 (ln(GSV))^2 \tag{11}$$

(d) Exponential function:

$$\sigma^0 = \beta_1 + \beta_2 sqrt(GSV) \tag{12}$$

(e) Water-Cloud analysis function:

$$\sigma^0 = \beta_1 + \beta_2 e^{(\beta_3 GSV)} \tag{13}$$

In addition, we also constructed a multi-variable regression model using three elements (HH, HV, and VV backscatter) on the diagonal of the covariance matrix, and compared it with the above five models to find a suitable model for forest GSV estimation and mapping.

$$ln(GSV) = a + b_1 \sigma_{HH}^0 + b_2 \left(\sigma_{HH}^0 \right)^2 + c_1 \sigma_{HV}^0 + c_2 \left(\sigma_{HV}^0 \right)^2 + d_1 \sigma_{VV}^0 + d_2 \left(\sigma_{VV}^0 \right)^2 \tag{14}$$

4. Results

4.1. Acquisition of Terrain Correction Factors

Before implementing terrain correction, the correction factors of each correction stage should be obtained. These correction factors could be divided into two categories: angular factors and parameter n, where the angular factors include the POA shift angle, projection angle, incidence angle, and local incidence angle. All of the angular factors are shown in Figure 4. Here, it is worth noting that the projection angle and local incidence angle of a single pixel are not complementary, especially in the terrain undulating regions. In order to show the correction parameters and correction effects, we chose one scene of data as an example to display the results. Here, the data acquired on 14 July 2016 was randomly selected.

Based on Equation (8), the distribution of correlation coefficients at different polarization channels can be obtained with a 0.01 interval. As shown in Figure 5, the different polarization channels are labeled with solid lines in different colors: HH polarization in red, HV polarization in blue, and VV polarization in green. The black dotted lines represent the position corresponding to the optimal n value. In order to effectively reduce the topographic effect, we have obtained the distribution of the correlation coefficients for different forest cover types, i.e., woodland (Figure 5a), shrubbery (Figure 5b), sparse woodland (Figure 5c), and other forest (Figure 5d).

From these figures, we can see that the variation trend of the correlation coefficients of different polarization channels is consistent for different forest cover types, increasing with the increase of parameter n values. After obtaining the correlation coefficient distribution, we can easily extract the optimum values of n by using Equation (8) in all four forest cover types.

In addition to the data acquired on 14 July 2016, the optimum n values of the remaining six scene data (16 June 2016, 30 June 2016, 25 August 2016, 22 September 2016, and 6 October 2016) are also calculated, and the results are shown in Table 2. From Table 2, it can be seen that the optimum n values of HV polarization is within the range of zero to one, while the HH and VV polarizations are greater than one.

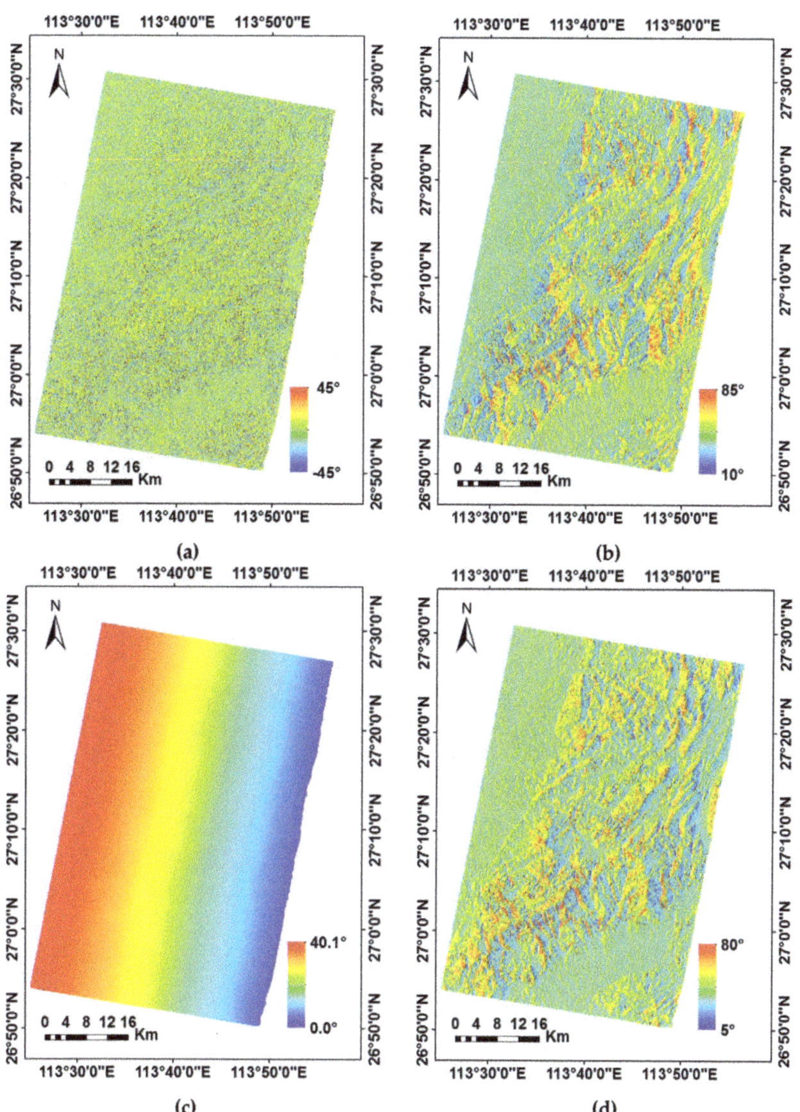

Figure 4. The geocoded angular factors for terrain correction. (**a**) Polarization orientation angle (POA) shift angle; (**b**) Project angle; (**c**) Incidence angle; and (**d**) Local incidence angle.

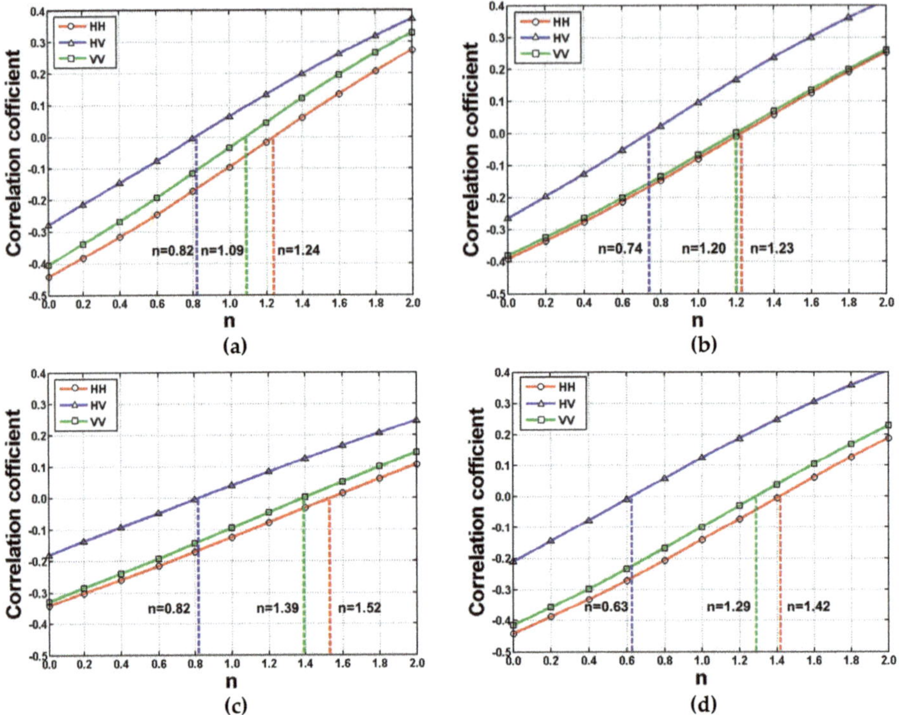

Figure 5. Distribution of correlation coefficients for various values of n and the positons of the optimum n values at four forest cover types (based on data obtained on 14 July 2016): (**a**) Woodland; (**b**) Shrubbery area; (**c**) Sparse woodland; and (**d**) Other forest.

Table 2. Results of the optimum n values. S-Woodland: sparse woodland, O-Forest: other forest.

	Data	Woodland	Shrubbery	S-Woodland	O-Forest
HH	16 June 2016	1.31	1.27	1.60	1.48
	30 June 2016	1.20	1.41	1.50	1.37
	14 July 2016	1.24	1.23	1.52	1.42
	11 August 2016	1.11	1.09	1.30	1.22
	25 August 2016	1.13	1.06	1.36	1.25
	22 September 2016	1.16	1.12	1.37	1.24
	6 October 2016	1.32	1.35	1.58	1.46
HV	16 June 2016	0.91	0.83	0.93	0.77
	30 June 2016	0.76	0.67	0.79	0.57
	14 July 2016	0.82	0.74	0.82	0.63
	11 August 2016	0.74	0.65	0.65	0.48
	25 August 2016	0.74	0.63	0.67	0.47
	22 September 2016	0.70	0.56	0.65	0.44
	6 October 2016	0.85	0.79	0.83	0.64
VV	16 June 2016	1.14	1.24	1.44	1.38
	30 June 2016	1.04	1.14	1.36	1.26
	14 July 2016	1.09	1.20	1.39	1.29
	11 August 2016	1.01	1.11	1.20	1.14
	25 August 2016	1.01	1.06	1.22	1.11
	22 September 2016	1.01	1.10	1.23	1.13
	6 October 2016	1.13	1.29	1.42	1.34

4.2. Results of Terrain Correction

Terrain correction of all the data was performed by using the correction factors obtained in the previous section. In the AVE correction stage, the forest area was divided into four cover types through using the land-use data product, and then the final correction results were merged for analysis. In this section, only the data results of 14 July 2016 were presented to analyze the effects of each correction stage. Figure 6 presents the backscatter coefficients' variation of the original data and each correction stage in different polarization channels. The horizontal axis and vertical axis are the longitude and latitude of the image, and the different colors represent the intensity of backscatter coefficients. Figure 6a1, b1, and c1 show the backscatter coefficients' distribution of different polarization channels in the original data, and the results after POA compensation are shown in Figure 6a2, b2, and c2. According to a visual inspection, no evident differences can be seen in the corresponding polarization channels. This means that the contribution of POA compensation to terrain correction is limited. That is because the impact of the azimuth slope is relatively weak compared to the distance direction for the full polarimetric data. Figure 6a3, b3, and c3 show the results of ESA correction. Obviously, the topographic effects have been improved in all of the polarization channels. However, there are still some topographic effects in high elevation areas, such as the ridge where local incidence angles are usually relatively large, which requires further correction through the AVE correction stage. As shown in Figure 6a4, b4, and c3, in the three polarization channels, the topographic effects of ridges have effectively been removed.

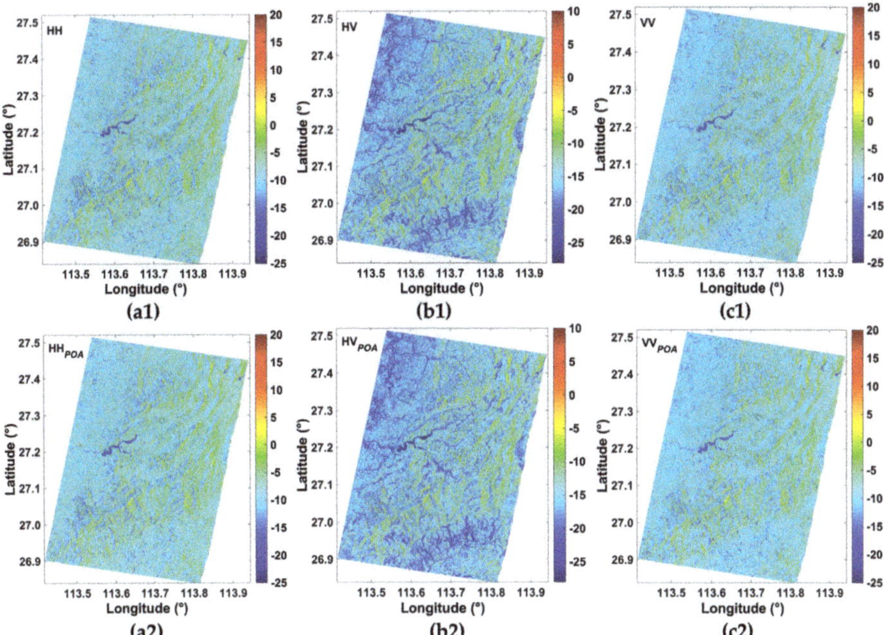

Figure 6. *Cont.*

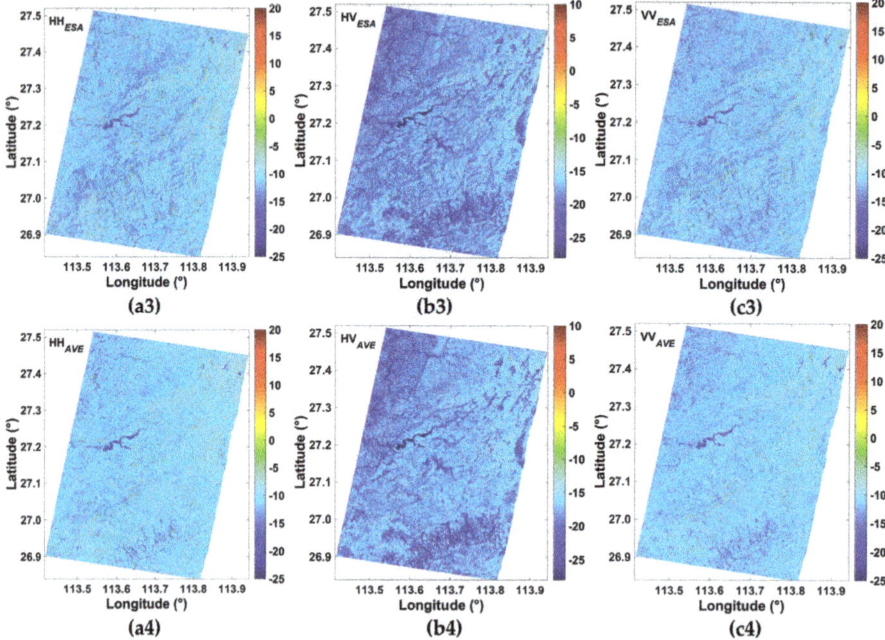

Figure 6. Backscatter coefficient variation of the original data and each correction stage in different polarization channels (HH, HV, and VV). Original data: **a1**, **b1**, **c1**; POA correction stage: **a2**, **b2**, **c2**; Effective scattering area (ESA) correction stage: **a3**, **b3**, **c3**; Angular variation effect (AVE) correction stage: **a4**, **b4**, **c4**.

To further illuminate the effects of terrain correction, we show the relationship between the local incidence angle and the backscatter coefficients of different polarization channels. The results are shown in Figure 7, where the red dashed line is the fitting curve of the backscatter coefficients and local incidence angle, and the different colors represent the density of points. We notice that there is a linear relationship between the backscatter coefficients and local incidence angle. The linear slope is relatively large in the POA compensation stage, but the linear slope becomes smaller following ESA correction and AVE correction. It indicates that POA compensation is limited in its ability to eliminate the impact of local complex terrain. In addition, the effect of the ESA correction stage is more considerable at low-local incidence angles than at high-local incidence angles, where it can effectively limit the overestimation of backscatter intensity. As shown in Figures 7a and 7d, in the range of 5° to 15°, the distribution of backscatter coefficients is from −20 dB to −5 dB at the ESA correction stage, which is much lower than that of the POA correction stage (−10 dB to 5 dB). However, in the range of 70° to 80°, the distribution range of backscatter coefficients does not change in the two correction stages, staying between −10 dB and −20 dB. In contrast, after AVE correction, the distribution of backscatter coefficients is from −15 dB to 0 dB, while it remains unchanged at low-local incidence angle areas. It indicates that the AVE correction method is more effective at high-local incidence angles than at low-local incidence angles, and it can limit the underestimation of backscatter intensity at high-local incidence angle areas.

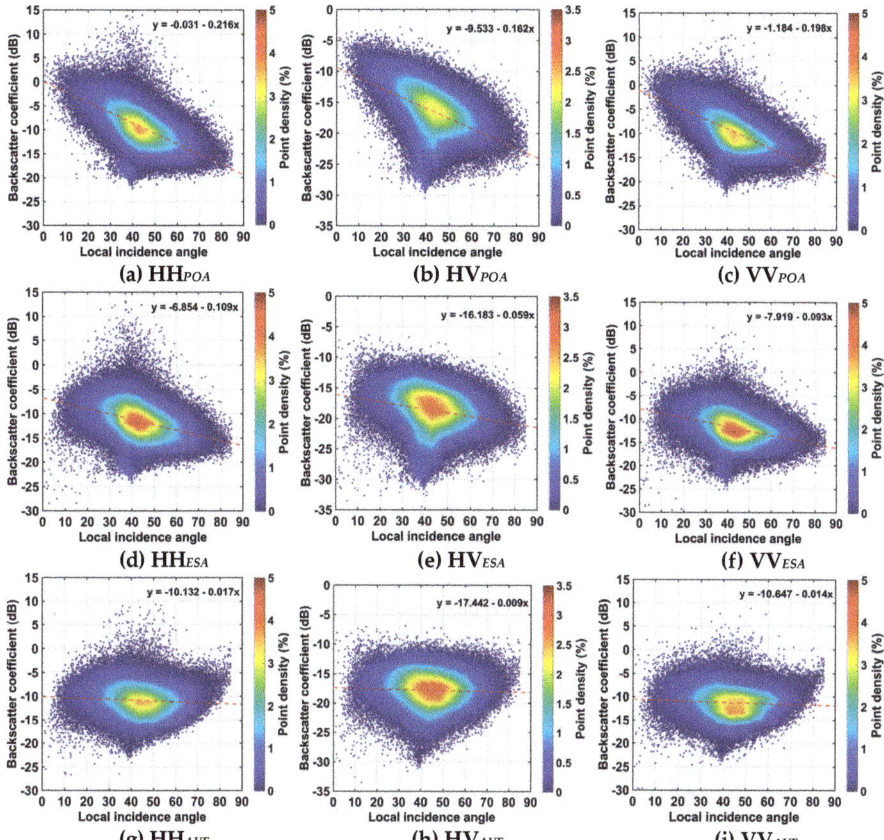

Figure 7. The relationship between backscatter coefficients of different polarization channels and local incidence angles at different correction stages: POA correction stage (HH: **a**, HV: **b**, VV: **c**); ESA correction stage (HH: **d**, HV: **e**, VV: **f**); AVE correction stage (HH: **g**, HV: **h**, VV: **i**).

4.3. Backscatter Sensitivity to Forest GSV

In this section, we analyze the sensitivity between forest GSV of all 60 plots and the individual polarization channel backscatter in seven scenes of PALSAR-2 data. As an example, Figure 8 shows the scatterplots for different polarizations on 14 July 2016, which describe the relationship between the forest GSV and backscatter coeffcents of the original and terrain-corrected data.

Compared with terrain correction, the dynamic range of these scatters is larger in the original data (Figure 8a–c) and shows low sensitivity to forest GSV. After POA compensation (Figure 8d–f), the correlation coefficient values are 0.129, 0.29, and 0.122 at HH, HV, and VV polarization, respectively, which indicate that the sensitivity has not been improved. After ESA correction (Figure 8g–i), the correlation coefficients are increased by 0.227, 0.27, and 0.24 at HH, HV, and VV polarization. The AVE correction stage also contributes to enhancing the sensitivity between forest GSV and backscatter where the correlation coefficients are 0.489, 0.643, and 0.473, respectively (Figure 8j–l). Clearly, the terrain correction can improve the sensibility between forest GSV and backscatter coefficients in this study area. However, we note that sample plots with too low GSV values (less than 37.06 m^3/ha) still have negative effects on the sensitivity between forest GSV and backscatter at the HH and VV channels.

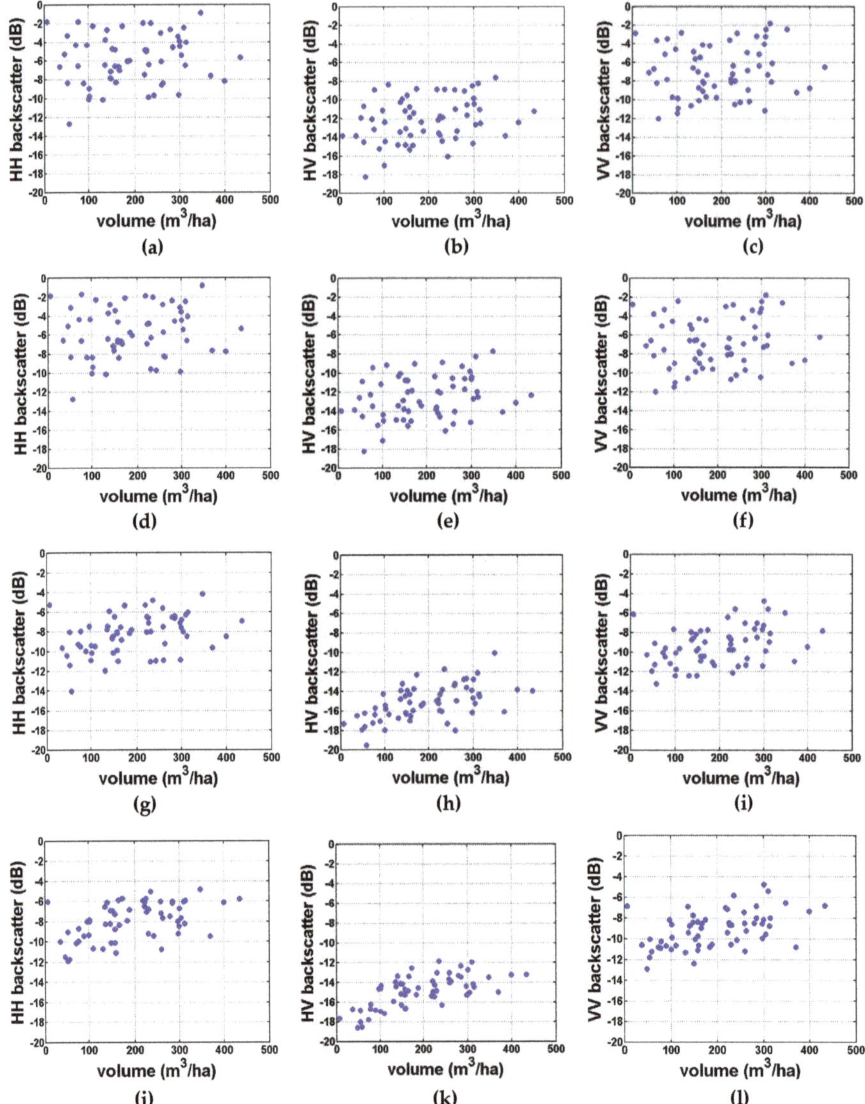

Figure 8. The relationship between forest growing stock volume (GSV) and backscatter coefficents of different polarzation channels. Original data: (**a**) HH, (**b**) HV, and (**c**) VV; After POA compensation: (**d**) HH, (**e**) HV, and (**f**) VV. After ESA correction: (**g**) HH, (**h**) HV, and (**i**) VV; After AVE correction: (**j**) HH, (**k**) HV, and (**l**) VV.

We also compare the correlations between forest GSV and backscatter coefficients of all the SAR data to select the most relevant data for GSV estimation. The results are summarized in Table 3. We find that the correlation coefficient values are different for data acquired on different dates, and the acquired data on 14 July 2016 are much higher than other data. Moreover, the HV backscattering intensities of each scene of PALSAR-2 data show stronger correlations with the GSV than with HH or VV. Therefore, the PolSAR data obtained on 14 July 2016 is selected for further analysis, and the element HV backscatter is used as an individual measurement for single-variable regression models.

Table 3. Correlation coefficients between GSV and backscatter coefficent of all data.

Acquisition Time	HH	HV	VV
16 June 2016	0.418	0.495	0.370
30 June 2016	0374	0.561	0.216
14 July 2016	0.489	0.643	0.473
11 August 2016	0.435	0.564	0.377
25 August 2016	0.469	0.563	0.428
22 September 2016	0.182	0.545	0.123
6 October 2016	0.381	0.487	0.349

4.4. GSV Estimation and Mapping

Based on the results mentioned in Section 4.3, the PolSAR data obtained on 14 July 2016 was used to estimate forest GSV. The three elements of the diagonal of covariance matrix generated by this data are used as variables of regression models; among them, the element HV backscatter is used as an individual measurement for the single-variable regression models. Table 4 provides the results of all the models' fitting based on the training sample dataset. The fitting curves generated by the single-variable models are shown in Figure 9. The decisive coefficients' R^2 values are 0.539 (Direct linear model, Figure 9a), 0.601 (Logarithmic model, Figure 9b), 0.603 (Quadratic model, Figure 9c), 0.579 (Exponential model, Figure 9d), and 0.612 (Water-Cloud analysis model, Figure 9e), respectively. The direct linear relationship (Figure 9a) between the backscatter coefficient and forest GSV is weak, but this phenomenon can be improved through the transformation of parameters, such as the natural logarithmic transformation of forest GSV (Figure 9b). In addition, the Water-Cloud analysis model is found to be the most reliable in the capacity of single-variable regression models, as it produces the highest coefficient of determination in the five models. Howeve, from Figure 9e, when the forest GSV is greater than 300 m^3/ha, the change of the fitting curve tends to be gentle, which may limit its ability regarding estimation in higher GSV areas.

Table 4. Summary of regression model results.

Model	Regression Equation	R^2
Direct linear	$\sigma^0_{HV} = -17.525 + 0.013GSV$	0.529
Logarithmic	$\sigma^0_{HV} = -26.608 + 2.296ln(GSV)$	0.601
Quadratic	$\sigma^0_{HV} = -28.155 + 2.941ln(GSV) - 0.066(ln(GSV))^2$	0.603
Exponential	$\sigma^0_{HV} = -19.914 + 0.377sqrt(GSV)$	0.579
Water-Cloud analysis	$\sigma^0_{HV} = -12.932 - 7.163\exp(-0.008GSV)$	0.612
Multi-variable	$ln(GSV) = -2.611 + 0.531\sigma^0_{HH} + 0.031(\sigma^0_{HH})^2 - 1.693\sigma^0_{HV} - 0.063(\sigma^0_{HV})^2 + 0.255\sigma^0_{VV} + 0.01(\sigma^0_{VV})^2$	0.674

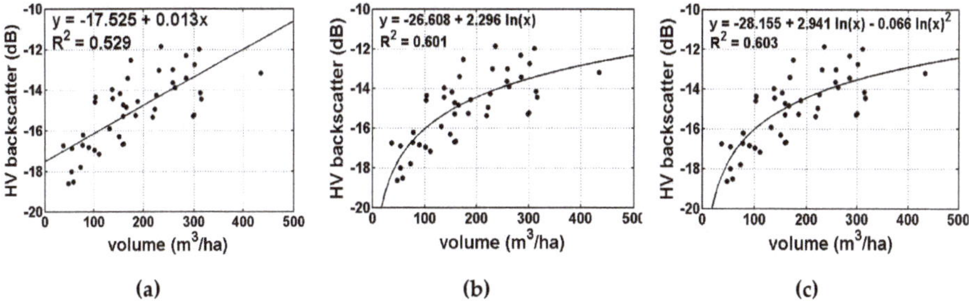

Figure 9. *Cont.*

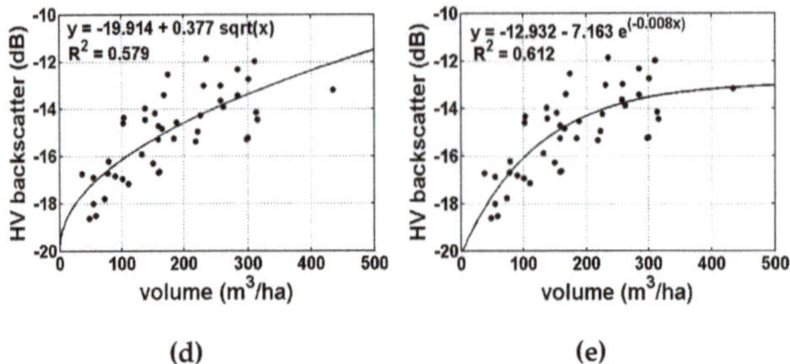

(d)   (e)

Figure 9. Fitting curves of the single-variable models: (**a**) Direct linear model; (**b**) Logarithmic model; (**c**) Quadratic model; (**d**) Exponential model; and (**e**) Water-Cloud analysis model.

Compared with the Water-Cloud analysis model, the established multi-variable model may have greater potential to provide useful GSV estimation. We validate this possibility based on the test sample dataset, and the results are displayed in Figure 10. The validated plot of the Water-Cloud analysis model is characterized by a correlation coefficient R^2 value of 0.417, whose root mean square error (RMSE) is 91.075 m^3/ha. For the multi-variable model, the correlation coefficient R^2 value is 0.630, whose RMSE is 70.965 m^3/ha. Obviously, the accuracy of the multi-variable model inversion is higher than that of the Water-Cloud analysis model inversion. Therefore, the multi-variable model is used to estimate the forest GSV for the whole study region. The results are shown in Figure 11. Figure 11a is the schematic diagram for the spatial distribution of the forest GSV at the pixel scale, which shows that the range of the estimated forest GSV is 0 to 450 m^3/ha. Figure 11b is the histogram of the GSV map. The mean and standard deviation of the GSV in the region are 135.759 m^3/ha and 47.255 m^3/ha. Furthermore, we also calculated the GSV of different land-cover types. The mean GSV of woodland, shrubbery, sparse woodland, and other forest were 137.701 m^3/ha, 130.541 m^3/ha, 125.991 m^3/ha, and 113.759 m^3/ha, respectively. The standard deviations were 45.906 m^3/ha, 42.172 m^3/ha, 56.274 m^3/ha, and 62.051 m^3/ha, respectively.

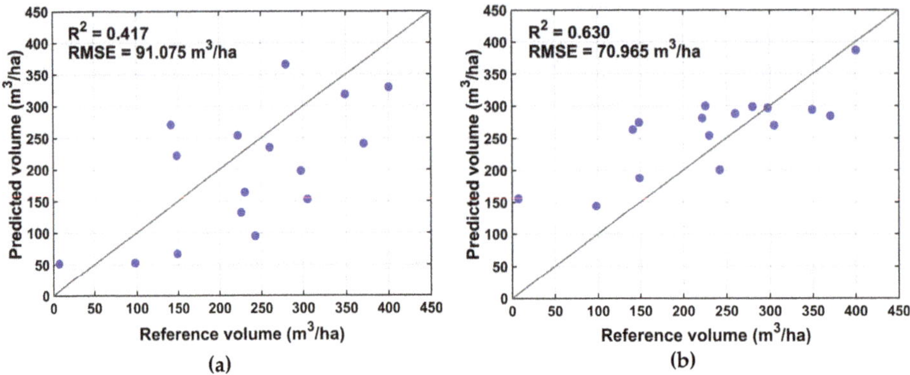

(a)   (b)

Figure 10. The relationship between predicated GSV and reference GSV using test sample dataset: (**a**) Water-Cloud analysis model; and (**b**) Multi-variable model.

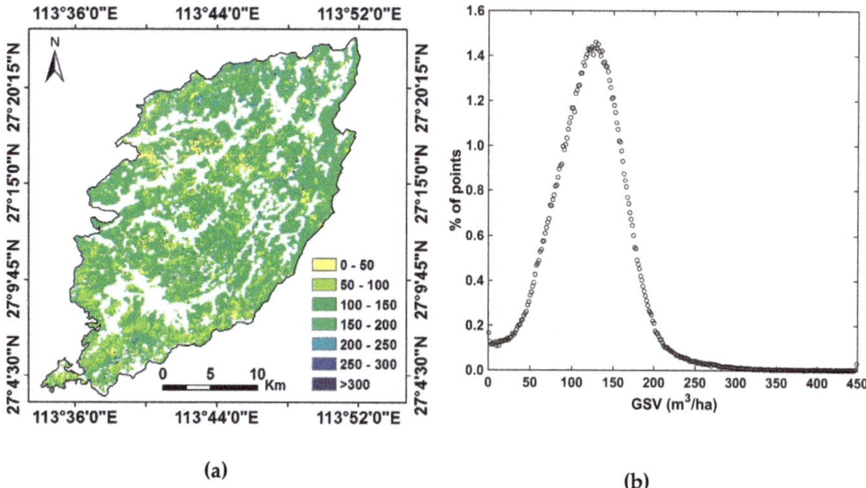

Figure 11. Forest GSV estimation results of the whole study region: (**a**) Spatial distribution of forest GSV at pixel scale; (**b**) Histogram of GSV map.

5. Discussion

PALSAR-2 L-band full polarimetric data has been widely used in the estimation of forest parameters. However, its use for the estimation of forest GSV presents a challenge in subtropical mountain areas, where the underlying topography is complex and diverse, seriously affecting the radiometric quality of SAR images [59]. We performed terrain correction through integrating three stages (POA, ESA, and AVE) to reduce the negative influence of topography on the full polarimetric data. In these three steps, the AVE correction step is based on a semi-empirical cosine model, which can be considered as a function of parameter n (Equation (6)). Therefore, the key to the effectiveness of the AVE correction is whether the value of n can be accurately obtained. A traditional way to obtain the value of n is to use an empirical value of one [60,61], which corrects the experimental data as a whole, and does not need to mask non-forest areas. However, it ignores the difference between polarization channels and the influence of forest features, which is often used for the terrain correction of single-polarimetric and dual-polarimetric data [39,62,63]. In this study, we calculated and evaluated the correlation results between the local incidence angle and the backscatter coefficient to generate the correction factor n of different polarization channels. It is adaptive and takes into account the difference between polarization channels. In addition, according to the results of Figure 5 and Table 2, we would like to stress the necessity of land-cover types for reducing the impact of microtopography in AVE correction, although it has many classification criteria.

In this paper, the SRTM DEM is used as an auxiliary data for SAR dataset geocoding and terrain correction. According to the results of Figures 7 and 8, the effects of removing terrain and improving sensitivity are obvious. However, the DEM data are digital surface models (i.e., DSM), not digital terrain models (i.e., DTM). Thus, the experimental process is carried out under the assumption that the fluctuation of the forest canopy top is consistent with that of the underlying topography. For coarse resolution SAR data, this assumption is not a serious limitation, because it is difficult to reflect the information of individual canopy fluctuation in DSM data with coarse resolution [50]. Therefore, in the premise of multi-look processing of SAR data, SRTM DEM (30 m or 90 m) can be used to assist terrain correction [50,61]. Based on the resolution of DEM data being far lower than the original PolSAR data, we used the projection angle method instead of the area integration method [48] in the ESA correction stage. Although the use of globally shared DEM products can reduce the impact of terrain, it is still worth looking forward to obtaining high-precision and high-resolution DTM data through PolInSAR technology.

Forest GSV has different sensitivity to PolSAR data backscatter coefficients from different dates, even for the same polarization channel (Table 3). This phenomenon may be related to external environmental conditions (e.g., moisture and wind speed variations). This is because irregular variation of the environment affects the interaction between the electromagnetic waves and vegetation components. Although multi-temporal SAR data with significant climatic difference have been used to assess the relationships between the backscatter coefficients and forest parameters [28,42], the specific impact of the external environment on PolSAR data in subtropical regions is still unknown, and should be further studied. In addition, from Table 3, we find that the cross (HV)-polarized backscatter intensity of each scene of PALSAR-2 data is more sensitive to forest GSV than co (HH, VV)-polarization in subtropical mountain areas. The most likely reason is that the cross-polarized backscatter mainly occurs from multiple scattering within the tree canopy, and is less affected by the external environment [13].

We use five single-variable models to establish relationships between the GSV and the backscatter coefficients of the HV polarization (Table 4 and Figure 9). The results suggest that the direct linear relationship between the backscatter coefficient and forest GSV is weak. However, this phenomenon can be improved through the transformation of parameters, such as natural logarithmic transformation of forest GSV. Through the contrast analysis of the fitting performance of the five models, the Water-Cloud analysis model is found to be the most reliable. However, the multi-variable model has greater potential to provide a useful estimation of GSV than the Water-Cloud model. In fact, the correlation coefficient R^2 values of the two models only differ by 0.062, but the former has a higher accuracy of GSV estimation than the latter in the model test. This indicates that co-polarization can also make a certain contribution in GSV estimation. Therefore, our study recommends using the multi-variable model to map the GSV in the study area.

6. Conclusions

This research investigated the capability of full polarized L-band backscattering for the estimation of forest GSV in a subtropical mountain region of eastern Hunan, China. However, it was challenging to estimate GSV due to complex topography of the region. In this paper, we proposed a strategy for fine terrain correction through integrating three stages (POA, ESA, and AVE) and taking into account the impact of land-cover types. In the AVE correction stage, we calculated and evaluated the correlation results between the local incidence angle and the backscatter coefficient to generate the correction factor n of different polarization channels. We found that the optimum n values of HV polarization were within the range of zero to one, while the HH and VV polarizations were greater than one. The results of terrain correction demonstrated that the terrain correction strategy effectively reduced the negative influence of topography and improved the sensitivity between the forest GSV and backscatter coefficients. The results also showed that the land-cover types were necessary data for reducing the impact of microtopography in AVE correction. In the three primary diagonal elements of the PolSAR covariance matrix, the cross-polarized backscatter was more sensitive to forest GSV than co-polarization, and could be used as a single variable for GSV estimation. To estimate and map the GSV of the study area, five single-variable models and a multi-variable model were built using field measurements and corrected PolSAR data. The multi-variable model that was constructed by combining three diagonal elements had greater potential to provide a useful estimation of GSV than the single-variable models, whose correlation coefficient value was 0.630 and RMSE was 70.965 m^3/ha. Therefore, our study recommended using the multi-variable model to map the GSV in the study area. The range of estimated forest GSV was 0 to 450 m^3/ha. The mean value and stander deviation were 135.759 m^3/ha and 47.255 m^3/ha, respectively. The study expands the application potential of PolSAR data in complex topographic areas; thus, it is helpful and valuable for the large-scale (e.g., national or global scale) estimation of forest parameters.

Author Contributions: H.Z. conceived the idea, performed the experiments, and wrote and revised the paper; J.Z. supervised the work and contributed some ideas; C.W. contributed some ideas, analyzed the experimental results, and revised the paper; H.L. and J.L. provided the field data; L.Z., H.F., and Z.L. contributed to the discussion of the results.

Funding: This research was funded by the National Natural Science Foundation of China (No. 41820104005, 41531068, 41842059 and 41671356), and the Innovation Foundation for Postgraduate of Central South University, China (No. 2017zzts179).

Acknowledgments: The authors would like to thank the Japan Aerospace Exploration Agency (JAXA) for providing PALSAR-2 data (3375).

Conflicts of Interest: The authors declare no conflict of interest.

References

1. Houghton, R.A. Aboveground forest biomass and the global carbon balance. *Glob. Chang. Biol.* **2005**, *11*, 945–958. [CrossRef]
2. Shvidenko, A.; Schepaschenko, D.; Nilsson, S.; Bouloui, Y. Semi-empirical models for assessing biological productivity of Northern Eurasian forests. *Ecol. Model.* **2007**, *204*, 163–179. [CrossRef]
3. Santoro, M.; Beaudoin, A.; Beer, C.; Cartus, O.; Fransson, J.E.S.; Hall, R.J.; Pathe, C.; Schmullius, C.; Schepaschenko, D.; Shvidenko, A.; et al. Forest growing stock volume of the northern hemisphere: Spatially explicit estimates for 2010 derived from Envisat ASAR. *Remote Sens. Environ.* **2015**, *168*, 316–334. [CrossRef]
4. Santoro, M.; Cartus, O. Research pathways of forest above-ground biomass estimation based on SAR backscatter and interferometric SAR observations. *Remote Sens.* **2018**, *10*, 608. [CrossRef]
5. Chowdhury, T.A.; Thiel, C.; Schmullius, C. Growing stock volume estimation from L-band ALOS PALSAR polarimetric coherence in Siberian forest. *Remote Sens. Environ.* **2014**, *155*, 129–144. [CrossRef]
6. Song, R.; Lin, H.; Wang, G.; Yan, E.; Ye, Z. Improving selection of spectral variables for vegetation classification of east dongting lake, China, Using a Gaofen-1 image. *Remote Sens.* **2018**, *10*, 50. [CrossRef]
7. Bilous, A.; Myroniuk, A.; Holiaka, D.; Bilous, S.; See, L.; Schepaschenko, D. Mapping growing stock volume and forest live biomass: A case study of the Polissya region of Ukraine. *Environ. Res. Lett.* **2017**, *12*, 105001. [CrossRef]
8. Huete, A.; Didan, K.; Miura, T.; Rodriguez, E.P.; Gao, X.; Ferreira, L.G. Overview of the radiometric and biophysical performance of the MODIS vegetation indices. *Remote Sens. Environ.* **2002**, *83*, 195–213. [CrossRef]
9. Nakaji, T.; Ide, R.; Takagi, K.; Kosugi, Y.; Ohkubo, S.; Nasahara, K.N.; Saigusa, N.; Oguma, H. Utility of spectral vegetation indices for estimation of light conversion efficiency in coniferous forests in Japan. *Agric. For. Meteorol.* **2008**, *148*, 776–787. [CrossRef]
10. Zheng, S.; Gao, C.; Dang, Y.; Xiang, H.; Zhao, J.; Zhang, Y.; Wang, X.; Guo, H. Retrieval of forest growing stock volume by two different methods using Landsat TM images. *Int. J. Remote Sens.* **2014**, *35*, 29–43. [CrossRef]
11. Chrysafis, I.; Mallinis, G.; Siachalou, S.; Patias, P. Assessing the relationships between growing stock volume and sentinel-2 imagery in a Mediterranean forest ecosystem. *Remote Sens. Lett.* **2017**, *8*, 508–517. [CrossRef]
12. Nichol, J.E.; Sarker, M.I.R. Improved biomass estimation using the texture parameters of two high-resolution optical sensors. *IEEE Trans. Geosci. Remote Sens.* **2011**, *49*, 930–948. [CrossRef]
13. Sinha, S.; Jeganathan, C.; Sharma, L.K.; Nathawat, M.S. A review of radar remote sensing for biomass estimation. *Int. J. Environ. Sci. Technol.* **2015**, *12*, 1779–1792. [CrossRef]
14. Donoghue, D.N.M.; Watt, P.J.; Cox, N.J.; Wilson, J. Remote Sensing of Species Mixtures in Conifer Plantations Using LiDAR Height and Intensity Data. *Remote Sens. Environ.* **2007**, *110*, 509–522. [CrossRef]
15. Cartus, O.; Kellndorfer, J.; Rombach, M.; Walker, W. Mapping canopy height and growing stock volume using airborne Lidar, ALOS PALSAR and Landsat ETM+. *Remote Sens.* **2012**, *4*, 3320–3345. [CrossRef]
16. Skowronski, N.S.; Clark, K.L.; Gallagher, M.; Birdsey, R.A.; Hom, J.L. Airborne laser scanner-assisted estimation of aboveground biomass change in a temperate oak-pine forest. *Remote Sens. Environ.* **2014**, *151*, 166–174. [CrossRef]
17. Lu, D.; Chen, Q.; Wang, G.; Li, G.; Moran, E. A survey of remote sensing- based aboveground biomass estimation methods in forest ecosystems. *Int. J. Digit. Earth* **2014**, *9*, 63–105. [CrossRef]

18. Le Toan, T.; Quegan, S.; Davidson, M.W.J.; Balzter, H.; Paillou, P.; Papathanassiou, K.; Plummer, S.; Rocca, F.; Saatchi, S.; Shugart, H.; et al. The BIOMASS Mission: Mapping global forest biomass to better understand the terrestrial carbon cycle. *Remote Sens. Environ.* **2011**, *115*, 2850–2860. [CrossRef]

19. Fu, H.; Zhu, J.; Wang, C.; Wang, H.; Zhao, R. Underlying topography estimation over forest areas using high-resolution P-band Single-baseline PolInSAR data. *Remote Sens.* **2017**, *9*, 363. [CrossRef]

20. Wang, C.; Cai, J.; Li, Z.; Mao, X.; Feng, G.; Wang, Q. Kinematic parameter inversion of the slumgullion landslide using the time series offset tracking method with UAVSAR data. *J. Geophys Res. Solid Earth* **2018**, *10*, 1029. [CrossRef]

21. Xie, Q.; Zhu, J.; Lopez-Sanchez, J.M.; Wang, C.; Fu, H. A modified general polarimetric model-based decomposition method with the simplified Neumann volume scattering model. *IEEE Geosci. Remote Sens. Lett.* **2018**, *15*, 1229–1233. [CrossRef]

22. Gao, H.; Wang, C.; Wang, G.; Zhu, J.; Tang, Y.; Shen, P.; Zhu, Z. A crop classification method integrating GF-3 PolSAR and Sentinel-2A optical data in the Dongting Lake Basin. *Sensors* **2018**, *18*, 3139. [CrossRef]

23. Sandberg, G.; Ulander, L.M.H.; Fransson, J.E.S.; Holmgren, J.; Le Toan, T. L-and P-band backscatter intensity for biomass retrieval in hemiboreal forest. *Remote Sens. Environ.* **2011**, *115*, 2874–2886. [CrossRef]

24. Mermoz, S.; Réjou-Méchain, M.; Villard, L.; Le Toan, T.; Rossi, V.; Gourlet-Fleury, S. Decrease of L-band SAR backscatter with biomass of dense forests. *Remote Sens. Environ.* **2015**, *159*, 307–317. [CrossRef]

25. Wu, C.; Wang, C.; Shen, P.; Zhu, J.; Fu, H.; Gao, H. Forest height estimation using PolInSAR optimal normal matrix constraint and cross-iteration method. *IEEE Geosci. Remote Sens. Lett.* **2019**. [CrossRef]

26. Ma, J.; Xiao, X.; Qin, Y.; Chen, B.; Hu, Y.; Li, X.; Zhao, B. Estimating aboveground biomass of broadleaf, needleleaf, and mixed forests in Northeastern China through analysis of 25-m ALOS/PALSAR mosaic data. *For. Ecol. Manag.* **2017**, *389*, 199–210. [CrossRef]

27. Treuhaft, R.N.; Goncalves, F.G.; Drake, J.B.; Chapman, B.D.; Santos, J.R.D.; Dutra, L.V.; Graca, P.M.L.A.; Purcell, G.H. Biomass estimation in a Tropical Wet forest using Fourier transforms of profiles from lidar or interferometric SAR. *Geophys Res. Lett.* **2010**, *37*, 225. [CrossRef]

28. Thiel, C.; Schmullius, C. The potential of ALOS PALSAR backscatter and InSAR coherence for forest growing stock volume estimation in Central Siberia. *Remote Sens. Environ.* **2016**, *173*, 258–273. [CrossRef]

29. Tebaldini, S. Algebraic synthesis of forest scenarios from multibaseline PolInSAR data. *IEEE Trans. Geosci. Remote Sens.* **2009**, *47*, 4132–4144. [CrossRef]

30. Cloude, S.R. Polarization coherence tomography. *Radio Sci.* **2006**, *41*, RS4017. [CrossRef]

31. Li, W.; Chen, E.; Li, Z.; Ke, Y.; Zhan, W. Forest aboveground biomass estimation using polarization coherence tomography and PolSAR segmentation. *Int. J. Remote Sens.* **2015**, *36*, 530–550. [CrossRef]

32. Zhang, H.; Wang, C.; Zhu, J.; Fu, H.; Xie, Q.; Shen, P. Forest above-ground biomass estimation using single-baseline polarization coherence tomography with P-band PolInSAR data. *Forests* **2018**, *9*, 163. [CrossRef]

33. Minh, D.H.T.; Le Toan, T.; Rocca, F.; Tebaldini, S.; Villard, L.; Réjou-Méchain, M.; Phillips, O.L.; Feldpausch, T.R.; Dubois-Fernandez, P.; Scipal, K.; et al. SAR tomography for the retrieval of forest biomass and height: Cross-validation at two tropical forest sites in French Guiana. *Remote Sens. Environ.* **2016**, *175*, 138–147. [CrossRef]

34. Peng, X.; Li, X.; Wang, C.; Fu, H.; Du, Y. A maximum likelihood based nonparametric iterative adaptive radar tomography and its application for estimating underlying topography and forest height. *Sensors* **2018**, *18*, 2459. [CrossRef]

35. Peng, X.; Wang, C.; Li, X.; Du, Y.; Fu, H.; Yang, Z.; Xie, Q. Three-Dimensional structure inversion of buildings with nonparametric iterative adaptive approach using SAR tomography. *Remote Sens.* **2018**, *10*, 1004. [CrossRef]

36. Lee, J.S.; Pottier, E. *Polarimetric Radar Imaging: From Basics to Applications*; CRC Press, Taylor & Francis Group: Boca Raton, FL, USA, 2009.

37. Chowdhury, T.A.; Thiel, C.; Schmullius, C.; Stelmaszczukgórska, M. Polarimetric parameters for growing stock volume estimation using ALOS PALSAR L-band data over Siberian forests. *Remote Sens.* **2013**, *5*, 5725–5726. [CrossRef]

38. Le Toan, T.; Quegan, S.; Woodward, I.; Lomas, M.; Delbart, N.; Picard, C. Relationg radar remote sensing of biomass to modeling of forest carbon budgets. *Clim. Chang.* **2004**, *76*, 379–402. [CrossRef]

39. Lucas, R.; Armston, J.; Fairfax, R.; Fensham, R.; Accad, A.; Carreiras, J.; Kelley, J.; Bunting, P.; Clewley, D.; Bray, S.; et al. An evaluation of the ALOS PALSAR L-band backscatter-Above ground biomass relationship Queensland, Australia: Impacts of surface moisrure condition and vegetation structure. *IEEE J. Sel. Top. Appl. Earth Obs. Remote Sens.* **2010**, *3*, 576–593. [CrossRef]

40. Wilhelm, S.; Hüttich, C.; Korets, M.; Schmullius, C. Large area mapping of boreal growing stock volume on an annual and multi-temporal level using PALSAR L-band backscatter mosaics. *Forests* **2014**, *5*, 1999–2015. [CrossRef]

41. Mermoz, S.; Le Toan, T.; Villard, L.; Réjou-méchain, M.; Seifert-Granzin, J. Biomass assessment in the Cameroon savanna using ALOS PALSAR data. *Remote Sens. Environ.* **2014**, *155*, 109–119. [CrossRef]

42. Urbazaev, M.; Thiel, C.; Mathieu, R.; Naidoo, L.; Levick, S.R.; Smit, I.P.J.; Asner, G.P.; Schmullius, C. Assessment of the mapping of fractional woody cover in southern African savannas using multi-temporal and polarimetric ALOS PALSAR L-band images. *Remote Sens. Environ.* **2015**, *166*, 138–153. [CrossRef]

43. Antropov, O.; Rauste, Y.; Häme, T.; Praks, J. Polarimetric ALOS PALSAR time series in mapping biomass of boreal forests. *Remote Sens.* **2017**, *9*, 999. [CrossRef]

44. Zhao, P.P.; Lu, D.S.; Wang, G.X.; Liu, L.J.; Li, D.Q.; Zhu, J.R.; Yu, S.Q. Forest aboveground biomass estimation in Zhejiang province using the integration of Landsat TM and ALOS PALSAR data. *Int. J. Appl. Earth Obs.* **2016**, *53*, 1–15. [CrossRef]

45. Bouvet, A.; Mermoz, S.; Le Toan, T.; Villard, L.; Mathieu, R.; Naidoo, L.; Asner, G.P. An above-ground biomass map of African savannahs and woodlands at 25m resolution derived from ALOS PALSAR. *Remote Sens. Environ.* **2018**, *206*, 156–173. [CrossRef]

46. Ningthoujam, R.K.; Joshi, P.K.; Roy, P.S. Retrieval of forest biomass for tropical deciduous mixed forest using ALOS PALSAR mosaic imagery and field plot data. *Int. J. Appl. Earth Obs. Geoinform.* **2018**, *69*, 206–216. [CrossRef]

47. Lee, J.S.; Ainsworth, T.L. The effect of orientation angle compensation on coherency matrix and polarimetric target decomposition. *IEEE Trans. Geosci. Remote Sens.* **2011**, *49*, 53–64. [CrossRef]

48. Small, D. Flattening gamma: Radiometric terrain correction for SAR imagery. *TEEE Trans. Geosci. Remote Sens.* **2011**, *49*, 3081–3093. [CrossRef]

49. Castel, T.; Beaudoin, A.; Stach, N.; Le Toan, T.; Durand, P. Sensitivity of space-borne SAR data to forest parameters over sloping terrain. Theory and experiment. *Int. J. Remote Sens.* **2001**, *22*, 2351–2376. [CrossRef]

50. Zhao, L.; Chen, E.; Li, Z.; Zhang, W.; Gu, X. Three-step semi-empirical radiometric terrain correction approach for PloSAR data applied to forested areas. *Remote Sens.* **2017**, *9*, 269. [CrossRef]

51. Deng, S.; Katoh, M.; Guan, Q.; Yin, N.; Li, M. Estimating forest aboveground biomass by combining ALOS PALSAR and WorldView-2 data: A case study at purple mountain national park, Nanjing, China. *Remote Sens.* **2014**, *6*, 7878–7910. [CrossRef]

52. Fang, J.; Liu, G.; Xu, S. Biomass and net production of forest vegetation in China. *Acta Ecol. Sin.* **1996**, *16*, 497–508.

53. Shimada, M.; Isoguchi, O.; Tadono, T.; Isono, K. PALSAR radiometric and geometric calibration. *IEEE Trans. Geosci. Reomte Sens.* **2009**, *47*, 3915–3932. [CrossRef]

54. Lee, J.S.; Schuler, D.L.; Ainsworth, T.L. Polarimetric SAR data compensation for terrain azimuth slope variation. *IEEE Trans. Geosci. Remote Sens.* **2000**, *38*, 2153–2163.

55. Ulander, L.M.H. Radiometric slope correcton of synthetic-aperture radar images. *IEEE Trans. Geosci. Rmote Sens.* **1996**, *34*, 1115–1122. [CrossRef]

56. Ulaby, F.T.; Moore, R.K.; Fung, A.K. Volume Scattering and Emission Theory. In *Microwave Remote Sensing Active and Passive*; Artech House: Norwood, MA, USA, 1982; Volume III.

57. Peregon, A.; Yamagata, Y. The use of ALOS/PALSAR backscatter to estimation above-ground forest biomass: A case study in Western Siberia. *Remote Sens. Environ.* **2013**, *137*, 139–146. [CrossRef]

58. Pham, T.D.; Yoshino, K. Aboveground biomass estimation of mangrove species using ALOS-2 PALSAR imagery in Hai Phong city, Vietnam. *J. Appl. Remote Sens.* **2017**, *11*, 026010. [CrossRef]

59. Villard, L.; Le Toan, T. Relating P-band SAR intensity to biomass for tropical dense forests in Hilly terrain: γ0 or t0. *IEEE J. Sel. Top. Appl. Earth Obs. Remote Sens.* **2015**, *8*, 214–223. [CrossRef]

60. Cartus, O.; Santoro, M.; Kellndorfer, J. Mapping forest aboveground biomass in the Northeastern United States with ALOS PALSAR dual-polarization L-band. *Remote Sens. Environ.* **2012**, *124*, 466–478. [CrossRef]

61. Attarchi, S.; Gloaguen, R. Improving the estimation of above ground biomass using dual polarimetric PALSAR and ETM+ data in the Hyrcanian Mountain forest (Iran). *Remote Sens.* **2014**, *6*, 3693–3715. [CrossRef]
62. Thiel, C.J.; Thiel, C.; Schmullius, C.C. Operational large-area forest monitoring in Siveria using ALOS PALSAR summer intensities and winter coherence. *IEEE Trans. Geosci. Remote Sens.* **2009**, *47*, 3993–4000. [CrossRef]
63. Kim, C. Quantataive analysis of relationship between ALOS PALSAR backscatter and forest stand volume. *J. Mar. Sci. Technol.* **2012**, *20*, 624–628.

© 2019 by the authors. Licensee MDPI, Basel, Switzerland. This article is an open access article distributed under the terms and conditions of the Creative Commons Attribution (CC BY) license (http://creativecommons.org/licenses/by/4.0/).

 *forests*

Article

Testing a New Ensemble Model Based on SVM and Random Forest in Forest Fire Susceptibility Assessment and Its Mapping in Serbia's Tara National Park

Ljubomir Gigović [1], Hamid Reza Pourghasemi [2,*], Siniša Drobnjak [3] and Shibiao Bai [4,*]

[1] Department of Geography, University of Defence, 11000 Belgrade, Serbia; gigoviclj@gmail.com
[2] Department of Natural Resources and Environmental Engineering, College of Agriculture, Shiraz University, Shiraz, Iran
[3] Military Geographical Institute, 11000 Belgrade, Serbia; sdrobnjak81@gmail.com
[4] College of Marine Sciences and Engineering, Nanjing Normal University, Nanjing 210023, China
* Correspondence: hr.pourghasemi@shirazu.ac.ir (H.R.P.); shibiaobai21@163.com (S.B.)

Received: 18 April 2019; Accepted: 7 May 2019; Published: 11 May 2019

Abstract: The main objectives of this paper are to demonstrate the results of an ensemble learning method based on prediction results of support vector machine and random forest methods using Bayesian average. In this study, we generated susceptibility maps of forest fire using supervised machine learning method (support vector machine—SVM) and its comparison with a versatile machine learning algorithm (random forest—RF) and their ensembles. In order to achieve this, first of all, a forest fire inventory map was constructed using Serbian historical forest fire database, Moderate Resolution Imaging Spectro radiometer (MODIS), Landsat 8 OLI and Worldview-2 satellite images, field surveys, and interpretation of aerial photo images. A total of 126 forest fire locations were identified and randomly divided by a random selection algorithm into two groups, including training (70%) and validation data sets (30%). Forest fire susceptibility maps were prepared using SVM, RF, and their ensemble models using the training dataset and 14 selected different conditioning factors. Finally, to explore the performance of the mentioned models we used the values for area under the curve (AUC) of receiver operating characteristics (ROC). The results depicted that the ensemble model had an AUC = 0.848, followed by the SVM model (AUC = 0.844), and RF model (AUC = 0.834). According to achieved AUC results, it can be deduced that SVM, RF, and their ensemble method had satisfactory performance. The study was applied in the Tara National Park (West Serbia), a region of about 191.7 sq. km distinguished by a very high forest density and a large number of forest fires.

Keywords: geographic information system; support vector machine; random forest; ensemble model; hazard mapping

---

1. Introduction

Forest fires (also called wildfires) represent the uncontrolled movements of fire along the forest surface and they are one of the most damaging natural disasters and forces [1]. According to Chuvieco [2] and Zheng [3], forest fires have become increasingly widespread, partly due to global warming; since summer periods have become hotter and drier than before, winds are getting stronger and the stability of the rainy periods is disturbed, but above all the changes are a result of human negligence and sometimes ulterior motives.

A forest fire turns out to be one of the most critical natural hazards in recent years, and results in a serious loss of human life and terrific damage to the ecological environment and human infrastructure [4]. Wildfires are natural causes for ecological change and a very destructive natural phenomenon the same as earthquakes, landslides, and floods. Therefore, desertification and deforestation are ones of the most important effects of wildfires [5].

Different methods and techniques for forest fire susceptibility mapping are introduced according to the literature and can be classified into three groups: Probabilistic, statistical, and machine learning methods [4].

Probabilistic (mechanistic) methods simulate and predict the possible behavior of forest fires using specific mathematical functions and equations [6]. For this reason, these methods have the ability to model and predict the behavior of fire in space and time. The most commonly used mechanistic forest fire models described in the literature are BEHAVE [7], FIRETEC [8], Fire station [9], and LANDIS-II [10].

Unlike probabilistic methods, the statistical method is a better way to model forest fires when the research field is large, in particular, the combination of remote sensing (RS) technology and geographic information systems (GIS). This is because the statistical method for modeling forest fires collects and processes a large number of spatial data with different scales and resolutions covering large areas. Furthermore, various statistical methods and techniques for forest fire modeling exist, such as logistical regression [2,11–13], Monte Carlo simulations [14], weights-of-evidence [15], logistic generalized additive model [16], evidential belief function [13], and geographically weighted regression [17].

Machine learning methods were proposed and introduced due to the critical accuracy of forest fire evaluation, such as the support vector machine [18,19], random forest [17,20], kernel logistic regression [4], maximum entropy [20], and artificial neural networks (ANN) [1,11,21]. Generally speaking, an evaluation assessment of the machine learning method is better than the statistical method [22]. Indeed, according to Tien Bui [4], due to multiple and complex interactions between conditioning and ignition factors for forest fires, it is still difficult to model and predict forest fires on a regional scale. The objective of this research is, therefore, to evaluate forest fire susceptibility maps using supervised and versatile machine learning algorithms and their ensemble and to compare their performance in the Tara National Park, Republic of Serbia. In this research, 126 forest fire occurrence locations have been identified from satellite images, aerial photo images, and extensive field surveys, and they constitute the basic content of the fire inventory database. Of these, 88 (70%) locations were indiscriminately identified as training data and the remaining 38 (30%) cases were used for confirmation goals. These training datasets and 14 different conditioning factors were used as input data for the application of machine learning algorithms in order to obtain wildfire susceptibility maps.

2. Materials and Methods

2.1. Study Area and Data

Forest area in the Republic of Serbia covers 27,200 sq. km, which is approximately 31.1% of the country area. The study area includes the whole of Serbia's Tara National Park, which approximately covers 191.7 sq. km between latitudes of 43°43′13″ to 44°01′09″ N, and longitudes of 19°13′51″ to 19°44′20″ E. It is in the west of the Republic of Serbia (Figure 1). The study area's altitude varies from 200 to 1591 m above mean sea level (m.s.l.). Tara National Park was founded in 1981. The Tara National Park and the Mokra Gora Nature Park were nominated as potential biosphere reserves by the UNESCO MAB Committee. Mount Tara belongs to the Dinaric Alps and is part of the Old Vlach Mountains of Serbia. It is situated in the far west of Serbia, bordering the Drina River and next to the state border. Mount Tara is a medium–high mountain with an average altitude of 1000–1200 m above mean sea level (m.s.l.). The highest peak is Kozji rid at 1591 meters of altitude.

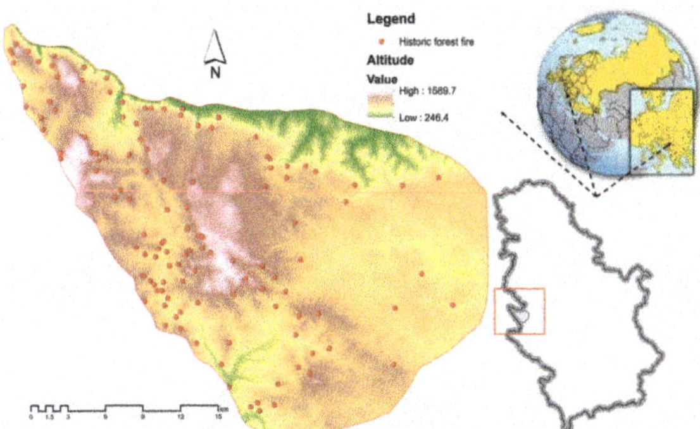

Figure 1. Location of the study area.

The Emerald Network program established Tara National Park as a primary butterfly area (PBA), an important bird area (IBA), and an important plant area (IPA). The Mount Tara area is a typical forest area covered by Silver Fir, European Beech, and European Spruce mixed forests (over 85% of forest area). The slope angles of the test area range from 0° to as much as 89°. The total annual rainfall ranges from 773 to 1038 mm/m^2, in different parts of the study region. The maximum rainfall is between March and June, based on records from the Republic Hydrometeorological Service of Serbia.

Producing forest fire inventory maps is an important step for forest fire susceptibility mapping. The best technique for collecting data on forest fire inventory maps is still unknown. The most common is an aggregation of data collected by a combination of remote sensing technology, geographic information systems, and field work. Therefore, in this study, historical reports, field surveys, high resolution Worldview-2 images, Landsat 8 OLI and MODIS satellite images, and aerial photo interpretation were applied to prepare a forest fire inventory map. The acquisition period of satellite images for the fire inventory map is between 2010 and 2016. The analyzed aerial photos are from 2015 and 2016, with a spatial resolution of 0.4 meters. Forest fire conditioning factor is another key topic and has been researched by a lot of scientists [13,17,19–22]. Hence, different layers, including altitude, aspect, slope degree, plan curvature, topographic wetness index (TWI), normalized difference vegetation index (NDVI), distance from rivers, distance from roads, distance from urban area, annual rainfall, land use/land cover, maximum annual temperature, wind power, and soil type, have been used to analyze the forest fire susceptibility.

Topography data and digital elevation models are among the most important conditions for forest fire sensitivity mapping [13]. In the literature, such as [23,24], the impacts of aspect, altitudes, degree of slope, and curvature have been widely reported. In the current study, a digital elevation model (DEM) with 20 m spatial resolution was developed using topography data contour lines. Conditioning factors such as altitude, aspect, slope degree, plan curvature, and TWI have been created using the mentioned DEM. The land use/land cover map was created using CORINE 2006 data, whereas soil texture is extracted from national soil data. The acronym CORINE stands for Co-ORdination of INformation on the Environment, an experimental programme of the Directorate-General for Environment, Nuclear Safety and Civil protection of the Commission of the European Communities. For assessment vegetation cover, the NDVI obtained from multispectral LANDSAT 8 OLI images. The NDVI index is obtained as the mean value from the average monthly values calculated for 2016. Distance from roads, distance from rivers, and distance from urban areas were prepared using a digital topographic database at scale 1:25,000 produced in the Serbian Military Geographical Institute. Maximum annual temperature, wind power, and annual rainfall were obtained using meteorological data from the Republic Hydro-Meteorological Service of Serbia. The detailed information of data sources for forest fire conditioning factors is shown in Table 1.

Table 1. Data sources and associated factor classes for forest fire susceptibility mapping.

Sub-Classification	Data Layers	Source of Data	GIS Data Type	Derived Map	Resolution
Fire Inventory Database	Historical forest Fire	Worldview-2 images, Landsat 8 OLI images, MODIS images, aerial photo, and National fire inventory database	Point	-	-
Topography	Elevation	DEM, contour lines with 20 m intervals	GRID	Elevation	20 m
	Slope	-	GRID	Slope degree	20 m
	Aspect	-	GRID	Aspect degree	20 m
	Curvature	-	GRID	Curvature	20 m
	TWI	-	GRID	TWI	20 m
Soil type	Soil	National soil data	Polygon	Soil	1:50,000
Land use/land cover	Land use	CORINE data	ARC/INFO GRID	Land use	30 m
NDVI	NDVI	Landsat 8 OLI images	ARC/INFO GRID	NDVI	30 m
Annual rainfall	Rainfall	Republic Hydro-Meteorological Service http://www.hidmet.gov.rs/index_eng.php	GRID	Precipitation map (mm/m^2)	1:50,000
Annual temperature	Max annual temperature	Republic Hydrometeorological Service http://www.hidmet.gov.rs/index_eng.php	GRID	Temperature map (°C)	20 m
Wind power	Wind power	Republic Hydrometeorological Service http://www.hidmet.gov.rs/index_eng.php	GRID	Wind power map (m/s)	20 m
River	Drainage network	MGI Digital topographic map http://www.vgi.mod.gov.rs/english/index_eng.html	Line	Distance from rivers (m)	1:25,000
Roads	Road network	MGI Digital topographic map http://www.vgi.mod.gov.rs/english/index_eng.html	Line	Distance from roads (m)	1:25,000
Urban areas	Urban areas	MGI Digital topographic map http://www.vgi.mod.gov.rs/english/index_eng.html	Polygon	Distance from urban areas (m)	1:25,000

2.2. Methods

The flowchart of the method used in the research is shown in Figure 2. In the first step, the data collection is presented, where all data are placed in the database. In the following, models of support vector machine, random forest, and their ensemble were applied. The validation of the constructed models was finally tested using receiver operating characteristic (ROC) curve.

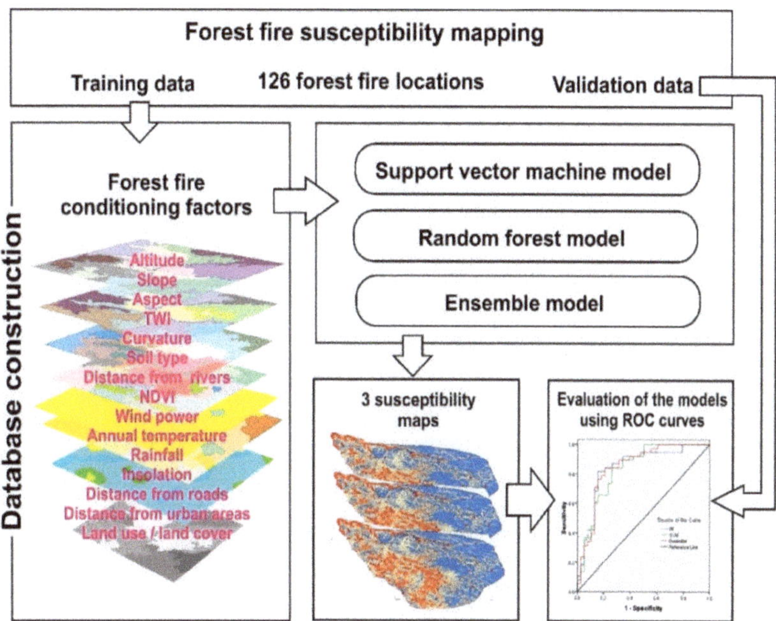

Figure 2. Flowchart of the forest fire susceptibility mapping.

3. Input Variables

3.1. Conditioning Factors

The selection of criteria for assessing the forest fire and its mapping is an important step in the analysis. To create a reliable forest fire susceptibility map, it is essential to identify forest fire conditioning factors [25]. Based on experts' opinions and longer field observations, this study adopted fifteen criteria that are an important cause of susceptibility to forest fires in the Tara National Park of Serbia. The selected criteria with a short description are given in Table 2, and they are shown in Figures 3–6.

Table 2. Conditioning factors for forest fire susceptibility model.

Category	Description
Topography (Figure 3)	Altitude is an important forest fire conditioning factor. An altitude map is prepared from the 20 × 20 m digital elevation model (1:25,000—scale with 20 m contour intervals).
	The slope is the gradient of the land expressed as percentages or angle and it has a great influence on fir behavior. Fires burn faster on a steeped slope due to convection column flame front proximately to new fuels. Slope influences the rate of speed and fire direction.
	Aspect is the direction in which a slope faces. It has an effect on the climate of the slope in terms of insolation, exposure of winds, etc. Therefore, the opposite aspect tends to retain more moisture supporting greenish and healthy vegetation.
	The curvature is defined as the change rate of slope gradient or aspect, usually in a particular direction. In addition, the curvature represents convergence or divergence of water level concurrently with an activity of downhill flow. Negative, zero, and positive curvature represent concave, flat, and convex, respectively.
	Topographic Wetness Index (TWI) describes the size of saturated areas of runoff generation and the effect of topography on the location. It is defined as [26]: TWI = ln (AS/tan β), where AS is the catchment area and β is the slope angle in degrees.
Environmental (Figure 4)	Soil type reflects the affect of textures and compositions of soil materials on fire occurrence. The soil map was constructed from the soil map of the state and was classified into fine-silt, course-loamy, fine-loamy, mixed-loamy, and skeletal-loamy.
	Distance from river was created using a topographical map and it was calculated based on the Euclidean distance method in ArcGIS 10.4 and were classified into (<100), (100–200), (200–500), (500–1000), (1000–2000), (2000–3000), (3000–4000), (4000-5000), and (>5000) meters classes
	Normalized Difference Vegetation Index (NDVI). The NDVI map was created using multispectral Landsat 8 OLI imagery showing the surface vegetation coverage and density in an image.
	Land use/land cover is considered as a factor in environmental protection. Data on land use/cover were taken on the basis of the Corine Land Cover 2006 (CLC2006) database, collected in the framework of the European Commission's CORINE (Coordination of Information on the Environment) programme.
Meteorological (Figure 5)	Wind power varies greatly, even at very short time scales (seconds to minutes). Two wind characteristics are used in wildfire susceptibility mapping: Wind speed and wind direction.
	Annual temperature is a basic weather factor and should be taken into account. The temperature influences the condition of forest fuel, as its main effect is to dry the fuel.
	Rainfall is the important effect that contributes to high fuel humidity and therefore is a negative indicator of the spread of fire. The scale was reversed to conform to the linear trend of other parameters. Annual rainfall values are divided into nine classes: (773.6–801.6, 801.7–831.6, 831.7–863.8, 863.9–895.9, 896–925.9, 926–950.8, 950.9–973.6, 973.7–998.5, 998.6–1037.9 mm/m^2)
Social (Figure 6)	Distance from roads was created using a topographical map, was calculated based on the Euclidean distance method in ArcGIS 10.4, and was classified into (<100), (100–200), (200-300), (300–500), (500–750), (750–1000), (1000–2000), (2000–3000), and (>3000) meters classes
	Distance from urban areas was created using a topographical map, was calculated based on the Euclidean distance method in ArcGIS 10.4, and was classified into (<1000), (1000–2000), (2000–3000), (3000–4000), (4000–5000), (5000–6000), (6000–7000), and (>7000) meters classes.

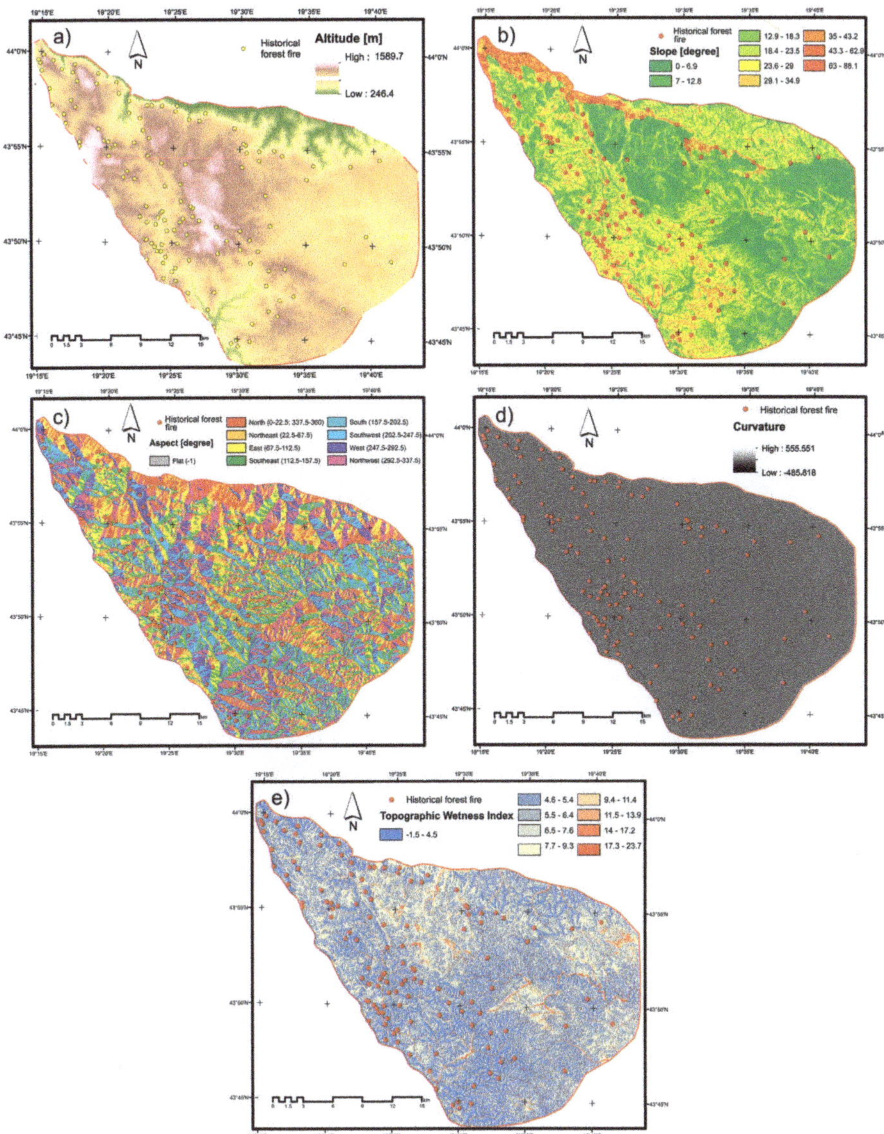

Figure 3. Topographical factors related to forest fire; (**a**) altitude, (**b**) slope degree, (**c**) aspect, (**d**) plan curvature, and (**e**) topographic wetness index (TWI).

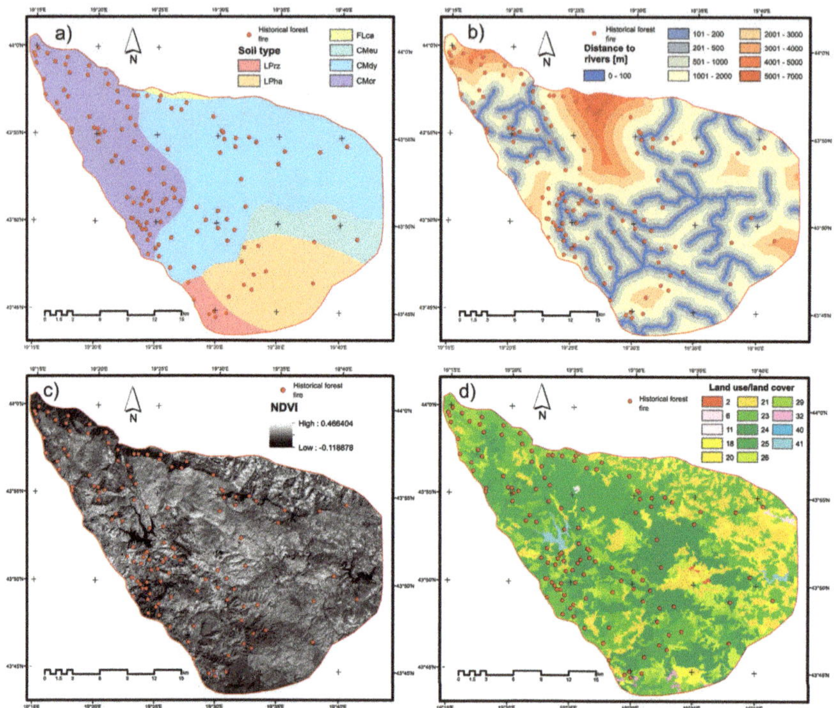

Figure 4. Environmental factors related to forest fire; (**a**) soil type, (**b**) distance from rivers, (**c**) normalized difference vegetation index (NDVI), and (**d**) land cover/land cover.

Weather patterns such as temperature, rainfall, and wind power are considered as principal factors that strongly affect forest fire behavior, in which the forest fire is more likely to occur under hot, windy, and dry weather conditions. For this study, the weather data in 2016 that were available at the Republic Hydrometeorological Service of Serbia were used, including average maximum annual climatic related data: Wind power, temperature, and the total sum of rainfall (Figure 5).

Table 3. Soil type classes.

Number	Code/Value	Description
1	Flca	Calcaric Fluvisol
2	CMcr	Chromic Cambisol
3	Cmdy	Dystric Cambisol
4	Cmeu	Eutric Cambisol
5	Lpha	Haplic Leptosol
6	LPrz	Rendzic Leptosol
7	Pldy	Dystric Planosol

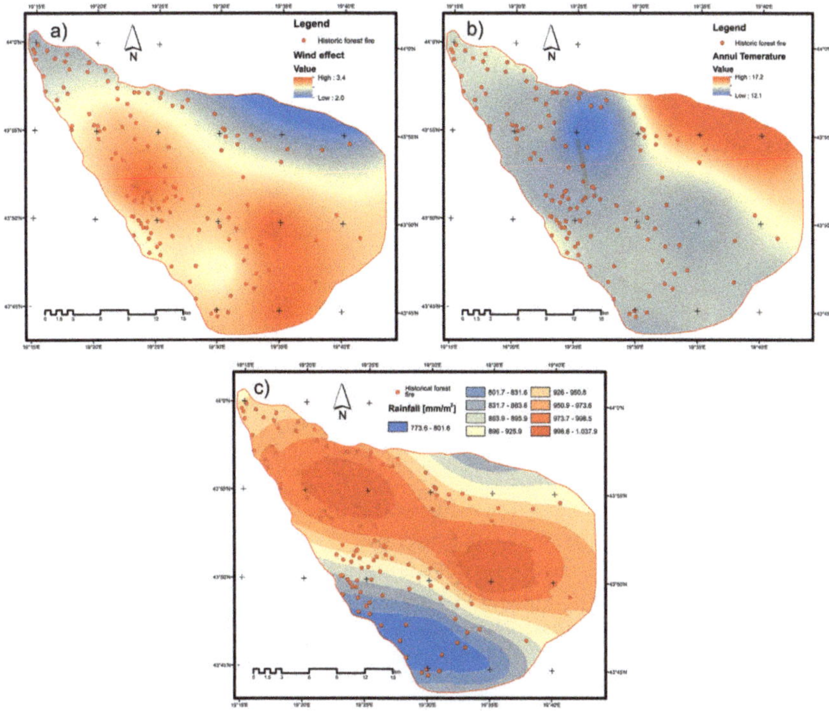

Figure 5. Meteorological factors related to a forest fire; (**a**) wind power, (**b**) maximum annual temperature, and (**c**) annual rainfall.

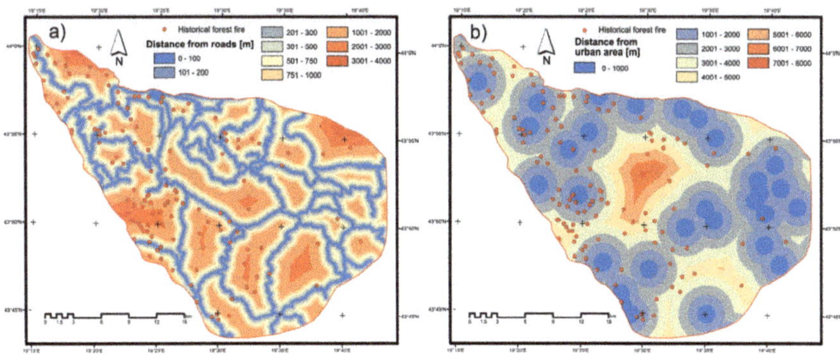

Figure 6. Social factors related to a forest fire; (**a**) distance from roads and (**b**) distance from urban areas.

In addition, a description of the soil types based on codes is shown in Table 3.

A full description of the land use/land cover conditioning factor (Figure 4d) based on codes is shown in Table 4.

Table 4. Land use/land cover.

Number	Code/Value	RGB	Code	Description
1	2	255,0,0	112	Discontinuous urban fabric
2	6	230,204,230	124	Airports
3	11	255,230,255	142	Sport and leisure facilities
4	12	255,255,168	211	Non-irrigated arable land
5	18	230,230,77	231	Pastures
6	20	255,230,77	242	Complex cultivation patterns
7	21	230,204,77	243	Land principally occupied by agriculture, with significant areas of natural vegetation
8	23	128,255,0	311	Broad-leaved forest
9	24	0,166,0	312	Coniferous forest
10	25	77,255,0	313	Mixed forest
11	26	204,242,77	321	Natural grasslands
12	29	166,242,0	324	Transitional woodland-shrub
13	32	204,255,204	333	Sparsely vegetated areas
14	40	0,204,242	511	Water courses
15	41	128,242,230	512	Water bodies

3.2. Multi-Collinearity Test

In the current research, the multi-collinearity test was used to avoid the occurrence of collinearity between the conditioning factors. Multi-collinearity is a phenomenon where one predictor variable can be predicted from the other predictor variables with an extensive degree of accuracy in a multiple regression model. To quantify the severity of multi-collinearity in an ensemble learning model, tolerance and variance inflation factor (VIF) was used. Variance inflation factor contributes to a measuring index that shows how much an estimated and collinearity effected regression coefficient is increased.

A tolerance value less than 0.2 indicates multi-collinearity between independent variables, and serious multi-collinearity occurs when the tolerance values are smaller than 0.1. If the VIF value exceeds 10, it is often regarded as a multi-collinearity indication [27,28]. The tolerance and VIF values in this study are estimated and shown in Table 5. The highest VIF and the lowest tolerance were 4.496 and 0.222, respectively, based on Table 5. There is, therefore, no multi-colinearity in current research between independent factors. In the meantime, insolation had a tolerance of less than 0.1 and was removed from the following analyses.

Table 5. Multi-collinearity test.

Model	Unstandardized Coefficients		Standardized Coefficients	T	Significant	Collinearity Statistics	
	B	Standard Error	Beta			Tolerance	VIF
(Constant)	1.674	1.726		0.970	0.334		
Aspect	0.004	0.014	0.021	0.325	0.746	0.953	1.049
Altitude	0.000	0.000	0.182	2.088	0.038	0.510	1.961
NDVI	0.122	0.671	0.014	0.182	0.856	0.663	1.508
Plan curvature	0.039	0.048	0.052	0.808	0.420	0.947	1.056
Rainfall	0.000	0.001	−0.056	−0.700	0.485	0.610	1.640
Distance from rivers	-3.372×10^{-5}	0.000	−0.072	−0.950	0.344	0.671	1.491
Distance from roads	1.013×10^{-5}	0.000	0.013	0.187	0.852	0.825	1.212
Soil type	0.007	0.037	0.016	0.184	0.854	0.541	1.850
Maximum annual temperature	−0.084	0.069	−0.155	−1.208	0.229	0.234	4.272
Distance from urban	6.978×10^{-6}	0.000	0.017	0.233	0.816	0.712	1.404
Wind power	−0.233	0.236	−0.130	−0.987	0.325	0.222	4.496
TWI	0.000	0.008	0.002	0.032	0.974	0.942	1.061
Slope	0.023	0.004	0.528	6.503	0.000	0.587	1.704

VIF = Variance Inflation Factor.

4. Training Data Selection

In order to collect data for forest fire database, we use Moderate Resolution Imaging Spectroradiometer (MODIS), Landsat 8 OLI and Worldview-2 satellite images, extensive field surveys, and aerial photo images. In this research, a total of 126 forest fire occurrence locations were identified. Locations of forest fires are mapped and analyzed as "points". These points refer to the points located on the center of gravity of the forest fire occurrence or centroids of the burned areas.

From a machine learning point of view of, mapping susceptibility to forest fire can be considered as a binary classification problem with two classes: Forest fire and non-forest fire. Forest fire points are coded as "1," while non-forest fire points are coded as "0" and the dependent variable is represented. For this analysis, all 126 forest fire locations were randomly divided by a random selection algorithm into two groups: Training 88 forest fire locations (70%) and validation data sets with remaining 38 forest fire hotspots (30%). The second validation dataset with the remaining 38 forest fires was used for the model validation and to confirm the prediction accuracy.

We need positive and negative examples of fire occurrence in order to build predictive models of forest fires. Positive examples were represented in the past by validation datasets of forest fire sites where we noticed the occurrence of the fire along with the date and time. The same quantity of non-forest fire points was randomly sampled from non-forest fire areas within the areas at least 15 km away from any positive example detected in timestamp ± 5 days and they represent negative examples.

5. Machine Learning Applications

5.1. Support Vector Machine

The support vector machine (SVM) is a widely used statistical machine learning algorithm proposed by Vapnik [29] based on the basic risk minimization principle. The support vector machine algorithm separates the classes with a final surface (called an optimal hyper-plane) that optimizes the margin among the classes in the dataset. The data points of these classes closest to the hyper-plane were originally called support vectors. The main objective of SVM statistical learning algorithms is not just to separate the two classes, but also to find an optimal hyper-plane separating the two classes (i.e., wildfires and no wildfires) and the training data set.

Training data are introduced by $\{x_i, y_i\}$, $i = 1, \ldots .. r$, $y_i = \{1, -1\}$, where r is a number of training samples and the training vector consists of two classes $y_i = 1$ for class α_1 and $y_i = -1$ for class α_2. If classes are linearly separable, it is possible to define at least one hyper-plane defined by vector w with bias b, which can separate the classes properly (training error is 0) according to Equation (1):

$$w{\cdot}x + b = 0 \tag{1}$$

To find such hyper-plane, w and b are estimated in the way that $y_i(w{\cdot}x_i + b) \geq 1$ for $y_i = 1$ (class α_1) and $y_i(w{\cdot}x_i + b) \geq -1$ for $y_i = -1$ (class α_2). These two can be associated based on Equation (2):

$$y_i(w{\cdot}x_i + b) - 1 \geq 0 \tag{2}$$

There are many hyper-plane systems that can be used to separate two classes, but there is only one optimal hyper-plane in n dimensions. The training points closest to the optimal hyper-plane and located at the two boundaries, given with $w{\cdot}x_i + b = \pm 1$, are called support vectors and the center of the margin is the optimal hyper-plane separation.

The optimal hyper-plane between two classes is defined by maximizing the gap between the nearest classes. Mathematically, this means that we want to differentiate the two classes by their maximum distance between support vectors. This distance is equal to $\frac{2}{||w||}$. This is expressed as follows:

$$\min \frac{1}{2} ||w||^2 \tag{3}$$

subject to the following constraints: $y_i(w{\cdot}x_i + b) \geq 1$, where, $|(|w|)|$ is the hyper-plane standard, b is a scalar base, and $(\cdot)$ denotes the scalar product. The cost function can be defined by using the Lagrangian multiplier as in Equation (4):

$$L = \frac{1}{2}\|w\|^2 - \sum_{i=1}^{r} a_i\left(y_i\left((w{\cdot}x_i) + b\right) - 1\right) \tag{4}$$

where, a_i is the Lagrangian multiplier.

For non-linearly separable classes, the constraints can be changed by introducing slack variables ξ_i [30].

Equation (3) becomes:

$$L = \min\frac{1}{2}\|w\|^2 + C\sum_{i=1}^{r}\xi_i \tag{5}$$

where C is the constant or penalty parameter that determines the correlation between training error and the complexity of the model [31].

In order to deal with the non-linearity of the classification or regression problem, the SVM classification approach introduced certain classes of functions called kernels $K\left(x_i, x_j\right) = \phi(x_i)\phi\left(x_j\right)$. The original input data can easily be transferred to high-dimensional function space with certain non-linear kernel functions. The most commonly used SVM classification kernels are a radial basis function (RBF), also known as Gaussian kernels, polynomial, linear, and sigmoid kernels [29].

In this study, the radial basis function (RBF) kernel is used to model forest fire using the SVM model [32]. Since the performance of the SVM model depends on the kernel width (γ) and the regularization constant (C), they should be carefully monitored. In this research, the R open source software "rminer" package [33] was used for support vector machine modeling and optimal parameters are provided. The tuning was done in a separate data set. Features of SVM applied for forest fire modeling are:

- SVM type applied for model: Radial Basis function.
- Hyper-parameter: sigma = 0.054
- Number of Support Vectors: 34
- Objective Function Value: −93.072
- Training error: 0.160

The best values for kernel width and regularization parameter of the SVM were obtained using the grid search method, and the optimal values were found as 0.125 and 7.95 for the kernel width and regularization parameters, respectively.

To conclude, the forest fire hazard index is symbolized with four classes using Natural Breaks classifications [31,32] and reclassified using the reclassify tool from Spatial Analyst Tools ArcGIS 10.4 software (ESRI, Redlands, California, CA, USA) release. Established on this, each cell is classified into four categories and receives a new value, low, moderate, high, or very high, representing the forest fire hazard index. The results of the forest fire susceptibility assessment using SVM model are given in Figure 7. In general, a low value is an area with the least probability of forest fire occurrence, while the very high value represents areas with the highest probability of forest fire susceptibility.

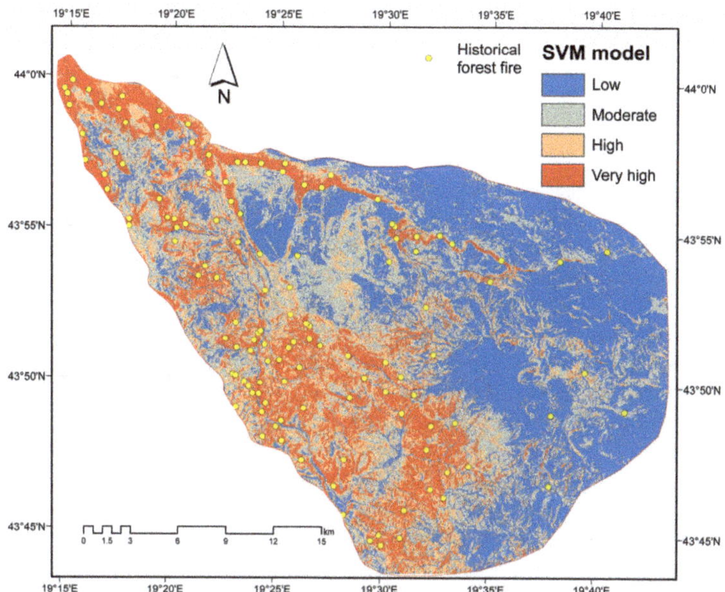

Figure 7. Forest fire susceptibility map using the support vector machine model.

5.2. Random Forests

The random forest (RF) algorithm is an influential method of collaborative learning developed for classification, regression, and unsupervised learning [33]. Moreover, the random forest method is widely used for data prediction and is suitable for high-dimensional non-linear modeling of forest fire susceptibility. The objective of RF is to identify the appropriate model for analyzing the relationship between independent variables and a dependent variable for weight determination for each factor. In this research, training data set forest fire locations (i.e., 88 forest fire locations) and 15 forest fire conditioning factors were used as dependent and independent variables.

The RF algorithm operates by building many classification trees during the training period [34] and the final output of the model generation process is the average value of the results of all classification trees [33].

In order to run the RF model, two main parameters of the random forest model must be defined a priori: The square root of the number of factors (m_{try}) and the number of trees to run the model (n_{tree}). The above parameters should be optimized to minimize the generalization error. In general, the model selects the best possible parameters for maximum accuracy [34].

Additionally, for tree learners, random forest training algorithm uses the regular technique of bagging or boot-strap aggregating. The RF method uses the Gini Index as a measure for the best split selection measuring the impurity of a given element in relation to the rest of the classes [35,36]. The Gini index is a measure of inequality of a distribution. The Gini index can be computed by summing the probability pi of a single class with label i being chosen multiplies by the probability $\sum_{k \neq i} p_k = 1 - p_i$ of a mistake in categorizing that class i. The Gini Index can be expressed as the following equation for a given training dataset T with j classes:

$$I_T(p) = \sum_{i=1}^{j} p_i \sum_{k \neq i} p_k = 1 - \sum_{i=1}^{j} p_i^2 \tag{6}$$

where, $i \in \{1, 2, ..., j\}$. Therefore, a decision tree is made to grow to its maximum depth by using a given combination of features.

In this research, the RF model was used to observe the link between forest fire conditioning factors and the occurrence of forest fire and to predict the susceptibility of a forest fire. In this study, we used the Random Forest package of R open source software [36] for RF modeling and then the final produced map was added to ArcGIS 10.4 to visualize the forest fire susceptibility maps using the Spatial Analyst Tools reclassification tool. The m_{try} parameter was regulated using the internal random forest function. In order to obtain the values of the study area's forest fire susceptibility index, the value of each wildfire environmental factor in each grid cell was calculated using a random forest model and the parameter configuration with the highest prediction accuracy was determined and set to $m_{try} = 5$. In addition, in this study, the number of trees (m_{tree}) in RF was fixed to 250 after a preliminary analysis and the number m of variables sampled at each node was selected to be 1. No calibration set is needed to tune the parameters. In addition, two types of error were calculated in this model: A mean decrease in accuracy and mean decrease in node impurity (mean decrease Gini). This different importance measure can be used for ranking variables and for variable selection.

The big advantage of the RF model is that it allows investigation of the variable importance (the contribution of each variable) measured by the mean reduction in prediction accuracy (Figure 8). Consequently, according to Peters [36], mean decrease in cross-validation and prediction accuracy assessment were used to examine the uncertainty propagation of conditioning factors for forest fire and to evaluate the whole random forest model.

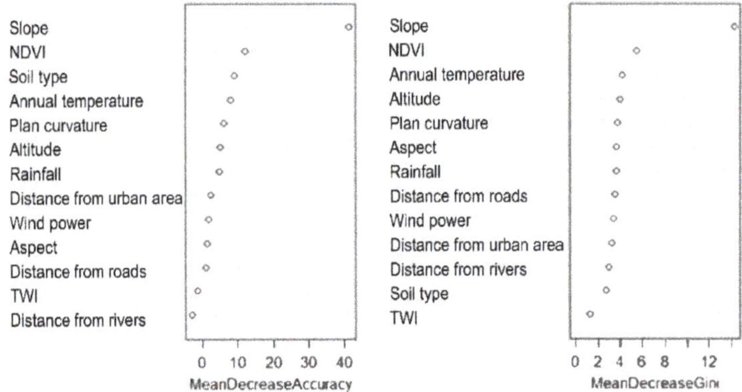

Figure 8. Mean decrease in prediction accuracy and mean decrease Gini index.

We can see from Figure 8 that the most important conditioning factor in wildfires modeling is the slope degree, followed by NDVI, soil type, and maximum annual temperature. Namely, the fire is usually climbs uphill more easily than it descends downhill. The higher inclination effects a faster spreading of the fire [17]. Moreover, the fire follows the direction of the surrounding wind, which usually blows uphill. In addition, the smoke and heat generated by the fire, are able to heat the fuel more than the fire itself.

Using a reclassification tool in the Spatial Analyst Tools ArcGIS 10.4 software, each final map cell is classified into four categories (low, moderate, high, and very high) representing the forest fire hazard index. The obtained results of the forest fire susceptibility assessment using the random forest model are given in Figure 9. A low value (blue color) is the areas with the least probability of forest fire occurrence, while the values of very high (red color) represent areas with the highest probability of forest fire hazard.

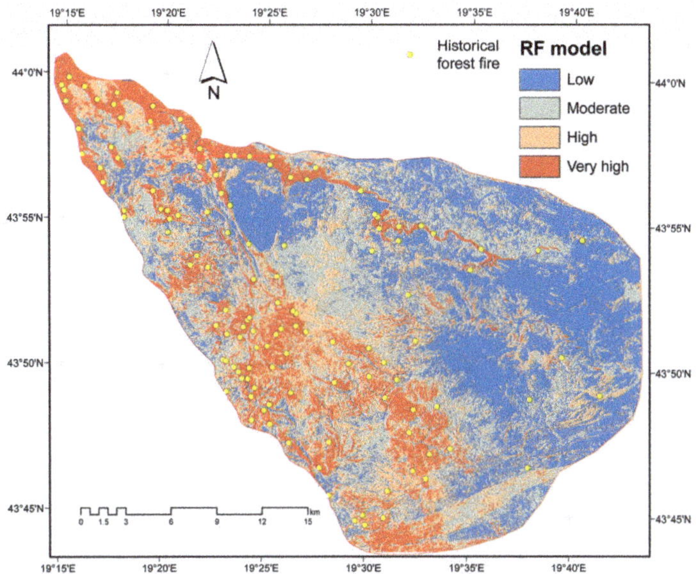

Figure 9. Forest fire susceptibility map using the random forest model.

5.3. Ensemble Modeling

Ensemble prediction is a learning algorithm that combines multiple model predictions [37] to reduce bias (boosting) and variance (bagging) or improve predictions (stacking). The Bayesian averaging is an original ensemble method, but the most popular methods for combining the predictions from different models are:

- Boosting, which is used to build multiple models (typically the same type) using previous chain model prediction errors.
- Bagging, which is used to create multiple models from different training dataset subsamples.
- Stacking, which is used to build multiple models and the supervisor model that best combines the predictions of the primary models.

In this research, we carefully combine mentioned machine learning models to get an ensemble model using Bayesian averaging [38,39] with efficient feature selection to address these issues and mitigate their effects on the defect classification performance. Multiple predictions are made for each data point in Bayesian averaging. In this method, we take an average of predictions from all the models and use it to make the final prediction. Bayesian averaging can be used for making predictions in regression problems or while calculating probabilities for classification problems. Along with efficient feature selection, a new ensemble learning algorithm is proposed to provide robustness to both data imbalance and feature redundancy.

The achieved results of the forest fire susceptibility assessment using ensemble model are given in Figure 10. A low value (blue color) is the areas with the lowest probability of forest fire, whereas the very high values (red color) are the areas with the highest risk of a forest fire.

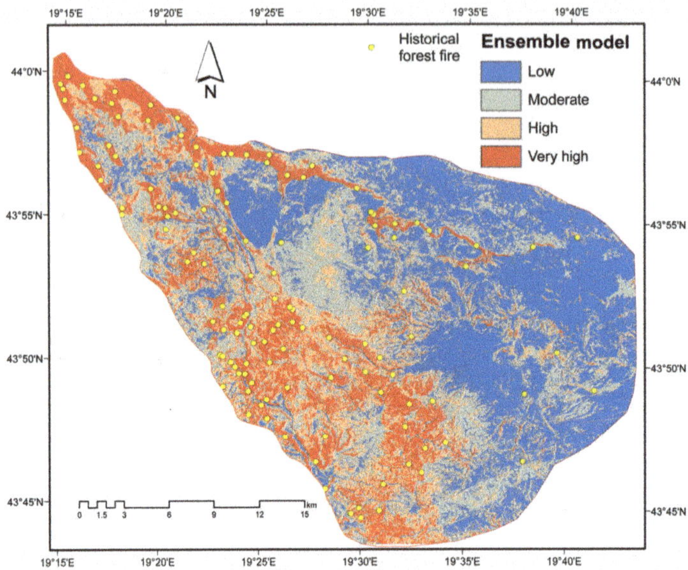

Figure 10. Forest fire susceptibility map using an ensemble method.

6. Validation

The validation of susceptibility maps for a forest fire is an important step in the modeling process. The capacity of support vector machine, random forests, and ensemble models was assessed using a non-dependent threshold approach: The operating characteristic of the receiver (ROC). The area under the curve (AUC) is a synthesized index calculated for ROC curves and it has been generally used in several types of research to assess the accuracy of the forest fire susceptibility map [40]. The AUC value is the probability that a positive event with the help of the test will be evaluated as positive. The ROC curves are generated by SPSS 17 software (IBM, New York, NY, USA) and represent the evolution of the proportion of genuine positive cases (also referred to as sensitivity) as a function of the proportion of false positive cases (corresponding to minus specificity). Graphic representation with a diagram of the pair (specificity, sensitivity) corresponds to the ROC curve for the numerous possible threshold values [41–44]. The ROC curves for SVM, RF, and Ensemble models are shown in Figure 11 and Table 6.

Table 6. The area under the curve.

Models	Area	Standard Error	Asymptotic Significant	Asymptotic 95% Confidence Interval	
				Lower Bound	Upper Bound
RF	0.844	0.047	0.001	0.751	0.937
SVM	0.834	0.047	0.001	0.743	0.926
Ensemble	0.848	0.046	0.001	0.758	0.938

According to validation results, all three forest fire susceptibility maps are considered to have the most acceptable and representable appearance (AUC > 0.8). In addition, both visual assessment and quantitative validation, using ROC curve, agreed that SVM, RF, and their ensemble models are the excellent performing model approaches with an AUC value shown in Table 6.

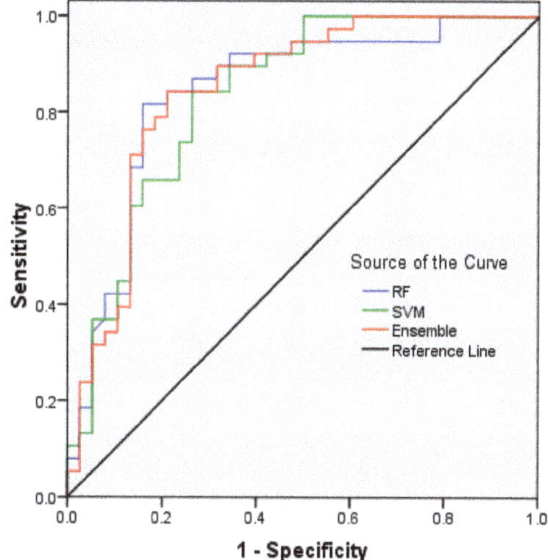

Figure 11. Receiver operating characteristics (ROC) curves for support vector machine (SVM), random forest (RF), and ensemble models.

7. Discussion

Machine learning algorithms specify computer-based tools that enable exploratory data and statistical examination to detect unknown patterns and relationships of dataset values in advance. In the current research, supervised and versatile machine learning algorithms and their ensemble were used to investigate the spatial relationship between the occurrence of forest fires and different environmental predictors [45]. The objective of this research was therefore to compare these statistical and decision tree-based regression models for forest fire mapping in the Tara National Park, Republic of Serbia. The results are presented and discussed in two important sections in the current research, including the importance of the conditioning factors and the performance of the models in the forest fire susceptibility mapping.

7.1. Importance of Conditioning Factors

Based on importance conditioning factor determination, the results of the current study showed that the most important conditioning factor in wildfire modeling is the slope angle, followed by NDVI, soil type, and the maximum annual temperature. Namely, the fire usually climbs uphill more easily than it descends downhill. The higher inclination effects faster spreading of the fire. On the other hand, the TWI and the distance from rivers were of the lowest importance in the occurrence of forest fires. In another study [46], slope, NDVI, and maximum annual temperature were reported to be more important in the occurrence of a forest fire, which is consistent with the results of the current research. In addition, another the research confirmed that NDVI [47], land use, soil type, and the annual temperature have a greater influence on the occurrence of the forest fire. In addition, researchers [48] found that NDVI, distance from urban areas, and distance from roads have the highest predictive values that indicate reasonable results in the forest fire susceptibility mapping.

7.2. Performance of the Used Models

The results show that the ensemble model had the highest AUC value (0.848), followed by the RF model (0.844) and the SVM model (0.834). The best performance in the current study had the ensemble method because that method combined the predictions of multiple different models together in order to decrease variance or bias and take into account advantages of both used machine learning methods.

In the ensemble method, we take and use an average of predictions from all models to make the ultimate prediction. The Bayesian average can be used to predict problems of regression or to calculate probabilities of classification issues. In addition to efficient feature selection, an ensemble learning algorithm is introduced to provide robustness for both data imbalance and feature redundancy [37,48].

According to the achieved results, the support vector machine has about the same accuracy as random forest method. SVM models have produced acceptable results in the mapping of susceptibility to the forest fire. The non-linear mapping is one of the greatest advantages of the SVM model. For each class of discrete covariates, a parametric model can, therefore, have different intercepts and coefficient values. Furthermore, the SVM model is not excessively influenced by noisy data and is not very likely to overfit. The SVM model has the advantage of complex, non-linear relationships and is highly noise-resistant [49]. On the other hand, the greatest weakness of the SVM method is the fact that testing different kernel combinations and model parameters requires finding the best model. In the meantime, the results obtained are very difficult to interpret because they are part of a complex black box model.

Due to their power, versatility, and ease of use, random forests are quickly becoming one of the most popular machine learning methods. The RF performance, in the current study, being better than the SVM model could be due to its ability to run on large datasets with a large number of predictors and its ability to handle thousands of input variables without variable deletion [19]. The random forest model uses regression trees to estimate an average of the dependent variable as the final prediction results in an internally unbiased estimation of the classification error. The RF algorithm has several advantages in relation to other machine learning methods. Firstly, RF method can handle noisy or missing data as well as categorical or continuous features; secondly, it does not require assumptions about the distribution of explanatory variables; and thirdly, it can deal with interactions and non-linearities between efficient factors [50,51]. These are major advantages that limit outlier generation, particularly when working with terrain variables with a high frequency of missing data [19].

The random forests method takes advantage of the high diversity between particular trees and operates by constructing many classification trees during the training period [52]. In addition, according to Catani [52], random forests method increases variety between the classification trees by randomly changing the predictive variable sets and by resampling the data with substitution over the various tree processes of induction [17]. The result of the model construction process is the average results of all trees, so cross validation is not necessary for this method. On the other hand, the biggest weakness of RF model is the fact that, unlike a decision tree, the model is not easily interpretable. In addition, the correct use of RF model might require some work to tune the model for the data.

8. Conclusions

Many countries have detailed programs for forest fire protection, which are based on prevention and fire-fighting measures. A fire detection system is one of the most important aspects of forest fire protection before the fire spreads over larger areas. Therefore, the main purpose of this paper is to demonstrate the results of an ensemble learning method using a Bayesian average based on predictive results from the support vector machine and random forest methods. In this paper, we modeled and predicted suitable locations for the outbreak of forest fires using machine learning algorithms. Regional forest fire modeling is a regular, nonlinear and complex issue that can not be easily assessed and predicted. In the current research, we attempted to compare the results of forest fire susceptibility maps using supervised and versatile machine learning algorithms (support vector machine and random forest) and their ensemble in the Tara National Park, Serbia. Based on the obtained area under the curve, all models had the most scientifically satisfactory reliability and could be used at the regional level for forest fire susceptibility mapping. The results depicted that the ensemble model using Bayesian average had the best performance. Finally, these maps can provide very useful information for fire managers, decision makers, and foresters to locate potential fire hazard areas spatially so that they can operate under conditions in fire prevention operations in the Tara

National Park of Serbia. Moreover, in national parks where the absolute priority is the preservation of natural features and endemic species, this kind of prevention from forest fires is justified and necessary. In addition, we believe that the results presented in this study make a substantial contribution to the literature on forest fire mapping.

Author Contributions: S.D. prepared the data layers, figures, and tables; H.R.P. and S.B. performed the experiments and analyses. L.G. supervised the research, finished the first draft of the manuscript, edited and reviewed the manuscript, and contributed to the model construction and verification.

Acknowledgments: This work supported research project 1.1.107/2018 "Possibilities of automatic extraction of vegetation data by a combination of satellite and aerial photogrammetric images" by the Ministry of Defence of the Republic of Serbia and research project VA/TT/3/17-19 "GIS Modeling of Risk Assessment of Disasters and Catastrophes in the Function of the Third Mission of the Army of Serbia" by the Ministry of Defense of the Republic of Serbia. This study was supported by 2015 Jiangsu provincial key R & D Program (Social Development) (BE2015704), the National Natural Science Foundation of China (No. 41877522)

Conflicts of Interest: The authors declare no conflicts of interest. The funders had no role in the design of the study, in the collection, analyses, or interpretation of data, in the writing of the manuscript, and in the decision to publish the results.

References

1. Satir, O.; Berberoglu, S.; Donmez, C. Mapping regional forest fire probability using artificial neural network model in a Mediterranean forest ecosystem. *Geomatics, Nat. Hazards Risk* **2016**, *7*, 1645–1658. [CrossRef]
2. Chuvieco, E.; Aguado, I.; Yebra, M.; Nieto, H.; Salas, J.; Martín, M.P.; Vilar, L.; Martínez, J.; Martín, S.; Ibarra, P.; et al. Development of a framework for fire risk assessment using remote sensing and geographic information system technologies. *Ecol. Model.* **2010**, *221*, 46–58. [CrossRef]
3. Zheng, Z.; Huang, W.; Li, S.; Zeng, Y. Forest fire spread simulating model using cellular automaton with extreme learning machine. *Ecol. Model.* **2017**, *348*, 33–43. [CrossRef]
4. Bui, D.T.; Le, K.-T.T.; Nguyen, V.C.; Le, H.D.; Revhaug, I. Tropical Forest Fire Susceptibility Mapping at the Cat Ba National Park Area, Hai Phong City, Vietnam, Using GIS-Based Kernel Logistic Regression. *Remote. Sens.* **2016**, *8*, 347.
5. Wegner, J.D.; Roscher, R.; Volpi, M.; Veronesi, F. Foreword to the Special Issue on Machine Learning for Geospatial Data Analysis 2018. Available online: https://www.mdpi.com/2220-9964/7/4/147 (accessed on 24 September 2018).
6. Pastor, E.; Zárate, L.; Planas, E.; Arnaldos, J. Mathematical models and calculation systems for the study of wildland fire behaviour. *Prog. Energy Combust. Sci.* **2003**, *29*, 139–153. [CrossRef]
7. Andrews, P.L. BEHAVE: Fire behavior prediction and fuel modeling system-BURN Subsystem, part 1. 1986. Available online: https://www.fs.usda.gov/treesearch/pubs/29612 (accessed on 27 September 2018).
8. Linn, R.; Reisner, J.; Colman, J.J.; Winterkamp, J. Studying wildfire behavior using FIRETEC. *Int. J. Wildl. Fire* **2002**, *11*, 233–246. [CrossRef]
9. Lopes, A.M.G.; Cruz, M.G.; Viegas, D.X. FireStation — an integrated software system for the numerical simulation of fire spread on complex topography. *Environ. Model. Softw.* **2002**, *17*, 269–285. [CrossRef]
10. Sturtevant, B.R.; Scheller, R.M.; Miranda, B.R.; Shinneman, D.; Syphard, A. Simulating dynamic and mixed-severity fire regimes: A process-based fire extension for LANDIS-II. *Ecol. Model.* **2009**, *220*, 3380–3393. [CrossRef]
11. Perestrello De Vasconcelos, M.J.; Sllva, S.; Tome, M.; Alvim, M.; Pereira, J.M. Spatial Prediction of Fire Ignition Probabilities: Comparing Logistic Regression and Neural Networks. *Photogramm. Eng. Remote Sens.* **2001**, *67*, 73–81.
12. Arndt, N.; Vacik, H.; Koch, V.; Arpacı, A.; Gossow, H. Modeling human-caused forest fire ignition for assessing forest fire danger in Austria. *iFor. - Biogeosci. For.* **2013**, *6*, 315–325. [CrossRef]
13. Pourghasemi, H.R. GIS-based forest fire susceptibility mapping in Iran: a comparison between evidential belief function and binary logistic regression models. *Scand. J. For. Res.* **2016**, *31*, 80–98. [CrossRef]
14. Conedera, M.; Torriani, D.; Neff, C.; Ricotta, C.; Bajocco, S.; Pezzatti, G.B. Using Monte Carlo simulations to estimate relative fire ignition danger in a low-to-medium fire-prone region. *Ecol. Manag.* **2011**, *261*, 2179–2187. [CrossRef]

15. Amatulli, G.; Pérez-Cabello, F.; De La Riva, J. Mapping lightning/human-caused wildfires occurrence under ignition point location uncertainty. *Ecol. Model.* **2007**, *200*, 321–333. [CrossRef]

16. Vilar, L.; Woolford, D.G.; Martell, D.L.; Martin, M.P. A model for predicting human-caused wildfire occurrence in the region of Madrid, Spain. *Int. J. Wildl. Fire* **2010**, *19*, 325–337. [CrossRef]

17. Oliveira, S.; Oehler, F.; San-Miguel-Ayanz, J.; Camia, A.; Pereira, J.M.; Pereira, J.M.C. Modeling spatial patterns of fire occurrence in Mediterranean Europe using Multiple Regression and Random Forest. *Ecol. Manag.* **2012**, *275*, 117–129. [CrossRef]

18. Sakr, G.E.; Elhajj, I.H.; Huijer, H.A.-S. Support Vector Machines to Define and Detect Agitation Transition. *IEEE Trans. Affect. Comput.* **2010**, *1*, 98–108. [CrossRef]

19. Pourtaghi, Z.S.; Pourghasemi, H.R.; Aretano, R.; Semeraro, T. Investigation of general indicators influencing on forest fire and its susceptibility modeling using different data mining techniques. *Ecol. Indic.* **2016**, *64*, 72–84. [CrossRef]

20. Arpaci, A.; Malowerschnig, B.; Sass, O.; Vacik, H. Using multi variate data mining techniques for estimating fire susceptibility of Tyrolean forests. *Appl. Geogr.* **2014**, *53*, 258–270. [CrossRef]

21. Maeda, E.E.; Formaggio, A.R.; Shimabukuro, Y.E.; Arcoverde, G.F.B.; Hansen, M.C. Predicting forest fire in the Brazilian Amazon using MODIS imagery and artificial neural networks. *Int. J. Appl. Earth Obs. Geoinformation* **2009**, *11*, 265–272. [CrossRef]

22. Massada, A.B.; Syphard, A.D.; Stewart, S.I.; Radeloff, V.C. Wildfire ignition-distribution modelling: A comparative study in the Huron–Manistee National Forest, Michigan, USA. *Int. J. Wildl. Fire* **2013**, *22*, 174. [CrossRef]

23. Renard, Q.; Pélissier, R.; Ramesh, B.R.; Kodandapani, N. Environmental susceptibility model for predicting forest fire occurrence in the Western Ghats of India. *Int. J. Wildl. Fire* **2012**, *21*, 368–379. [CrossRef]

24. Adab, H.; Kanniah, K.D.; Solaimani, K. Modeling forest fire risk in the northeast of Iran using remote sensing and GIS techniques. *Nat. Hazards* **2013**, *65*, 1723–1743. [CrossRef]

25. Gigović, L.; Pamučar, D.; Bajić, Z.; Drobnjak, S. Application of GIS-Interval Rough AHP Methodology for Flood Hazard Mapping in Urban Areas. *Water* **2017**, *9*, 360. [CrossRef]

26. Moore, I.D.; Grayson, R.B.; Ladson, A.R. Digital terrain modelling: A review of hydrological, geomorphological, and biological applications. *Hydrol. Process.* **1991**, *5*, 3–30. [CrossRef]

27. O'Brien, R.M. A Caution Regarding Rules of Thumb for Variance Inflation Factors. *Qual. Quant.* **2007**, *41*, 673–690. [CrossRef]

28. Pourghasemi, H.; Beheshtirad, M.; Pradhan, B. A comparative assessment of prediction capabilities of modified analytical hierarchy process (M-AHP) and Mamdani fuzzy logic models using Netcad-GIS for forest fire susceptibility mapping. *Geomatics Nat. Hazards Risk* **2016**, *7*, 861–885. [CrossRef]

29. Vapnik, V.N. *The Nature of Statistical Learning Theory*; Springer Science & Business Media: Berlin, Germany, 1995.

30. Foody, G.; Mathur, A. A relative evaluation of multiclass image classification by support vector machines. *IEEE Trans. Geosci. Remote Sens.* **2004**, *42*, 1335–1343. [CrossRef]

31. Schölkopf, B.; Smola, A.J.; Williamson, R.C.; Bartlett, P.L. New support vector algorithms. *Neural Comput.* **2000**, *12*, 1207–1245. [CrossRef]

32. Lee, S.; Hong, S.-M.; Jung, H.-S. A Support Vector Machine for Landslide Susceptibility Mapping in Gangwon Province, Korea. *Sustainability* **2017**, *9*, 48. [CrossRef]

33. Liaw, A.; Wiener, M. Classification and regression by randomForest. *R News* **2002**, *2*, 18–22.

34. Breiman, L. Random Forests. *Mach. Learn.* **2001**, *45*, 5–32. [CrossRef]

35. Cutler, D.R.; Edwards Jr, T.C.; Beard, K.H.; Cutler, A.; Hess, K.T.; Gibson, J.; Lawler, J.J. Random forests for classification in ecology. *Ecology* **2007**, *88*, 2783–2792. [CrossRef]

36. Breiman, L.; Cutler, A. Random forests — Classification description: Random forests. Available online: http//stat-www.berkeley.edu/users/breiman/RandomForests/cf_home.html (accessed on 28 September 2018).

37. Kuhnert, P.M.; Henderson, A.; Bartley, R.; Herr, A. Incorporating uncertainty in gully erosion calculations using the random forests modelling approach. *Environmetrics* **2010**, *21*, 493–509. [CrossRef]

38. McKay, G.; Harris, J.R. Comparison of the data-driven Random Forests model and a knowledge-driven method for mineral prospectivity mapping: a case study for gold deposits around the Huritz Group and Nueltin Suite, Nunavut, Canada. *Nat. Resour. Res.* **2016**, *25*, 125–143. [CrossRef]

39. Dieterich, T.G. Ensemble Methods in Machine Learning. In *Proceedings of the ≪UML≫ 2001—The Unified Modeling Language. Modeling Languages*; Springer Nature: Berlin, Germany, 2000; pp. 1–15.

40. Brownlee, J. Machine learning mastery. Available online: http//machinelearningmastery.com (accessed on 25 September 2018).

41. Hoang, N.-D.; Bui, D.T. A Novel Relevance Vector Machine Classifier with Cuckoo Search Optimization for Spatial Prediction of Landslides. *J. Comput. Civ. Eng.* **2016**, *30*, 4016001. [CrossRef]

42. Cortez, P. Data Mining with Neural Networks and Support Vector Machines Using the R/rminer Tool. In *Proceedings of the ≪UML≫ 2001—The Unified Modeling Language. Modeling Languages, Concepts, and Tools*; Springer Nature: Berlin, Germany, 2010; pp. 572–583.

43. Chen, W.; Pourghasemi, H.R.; Naghibi, S.A. A comparative study of landslide susceptibility maps produced using support vector machine with different kernel functions and entropy data mining models in China. *Bull. Eng. Geol. Environ.* **2018**, *77*, 647–664. [CrossRef]

44. Rahmati, O.; Tahmasebipour, N.; Haghizadeh, A.; Pourghasemi, H.R.; Feizizadeh, B. Evaluation of different machine learning models for predicting and mapping the susceptibility of gully erosion. *Geomorphology* **2017**, *298*, 118–137. [CrossRef]

45. Peters, J.; De Baets, B.; Verhoest, N.E.C.; Samson, R.; Degroeve, S.; De Becker, P.; Huybrechts, W. Random forests as a tool for ecohydrological distribution modelling. *Ecol. Model.* **2007**, *207*, 304–318. [CrossRef]

46. Hong, H.; Naghibi, S.A.; Dashtpagerdi, M.M.; Pourghasemi, H.R.; Chen, W. A comparative assessment between linear and quadratic discriminant analyses (LDA-QDA) with frequency ratio and weights-of-evidence models for forest fire susceptibility mapping in China. *Arab. J. Geosci.* **2017**, *10*, 1723. [CrossRef]

47. Pourtaghi, Z.S.; Pourghasemi, H.R.; Rossi, M. Forest fire susceptibility mapping in the Minudasht forests, Golestan province, Iran. *Environ. Earth Sci.* **2014**, *73*, 1515–1533. [CrossRef]

48. Bui, D.T.; Bui, Q.-T.; Nguyen, Q.-P.; Pradhan, B.; Nampak, H.; Trinh, P.T. A hybrid artificial intelligence approach using GIS-based neural-fuzzy inference system and particle swarm optimization for forest fire susceptibility modeling at a tropical area. *Agric. Meteorol.* **2017**, *233*, 32–44.

49. Ballabio, C.; Sterlacchini, S. Support Vector Machines for Landslide Susceptibility Mapping: The Staffora River Basin Case Study, Italy. *Math. For. Geosci.* **2012**, *44*, 47–70. [CrossRef]

50. Aertsen, W.; Kint, V.; Van Orshoven, J.; Özkan, K.; Muys, B. Comparison and ranking of different modelling techniques for prediction of site index in Mediterranean mountain forests. *Ecol. Model.* **2010**, *221*, 1119–1130. [CrossRef]

51. Aertsen, W.; Kint, V.; Van Orshoven, J.; Muys, B. Evaluation of modelling techniques for forest site productivity prediction in contrasting ecoregions using stochastic multicriteria acceptability analysis (SMAA). *Environ. Model. Softw.* **2011**, *26*, 929–937. [CrossRef]

52. Catani, F.; Lagomarsino, D.; Segoni, S.; Tofani, V. Landslide susceptibility estimation by random forests technique: sensitivity and scaling issues. *Nat. Hazards Earth Syst. Sci.* **2013**, *13*, 2815–2831. [CrossRef]

© 2019 by the authors. Licensee MDPI, Basel, Switzerland. This article is an open access article distributed under the terms and conditions of the Creative Commons Attribution (CC BY) license (http://creativecommons.org/licenses/by/4.0/).

 *forests*

Article

Tree Biomass Equations from Terrestrial LiDAR: A Case Study in Guyana

Alvaro Lau [1,2,*], Kim Calders [3], Harm Bartholomeus [1], Christopher Martius [4], Pasi Raumonen [5], Martin Herold [1], Matheus Vicari [6], Hansrajie Sukhdeo [7], Jeremy Singh [7] and Rosa C. Goodman [8]

[1] Laboratory of Geo-Information Science and Remote Sensing, Wageningen University & Research, Droevendaalsesteeg 3, 6708 PB Wageningen, The Netherlands; harm.bartholomeus@wur.nl (H.B.); martin.herold@wur.nl (M.H.)

[2] Center for International Forestry Research (CIFOR), Jalan CIFOR, Situ Gede, Bogor Barat 16115, Indonesia

[3] CAVElab-Computational & Applied Vegetation Ecology, Ghent University, Coupure Links 653, 9000 Gent, Belgium; kim.calders@ugent.be

[4] Center for International Forestry Research (CIFOR) Germany, Charles-de-Gaulle-Strasse 5, 53113 Bonn, Germany; c.martius@cgiar.org

[5] Computing Sciences, Tampere University, Korkeakoulunkatu 7, 33720 Tampere, Finland; pasi.raumonen@tuni.fi

[6] Department of Geography, University College London, Gower Street, London WC1E 6BT, UK; matheus.vicari.15@ucl.ac.uk

[7] Guyana Forestry Commission (GFC), 1 Water Street, Kingston, Georgetown, Guyana; hans.sukhdeo@gmail.com (H.S.); jeremy_singh45@yahoo.com (J.S.)

[8] Department of Forest Ecology and Management, Swedish University of Agricultural Sciences (SLU), Skogsmarksgränd, 901 83 Umeå, Sweden; rosa.goodman@slu.se

* Correspondence: alvaro.lausarmiento@wur.nl; Tel.: +31-317-481-937

Received: 1 May 2019; Accepted: 20 June 2019; Published: 25 June 2019

Abstract: Large uncertainties in tree and forest carbon estimates weaken national efforts to accurately estimate aboveground biomass (AGB) for their national monitoring, measurement, reporting and verification system. Allometric equations to estimate biomass have improved, but remain limited. They rely on destructive sampling; large trees are under-represented in the data used to create them; and they cannot always be applied to different regions. These factors lead to uncertainties and systematic errors in biomass estimations. We developed allometric models to estimate tree AGB in Guyana. These models were based on tree attributes (diameter, height, crown diameter) obtained from terrestrial laser scanning (TLS) point clouds from 72 tropical trees and wood density. We validated our methods and models with data from 26 additional destructively harvested trees. We found that our best TLS-derived allometric models included crown diameter, provided more accurate AGB estimates ($R^2 = 0.92$–0.93) than traditional pantropical models ($R^2 = 0.85$–0.89), and were especially accurate for large trees (diameter > 70 cm). The assessed pantropical models underestimated AGB by 4 to 13 %. Nevertheless, one pantropical model (Chave et al. 2005 without height) consistently performed best among the pantropical models tested ($R^2 = 0.89$) and predicted AGB accurately across all size classes—which but for this could not be known without destructive or TLS-derived validation data. Our methods also demonstrate that tree height is difficult to measure in situ, and the inclusion of height in allometric models consistently worsened AGB estimates. We determined that TLS-derived AGB estimates were unbiased. Our approach advances methods to be able to develop, test, and choose allometric models without the need to harvest trees.

Keywords: 3D tree modelling; aboveground biomass estimation; destructive sampling; Guyana; LiDAR; local tree allometry; model evaluation; quantitative structural model

---

1. Introduction

Guyana has approximately 18.3 million hectares of forests with a relatively low deforestation rate (between 0.1 and 0.3% per year), but is expected to increase in the future [1]. A cooperation between the Governments of Norway and Guyana expresses their willingness to provide a replicable model on how REDD+ can align the national forest countries' objectives with the world's need to combat climate change [1]. For that, Guyana is one of the first countries to establish a national program for reducing emissions from deforestation and degradation (REDD+; [2]). Guyana's REDD+ activities include the design and implementation of a national monitoring, measurement, reporting and verification (MMRV) system, which should be able to assess and reduce aboveground biomass (AGB) uncertainties within the country's capacities and capabilities [3].

AGB is typically estimated with allometric models built from empirical data. The applicability of any allometric model is thus largely dependent on the data used for its development and can produce systematic over- or under-estimations of the true AGB when applied to other geographic regions, species, or tree sizes where little or no data were included [4–6]. Since the performance of a country's MMRV program will be based on the quantification of emission reduction [6,7], Guyana seeks to test the accuracy of pantropical models and develop a country-specific allometric model.

Current AGB allometric models in tropical forests are commonly based on diameter at breast height (D; which can be measured in the field) and wood density (WD; from existing databases such as Global Wood Density Database [8,9]). In recent years, other tree biophysical attributes have been included such as height (H; [10]) and crown diameter (CD; [5]); and regional trends in height [10] crown width [5] and climate variability (E; [11]). However, due to the difficulty of measuring tree heights, the pantropical allometric models developed by Chave et al. [12] in 2005 are still widely used because they only require tree diameter and species [13].

An accurate biomass estimation of large trees is particularly important for both forest biomass [14] and forest biomass estimates [5]. Large trees account for around 75% of total forest AGB variation [14–16], and the uncertainty of tree biomass estimates increases with size [5,17,18]. Despite their relevance, large trees make up only 7% of available tropical biomass data (as of 2014; [11]), and the lack of inclusion of large tree biomass data in the development of allometric models is increasingly viewed as problematic [19–22]. Large trees explain over 75% of variation in total forest biomass [14–16]) and can predict the plot-level AGB. However, AGB error increases greatly with increasing tree size, and pantropical allometric models often underestimate large tree biomass [5,17,18].

Terrestrial Light Detection And Ranging (LiDAR), also known as terrestrial laser scanning (TLS), has been proven to be a valuable tool to assess the woody structure of trees [18,23,24]. TLS data provide high level of three-dimensional detail of forest and tree structure, which allow extrapolation to broader scales, or national scales using remote sensing systems [25]. Several studies have successfully taken TLS from its original utility—precision surveying applications—to tropical forests [18,22,23,26–29] and extracted tree attributes such as tree diameter [17], height [30,31], and crown width [32]. In combination with quantitative structure modelling (e.g., *TreeQSM* [33] or *SimpleTree* [34]), 3D tree point clouds were used to infer attributes such as total volume, AGB [17,18,22,26,30], AGB change [35], and tree species [36], as well as ecological questions such as tree mechanics, branching architecture, and surface area scaling [24].

Tropical countries seeking to participate in REDD+ that do not possess their own tree biomass database might find TLS-driven methodology a resourceful alternative. We have earlier used TLS to evaluate the accuracy of existing allometric models [17,18], and now we assess the potential of TLS and *TreeQSM*-method to develop allometric models to estimate AGB in forest trees of Guyana. We produced a unique tropical tree mass dataset of traditional inventory data and TLS scans of 98 tropical trees; 26 of these trees were destructively harvested, re-measured, and weighed as validation data. The objectives of this study are: (i) to model tree volume and estimate tree AGB from *TreeQSM* models of tree point clouds; (ii) to build allometric models based on TLS measurements for Guyana; and (iii) to evaluate the performance of these TLS-derived data and allometric models against destructively-harvested reference data and estimates from pantropical models.

2. Materials and Methods

2.1. Study Area

Field work was conducted January–February 2017 inside an active logging concession near the Berbice river in the East Berbice-Corentyne Region of Guyana (4.48 to 4.56 lat and −58.22 to −58.15 long; Figure 1). Located at 106 masl, the study area is a mixture of white sand plateu and mixed forest [37]. The region is dominated by evergreen trees [37], with an an average precipitation of 3829 mm yr^{-1}, an average temperature between 22.5 and 30.5 °C, and an average humidity of 86% [38].

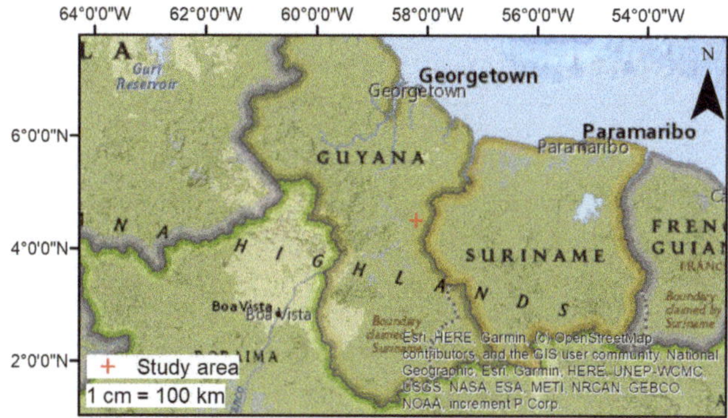

Figure 1. Map of the study area in Guyana (cross).

2.2. Tree Selection and Data Collection

2.2.1. Tree Inventory

An exploratory survey of the area was performed as a guide to sample the species composition of the forest. We grouped our trees into five diameter (D) size classes ($10 \leq D < 30, 30 \leq D < 50,\ldots$, $D \geq 90$ cm) and inventoried 15 to 23 trees per size class. A total of 106 trees were inventoried and scanned in 37 plots with TLS across a large range of tree sizes (D 11.2 to 149.8 cm), families (26), and species (50). Of these, 26 were destructively sampled and other 8 were discarded due to poor point cloud quality. We measured D and point of measurement (POM) of D, total H, CD, and recorded species, stem damage, and any irregularities. Total H and CD were measured with a Nikon Forestry-Pro hypsometer. Total H was measured from the base of the tree to the top [39] and CD was measured as the mean of the projected tree edge N-S and E-W [40]. An experienced local taxonomist matched reported local names with scientific names (Supplementary Material S.1).

2.2.2. TLS Data Acquisition

All TLS datasets were acquired using a RIEGL VZ-400 3D terrestrial laser scanner (RIEGL Laser Measurement Systems GmbH, Horn, Austria). We scanned at each position with a resolution of 0.04°, following the suggestion by Wilkes et al. [23]. The TLS data acquisition and plot sampling design can be found in Supplementary Material S.2.

From our full field inventory (a total of 106 trees), 26 trees were destructively sampled trees and removed from the TLS dataset to serve as validation data. We inspected the point clouds of the remaining 80 trees and discarded 8 trees whose point clouds were poor due to understory occlusion. Thus, 72 trees were used in our TLS database to build allometric models.

2.2.3. Destructive Harvesting and Fresh Mass Sampling

Twenty-six trees were selected and destructively sampled to serve as validation data. This selection was based on diameter class, species, and wood density to maximize the number of species sampled and avoid selection biases; all other characteristics (e.g., commercial value, trunk or crown form, hollowness, structural damage or any other irregularities) were ignored—following Goodman et al. [5]. After felling each tree, we re-measured D and H (denoted with the prefix post-harvest) and weighed each part in situ. Each part was separated and weighed. The fresh mass was measured directly in the field. Larger and non-irregular stems and very large branches were measured through volume estimation. We measured length, top diameter, and bottom diameter of any hollow sections.

We collected three wood samples from each part of each tree to estimate water content. Samples were weighed immediately in the field, labelled, and air dried during the field campaign. Samples from the bole, stump, and large branches (when volume was measured) were also used to determine wood density. For this, measured fresh volume of each wood sample by water displacement. Detailed information of the fresh mass sampling procedure can be found in Supplementary Material S.3.

2.2.4. Laboratory Analysis

We transported all wood samples to the laboratory at the Guyana Forestry Commission (Georgetown, Guyana) for species identification, drying, and storage. Wood samples were oven dried (101 to 105 °C until they reached a constant mass)—as in Williamson and Wiemann [41]—and re-weighed. Detailed explanation of the laboratory work can be found in the Supplementary Material S.3. Wood density was calculated as dry mass per fresh volume; and dry mass fraction (*dmf*) was the ratio dry to fresh mass.

2.3. Diameter, Tree Height and Crown Diameter from TLS Data

TLS-derived D was calculated from cross sectional point clouds (6 cm width) taken at every 10 cm on the Z-axis up to 6 m height. Least square circle approach was used to fit circles in each cross sectional point cloud as in Calders et al. [17]. We automatically determined POM by analysing the angle between two consecutive diameters, starting from the bottom. The first angle within 1° of 90° (i.e., vertical) was considered as the POM. Total height was estimated as the distance between the maximum and the minimum point in the Z-axis from each tree point cloud. Crown diameter was estimated as the average of two horizontal distances between the maximum and the minimum point in the X- and Y-axis from each tree point cloud.

2.4. Tree Volume and Biomass from TLS Data

We estimated tree wood volume from 3D quantitative structure models (*TreeQSM* version 2.0; [17,18,33]) from our reconstructed 72 trees. *WD* values were assigned to each species or genus according to Global Wood Density Database [8,9]. To obtain tree volume, we had two main components: (i) semi-automatic individual tree extraction from TLS plots (Figure 2a–c), and (ii) 3D reconstruction of QSMs for individual extracted trees (Figure 2d and Supplementary Material S.4).

Once we extracted the trees, we reconstructed their volume with cylinder features using the automated framework presented in [17] to optimize QSMs. We optimized cover patch size (*d*) by reconstructing the volume using a *d* range from 0.02 to 0.09 m with a 0.005 m increment and a minimum number of points per cover patch *nmin* of 4. The optimization process returned the most suitable *d* for each tree based on least square fit process and 20 models were reconstructed on average. The heuristic decision to accept/reject was taken based on analyst's experience and judgement (Figure 2d; [17,18]).

For 43 trees with large buttresses (selected by POM $\neq$ 1.3 m, visual inspection, and author's expertise), a triangular mesh was used for the volume modelling in the bottom part of the stem rather than cylinders (Figure 3; see Disney et al. [22]). The volume of the mesh replaced the volume of the cylinders on the total tree volume estimation.

Figure 2. (**a**) *Vitex spp.* tree point cloud (TLS-derived $H = 51.8$ m and TLS-derived $D = 114.6$ cm with POM at 5.3 m), (**b**) down-sampled tree point cloud (0.026 m point spacing, as in Calders et al. [42]), (**c**) soft tissues (green) and hardwood (black) separated point cloud [43], and (**d**) *TreeQSM* modelled after the hardwood point cloud.

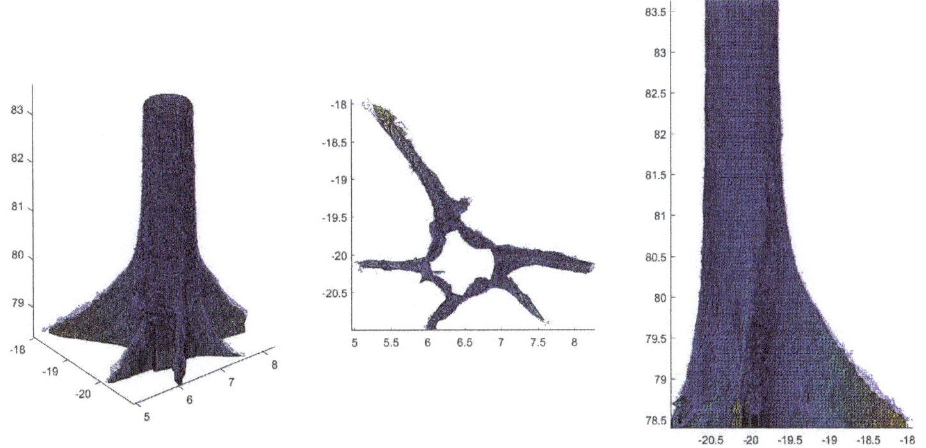

Figure 3. Buttresses modelling of a *Hymenolobium flavum* tree. The bottom part of the stem was modelled with a triangular mesh instead of cylinders. The mesh volume replaced the volume of the cylinders.

Finally, to estimate AGB, we multiplied the total tree volume with the corresponding wood density. As an indicator of the reconstruction accuracy of the *TreeQSM*, root mean square error (RMSE) was calculated to measure the difference between the reference and modelled AGB; R^2 was used to judge the fit of the *TreeQSM* models; concordance correlation coefficient (CCC; Lin [44]) was calculated to compute the agreement on a continuous measure obtained by two methods; and the coefficient of variation of RMSE (CV RMSE) was calculate to measure the difference between our TLS-derived AGB and reference AGB [17,18].

2.5. TLS-Derived Allometric Models

We examined five model forms (Table 3) based on previous forms developed by Chave et al. [12] and Goodman et al. [5] using TLS-derived attributes (D, H, and CD) and WD to test the relevance of these attributes to predict AGB. To build the allometric models, all data were transformed to the natural logarithm to comply with allometric theory and meet the assumptions of linear regression. The models were built using least-squares linear regression and the attributes were removed using backwards stepwise regression to produce a minimum adequate model in the statistical program R [45]. We tested the assumptions of the final five models. We tested for normal distribution of the residuals using Q-Q plots and Anderson-Darling test to assess independence, and plots of residuals against fitted values to assess homogeneous variance and linearity.

We used a paired t-test to analyse whether the TLS-derived attributes were different than our post-harvested field measurements. If TLS and field-based measurements differed significantly, a calibration factor was applied to the input attribute.

We also tested whether there were significant differences between models built using field measurements or TLS-derived data. We built models using both sets of data with data source as a dummy variable. When the dummy variable was significant, we applied its attribute estimate (Reference Dummy Variable Corrector; RDVC) to the corresponding model form to modify the TLS built models.

2.6. Tree Aboveground Biomass Estimation from Pantropical Allometric Models

Structural (D and H) and WD data from the 26 harvested trees were used to estimate AGB from pantropical allometric models (Table 1). We estimated AGB and error using the most widely used pantropical allometric models (Eqn. Ch05.II.3, Ch05.I.5, Cha14.H and Ch14.E) for moist forests [11,12] and an updated version of Ch14.E. This revised version (Eqn. Rj17.E) is a direct model fit equation, while the original equation Ch14.E was obtained by merging two equations [46].

Table 1. Pantropical models from [11,12,46] included diameter at breast height (D, in cm), specie-specific wood density values according to the GWDD (WD, in $g\,cm^{-3}$ or $kg\,m^{-3}$), total height (H, in m), the environmental stress (E, calculated from the GPS average location of each tree [1]) to estimate aboveground biomass (AGB, in kg dry mass) and ε is the model error.

Model	Form $AGB =$
Ch05.II.3	$WD \cdot exp(-1.499 + 2.1481 \cdot ln(D) + 0.207 \cdot ln(D)^2 - 0.0281 \cdot ln(D)^3) + \varepsilon$
Ch05.I.5	$0.0509 \cdot WD \cdot D^2 \cdot H + \varepsilon$
Ch14.H	$0.0673 \cdot (WD \cdot D^2 \cdot H\cdot)^{0.976} + \varepsilon$
Ch14.E	$exp(-1.803 - 0.976 \cdot E + 0.976 \cdot ln(WD) + 2.673 \cdot ln(D) - 0.0299 \cdot ln(D^2)) + \varepsilon$
Rj17.E	$exp(-2.024 - 0.896 \cdot E + 0.920 \cdot ln(WD) + 2.795 \cdot ln(D) - 0.0461 \cdot ln(D^2)) + \varepsilon$

[1] http://chave.ups-tlse.fr/pantropical_allometry.htm.

2.7. Assessment of Allometric Models

We made two types of assessment. First, we evaluated the log-transformed models based on the fit of the data used to build the models. For that, we evaluated the models by using a penalized likelihood criterion on the number of attributes: adjusted R-square (R^2), corrected Akaike's information criterion (AICc), the coefficient of variation of RMSE ($CV\,RMSE$), and residual standard error (RSE). Goodness of fit is indicated by high R^2 and low AICc, CV RMSE, and RSE. To estimate AGB values (in Mg) a correction factor ($CF = exp[RSE^2/2]$) was applied to backtransform predicted values and remove bias from the log-transformed data [47].

Second, we assessed the ability of our allometric models to predict AGB (Table 3) and compared them to five pantropical allometric models (Table 1). We validated these AGB estimates with field-measured reference AGB using the metrics listed below: model error (Equation (1)) and relative error (Equation (2)).

$$AGB_{error}(\text{Mg}) = AGB_{est} - AGB_{ref},$$

(1)

$$AGB_{relative\ error}(\%) = \frac{AGB_{est} - AGB_{ref}}{AGB_{ref}} \cdot 100 \qquad (2)$$

where AGB_{est} is AGB predicted by each model, AGB_{ref} is the AGB calculated from our destructive sampling, $mean_{error}$ is the average of AGB_{error} for all 26 trees, and $mean_{ref}$ is the average of AGB_{ref}. We calculated these metrics for our entire harvested tree dataset ($n = 26$) and subsequently split this data into small ($D \leq 70$ cm; $n = 17$) and large ($D > 70$ cm; $n = 9$) trees. We used both assessments to identify our "best" model(s).

3. Results

3.1. Tree Attributes and Estimated Biomass

A total of 72 trees were used to build allometric models based on TLS-derived data (Table 2 and Supplementary Material S.5). Six harvested trees had hollow sections in the bole.

Table 2. Pre- and post-harvested field measured attributes, and terrestrial laser scanning (TLS)-derived attributes range for: diameter (*D*), height (*H*), crown diameter (*CD*), wood density (*WD*; see note below) and aboveground biomass (AGB).

Attributes	Allometric Model Dataset ($n = 72$)		Validation Dataset ($n = 26$)	
	Measured$_{pre}$	TLS-Derived	Measured$_{post}$	TLS-Derived
Diameter (cm)	12.9 – 134.0	13.3 – 126.2	16.7 – 128.7	16.7 – 130.2
Tree height (m)	14 – 43.0	16.9 – 51.8	16.4 – 51.6	16.6 – 49.1
Crown diameter (m)	4.4 – 42.6	2.5 – 42.9	3.4 – 30.8 $_{pre}$	4.6 – 30.2
WD (g cm^{-3})	0.4 – 1.0	0.4 – 1.0	0.4 – 0.9	0.4 – 1.0
AGB (Mg)	NA	0.2 – 28.5	0.9 – 21.8	0.2 – 27.4

Note: Wood density was taken from direct measurements for measured trees in reference dataset and from the GWDD for all others.

Our study found a systematic difference between the three measurements for *D* in our dataset of 26 validation trees (*p*-value < 0.05; Table 2; Figure A1a,c of Appendix A): mean values were 59.4, 57.7, and 55.2 cm for pre-harvest, post-harvest, and TLS-derived diameters, respectively. Using post-harvest field measurements as the reference data, we calibrated the TLS-derived *D* estimate (Figure A1c). Neglecting to adjust the systematically lower TLS-derived *D* would result in a systematic overestimation of AGB when applying TLS-derived allometric models with field measurements diameters. The TLS-derived attributes (*H* and *CD*) and AGB were not significantly different from our reference data (*p*-value > 0.05; Figure A1d–f). However, post-harvest *H* measurements were significantly taller than pre-harvest *H* measurements taken on standing trees ($p < 0.05$; $n = 26$; Figure A1b). *WD* values from our measurements and the GWDD were similar (our values were 1 % greater; Supplementary Material S.6). Finally, there was no systematic difference between AGB estimated from *TreeQSM* and our destructively-sampled reference data (Figure A1f).

3.2. Allometric Models Using TLS-Derived Measurements

We created five allometric models using combinations of the TLS-derived attributes *D*, *H* and *CD* and *WD* from the GWDD (Table 3). For models using *CD* (m4 and m5) we found that the residuals were not normally distributed using Anderson-Darling test. Upon further visual inspection of normal quantile-quantile plots, we considered the residuals to be reasonably normally distributed and that these models are reliable (Supplementary Material S.7). For model forms m2 and m3, models built using field and TLS-derived data were significantly different ($p < 0.05$), thus we applied RDVC to the TLS models (Table 3).

Table 3. Models description for the TLS-derived aboveground biomass estimations including diameter (D), wood density (WD), height (H), crown diameter (CD), Reference Dummy Variable Corrector ($RDVC$) and associated statistical parameters based on 72 trees.

Model	Type	Form	a	b	c	d	e	RDVC	df	RSE	adj-R^2	AICc
m1	D	$ln(AGB) = a + b \cdot ln(D) + \varepsilon$	0.6788	1.9337	...	...	...	...	70	0.360	0.90	61.52
m2	D.WD	$ln(AGB) = a + b \cdot ln(D) + c \cdot ln(WD) + RDVC + \varepsilon$	0.6765	2.0246	1.0932	...	...	−0.1968	69	0.274	0.94	23.61
m3	D.WD.H	$ln(AGB) = b \cdot ln(D) + c \cdot ln(WD) + d \cdot ln(H) + RDVC + \varepsilon$	...	1.9091	1.0978	0.3224	...	−0.2138	69	0.266	NA	19.48
m4	D.WD.H.CD	$ln(AGB) = b \cdot ln(D) + c \cdot ln(WD) + d \cdot ln(H) + e \cdot ln(CD) + \varepsilon$	...	1.7282	0.2603	1.1522	0.3698	...	68	0.240	NA	6.23
m5	D.WD.CD	$ln(AGB) = a + b \cdot ln(D) + c \cdot ln(WD) + e \cdot ln(CD) + \varepsilon$	0.5366	1.8124	1.1512	...	0.3878	...	68	0.246	0.96	9.28

Notes: df are degrees of freedom of the model, RSE is residual standard error of estimates, R^2 is adjusted R^2, AICc is the corrected Akaike's information criterion and NA is not applicable. In the models where the intercept was removed, R² was not calculated.

3.3. Evaluation of Allometric Models

We assessed how well the five TLS-derived allometric models developed in this study and five pantropical allometric models estimate AGB of our destructively-harvested reference data on the original scale (in Mg; Figure 4 and Table 4). Results were similar to the assessment done on the log-transformed scale, and our models including CD estimated AGB better than all other models. On the original scale (Table 4), AGB estimates from TLS-derived allometric models were slightly better (R^2 0.87–0.93; CCC 0.89–0.96) than the pantropical models assessed (R^2 0.85–0.89; CCC 0.92–0.94). However, *TreeQSM* models m4 and m5 outperformed the pantropical models (CCC = 0.96, R^2 = 0.92, CV RMSE = 33 %). Pantropical models showed slightly lower level of agreement (CCC = 0.92–0.94); with a R^2 ranging 0.85–0.89 and CV RMSE ranging 37–44 %.

Table 4. Summary of AGB estimates from TLS-derived and pantropical allometric models—R^2, root mean square error (RMSE), concordance correlation coefficient (CCC), sum of errors (sum, mean, standard deviation (SD)), mean percent error and relative error (n = 26). Models are arranged based on the statistical parameters from the best model to the worst model.

Model	Type	R^2	RMSE	CCC	Error (Mg)			Relative Error (%)	
					Sum	Mean	SD	Mean	SD
m5	D.WD.CD	0.93	1.91	0.96	1.03	0.04	1.95	28.25	61.35
m4	D.WD.H.CD	0.92	1.99	0.96	1.36	0.05	2.03	28.33	57.91
Ch05.II.3	WD.D.D².D³	0.89	2.32	0.94	−7.49	−0.29	2.35	5.54	48.26
Ch05.I.5	D².WD.H	0.85	2.75	0.92	−9.86	−0.38	2.78	7.35	41.98
Ch14.H	(D².WD.H)	0.85	2.67	0.92	−11.89	−0.46	2.69	9.59	43.31
m1	D	0.87	2.52	0.93	13.65	0.53	2.51	68.01	105.95
Rj17.E	D.WD.E	0.88	2.43	0.93	−19.28	−0.74	2.36	−1.62	44.04
Ch14.E	D.WD.E	0.88	2.49	0.93	−21.56	−0.83	2.39	−3.62	43.52
m3	D.WD.H	0.88	2.92	0.89	−32.11	−1.23	2.69	14.80	60.97
m2	D.WD	0.89	2.96	0.92	32.21	1.24	2.74	73.80	98.95

Our TLS-derived allometric models with CD performed better than all other models in terms of R^2, RMSE, CCC, and absolute errors (Table 4 and Figure 4). The pantropical models tended to underestimate AGB but had lower relative errors (Table 4 and Figure 4). In contrast to the model evaluation with the 72 TLS-trees on the logarithmic scale, m5 (without H) performed slightly better than m4 (with H). By several metrics, model m2 and m3 performed the worst of all models evaluated. Adding WD and H to model m1 (with D only) did not improve estimates. In fact, adding these attributes increased overall error. However, adding CD to models with WD and H improved estimates by all metrics. Adding H to models never improved their accuracy. In both cases, models m5 vs. m4 and m3 vs m2, the model without H performed better than the counterpart model with H.

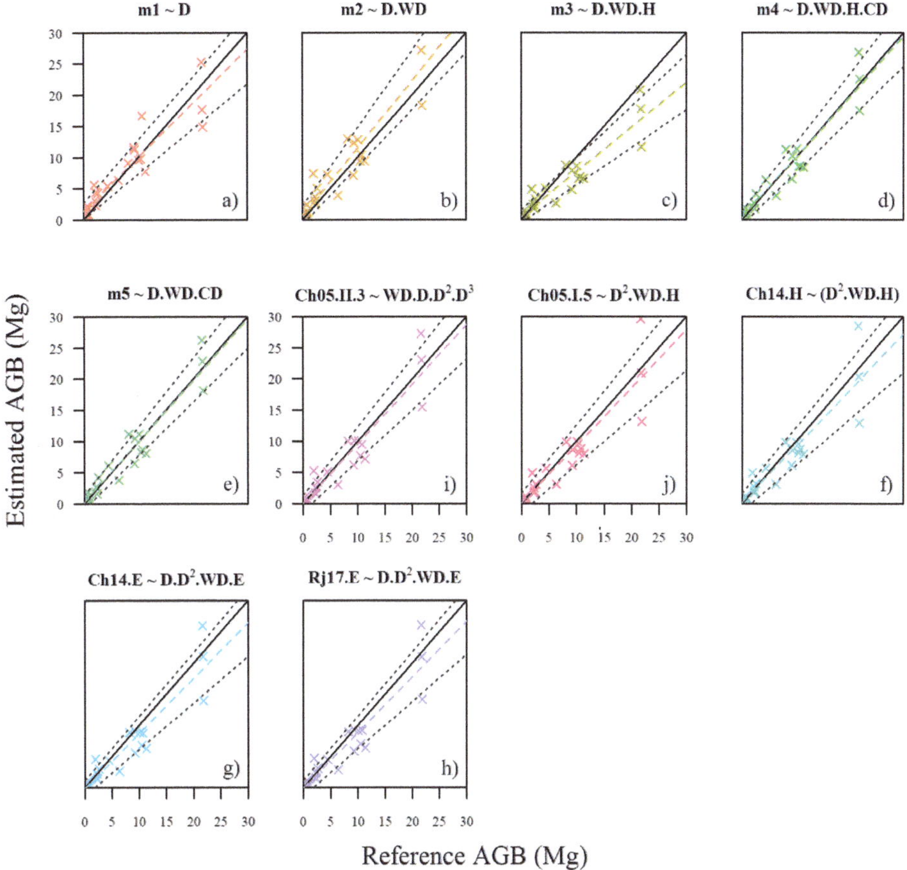

Figure 4. Relationship between reference AGB (harvested trees; $n = 26$) and AGB estimated by TLS-derived and pantropical allometric models. Black solid line is 1:1 relationship; dashed coloured lines depict linear fit; and dotted grey lines indicate 95 % confidence interval for the linear fit.

Among the pantropical models, Ch05.II.3 with just D and WD, had the highest R^2 and CCC and the lowest RMSE, but it slightly underestimated AGB (Table 4 and Figure 4). The two models including the "environmental stress" attribute (E; Ch14.E and Rj17.E) produced the largest underestimates of AGB in absolute terms but lowest relative error (Table 4). Pantropical models including H (Ch05.I.5 and Ch14.H) also underestimated AGB and were, by several metrics, the least accurate among all models evaluated (lowest R^2, lowest CCC, and highest RMSE).

Because allometric models often over- or underestimate AGB differently systematically with tree size, we also assessed the performance of the TLS-derived and pantropical models for small trees ($D \leq 70$ cm; $n = 17$) and large trees ($D > 70$ cm; $n = 9$) separately (Figure A2). The inclusion of CD in the models reduced error in AGB estimates for both size classes, but the effect was much more substantial in the large trees. The mean error of TLS-derived models m4 and m5 is very close to zero for both small and large trees, but mean relative error is very high for small trees (44 %). The inclusion of H in TLS-derived models improved AGB estimations for small trees but reduced the accuracy of AGB estimates for large trees when compared to their counterpart model without H. All pantropical models underestimated AGB, and the underestimation is greater for large trees (Table A1).

For the most accurate AGB estimates in Guyanese forests, we recommend model m5 (Equation (3)), especially for large trees. AGB is measured in kg dry mass, D in cm, H and CD in m and WD in $g\,cm^{-3}$. The back-transformation correction factor has been incorporated:

$$AGB = exp(0.5669 + 1.8124 \cdot ln(D) + 1.1512 \cdot ln(WD) + 0.3877 \cdot ln(CD)) \tag{3}$$

4. Discussion

4.1. Developing Allometric Models from TLS-Derived Attributes

This study presents the first assessment of the potential of TLS and *TreeQSM* to develop TLS-derived allometric models to estimate AGB for trees in Guyana and takes Guyana one step closer to its aim of developing a national MMRV system. Previous work by Gonzalez de Tanago et al. [18] tested TLS-derived tree volume estimates against field-based volume estimates and then converted to AGB wood density values from the Global Wood Density Database [8]. Here, we weighed trees directly to build our reference data and tested the effects of tree hollowness on the accuracy of TLS-QSM AGB estimates. We found that TLS-QSM AGB estimates were not biased, even with the presence of hollow and irregularly shaped trees.

This is one of the first studies to have explicitly used TLS-derived attributes and wood density to develop allometric models to estimate AGB of tropical trees. In a similar study, Stovall et al. [29] reconstructed 329 trees in Virginia, USA and found that TLS-derived allometric models predicted AGB better than national models. Another study, Momo Takoudjou et al. [26] calibrated an allometric model from Chave et al. [11] using TLS tree point clouds from a semi-deciduous forest in Cameroon and yielded a good fit (R^2 = 0.95). Both studies [26,29] used TLS-derived attributes (D and H) to predict AGB with accurate results. While one reconstructed AGB from coniferous trees [29], the other one was applied to tropical forest trees in Cameroon. Our results showed better results for TLS-derived allometric models that include CD compared to the pantropical models tested [11,12,46], which included trees from the same region (French Guiana, $n \approx 390$ trees). In the absence of CD data, the oldest and simplest of the pantropical models (Ch05.II.3) provides the most accurate AGB estimates in this region. This is one of the first studies to have explicitly used TLS-derived attributes and wood density to develop allometric models to estimate AGB of tropical trees.

4.2. Choosing the Adequate Tree Attributes for Allometric Models

Recent studies have highlighted the relevance of including crown dimension in allometric models [5,15,48]. Models including CD had the highest R^2 in Goodman et al. [5] and Jucker et al. [48] and lower bias [48]. These authors had independently agreed that CD improves tropical tree biomass estimates, especially for large trees (D > around 100 cm in Goodman et al. [5] and Ploton et al. [15]; and $\geq$10 Mg for Jucker et al. [48]). From our results, we found that allometric models using CD provide better AGB estimates for trees in Guyana, especially for large D size classes (Figure A2). While crown diameter is generally not collected in traditional forest inventories; UAV (unmanned aerial vehicles) with structure for motion algorithms allows you to delineate tree crowns [49]; increasing the potentiality of using crown models in allometric models.

Our results agree with Goodman et al. [5] that allometric models with H performed worse underestimating AGB than their counterparts without H in this region too. Goodman et al. [5] also found that models including H underestimated AGB and suggested the inclusion of CD instead of H in allometric models. Our results contrast those of Feldpausch et al. [10], in which their models with H performed better estimating AGB than models without H, and agree with Goodman et al. [5], which found that pantropical models that include H tend to be systematically inaccurate when applied to other locations.

We found significant differences in the pre-harvest, post-harvest, and TLS-derived D and H values, demonstrating the difficulty and ambiguity of measuring the diameter and height of tropical trees, as seen in Table 2. While it is nearly impossible to say what "true" and repeatable diameter above buttresses and total H are, TLS offers new insights. For example, even with rigorous protocols, determining the top of buttresses is a judgement call, and our data show variation in measurements on standing and felled trees. Pre-harvest D measurements were significantly higher, reflecting the difficulty of measuring high above the ground; and post-harvest measurements were much more conservative. Our novel method of determining D from TLS point clouds was perhaps too conservative but probably more repeatable. It is important to note that conservative D measurements yield conservative AGB estimates from allometric equations, but the opposite is true when building allometric models.

Measuring tree height in tropical forests is notoriously difficult, especially for trees above forest canopy. In Hunter et al. [50], the precision of repeated height measurements from the ground ranged from 3 to 20% of the total height, which resulted in 16% mean error of AGB estimates. In our study we did not record repeated measurements, but our pre-harvest height measurements were on average 10% lower and TLS-derived measurements were 2% lower than our reference (post-harvest) heights. Pre-harvest measurements had a greater variation and higher underestimation than our TLS-derived height. A novel alternative for would be to use a UAV-LS (UAV laser scanning) to estimate tree height and crown diameter. Brede et al. [51] were able to derived D and canopy height models using a RIEGL RiCOPTER with VUX$^{\circledR}$ -1UAV and compared those with a RIEGL VZ-400 TLS with high agreement for DBH (correlation coefficient of 0.98 and RMSE of 4.24 cm). However their study area was composed mostly by European Beech (*Fagus sylvatica*) and future work is needed to explore UAV-LS techniques in tropical forests.

4.3. Local or Pantropical Allometric Models?

We also estimated AGB for our trees using five well-known pantropical models (Table 1), showing that all five models underestimated AGB. Our results contrast with some studies and support others that used pantropical models for local studies. In Colombia, Alvarez et al. [4] found that using Chave et al. [12] moist pantropical equations (with and without H) all overestimated AGB (with relative errors up to 52.8 %); while Gonzalez de Tanago et al. [18] found that AGB was underestimated 15.2 to 35.7 % when compared to estimated AGB from *TreeQSM* models in Guyana, Indonesia, and Peru. As in Alvarez et al. [4], Kuyah et al. [52] found that AGB in Kenya was overestimated in 22 % using Chave et al. [12] moist forest equation (with H) and suggested that overestimations were due to the dominance of small trees ($D < 10$ cm) and lack of larger trees in their plots. We theorized that with more trees scanned, we could understand the reasons for these differences. When disaggregating by diameter size, we found that pantropical models tended to underestimate small and large trees (Figure A2). This matches Chave et al. [12], whose models tended to underestimate small trees. Our results are also aligned with Chave et al. [11], insofar as their models tended to be fairly accurate with medium size trees, and underestimate larger trees. In this study, our TLS-derived allometric models ($n = 72$) were based on 32 different species and 18 different families. Just one ha of tropical forest can have 300 different species [53]; thus, we aimed to cover all different tree species in our sample, with a more precise range of wood density and avoid bias by selecting few species.

Interestingly, our study also showed that locally developed allometric models are not always better than pantropical allometric models. We found out that the pantropical allometric model Ch05.II.3 (without H) consistently performed best ($R^2 = 0.89$) among the other pantropical allometric models assessed in this study ($R^2 = 0.85$–0.88) and even better than some of the TLS-derived allometric models developed (m1, m2, and m3). Thus, our approach can be used also to select the most appropriate allometric model available without the need of destructively harvested validation data. Moreover, our best TLS-derived allometric models (m4 and m5) estimated the AGB of large trees ($D > 70$ cm) very accurately, with mean AGB overestimate of just 1 %, while the pantropical models assessed yielded mean underestimates between -4% and -24% (Table A1). This is of much relevance due to the significance of large trees to total forest biomass [14–16] and the fact that pantropical allometric models systematically underestimated large tree biomass [5,17,18].

4.4. Challenges and Outlook

Scanning tropical trees in situ remains challenging, not only because of the harsh environment but also because the novel sampling design we developed for this study (Supplementary Material S.2.3). In our study, we increased the scanning resolution from $0.06°$ to $0.04°$ (Supplementary Material S.2.1) as suggested by Wilkes et al. [23] to balance trade-off between accuracy and scanning time requirements. Gonzalez de Tanago et al. [18] and Lau et al. [54] pointed out that low-density point clouds created unrealistic branching reconstructions (i.e., where cylinders were constructed due to low point cloud density rather than actual branching). Still, we discarded 8 tree point clouds due to under-story occlusion. Occlusion by dense vegetation might reduce TLS acquisition range and future TLS campaigns in tropics will cope with occlusion; either increasing their sampling rate, using a different type of TLS (single or multiple returns), or increasing their scanning resolution [17,23]. Because a detailed analysis of quality of tree point clouds is usually undertaken after the fieldwork, we suggest scanning 10% more trees than the desired sample size in case some trees need to be removed due to poor quality.

Our methodology provided unbiased AGB estimates regardless of tree structure, even with partially hollow and irregular stems. However, we suggest that this outcome is further tested with destructively harvested data from other forests. In Guyana, developing a national monitoring system based on the now known most appropriate pantropical model or their own national model could contribute to more accurate biomass estimates for REDD+ MMRV. Our results demonstrated that TLS-derived AGB estimates can be used as a decision-making tool in MMRV selection of an adequate pantropical allometric model, and in case TLS-derived allometric models using CD are out of scope, Ch05.II.3 would be an adequate model (on average conservative and reasonably accurate).

Being able to obtain TLS-derived AGB estimates without destructively harvesting trees is also environmentally desirable. We are able to quantify AGB for trees that would be illegal, expensive, impractical, or simply unnecessary to harvest. We are aware that our methods and analyses require expensive equipment and expert knowledge, but the process is much faster, less labour intensive, and more environmentally sustainable than destructive harvesting, especially for large trees. Research has already begun to semi-automatize the modelling processes by separating individual trees [55] and segregating wood from soft tissues [22].

With the advances on tree segmentation algorithms [42,56], the modelling of trees is being semi-automatized. Tree segmentation algorithms would allow us to segment and estimate AGB using *TreeQSM* for entire TLS scanned forest plots. While this is a case study of creating site-specific allometric models for Guyana, we showed that TLS-derived allometric models (including CD) can be an unbiased estimator of AGB, even in a logged forest where a high proportion of the remaining trees were damaged and hollow. Our method's potential to rapidly produce large, unbiased calibration tree datasets is of great significance to remote sensing missions, which rely on field data for their calibration. With TLS and 3D modelling, plot-level AGB could be estimated directly, and these plots could be used as calibration sites for upcoming satellite missions. However, this is a major undertaking. As mentioned by Disney et al. [57], data acquisition standardization, automatization of processing, and more accurate 3D reconstructions are needed first to narrow the bridge between TLS and remote sensing space missions. We strongly encourage other studies to expand upon our findings to determine whether TLS always provides unbiased AGB estimates and could replace destructive sampling in the future.

5. Conclusions

We advanced TLS methods to estimate tree metrics and explored the accuracy of field and TLS-derived data to develop local allometric models for Guyana. We showed that AGB from *TreeQSM* estimations were not biased. Moreover, we found that allometric models can be built from TLS-derived tree volume and wood density, even with the occurrence of hollow and irregular stems. We demonstrated that tree AGB estimates from TLS-derived allometric models including crown diameter (models m4 and m5) provide better agreement with reference data than AGB estimates from pantropical allometric models, especially for large trees ($D > 70$ cm). AGB estimates from TLS-derived allometric models and pantropical models including H provided poor agreement with reference data when compared to their counterpart without H. The simplest pantropical model (Ch05.II.3 [12] with only D and WD) provided very good estimates of our data. While our results are based on 72 TLS scanned trees, a small number of trees compared to other studies, our new approach can be further applied for developing allometric models without the need to harvest vast numbers of trees. Further, this new approach can be used to test and choose existing allometric models calibration remote sensing metrics to forest biomass, and improve the future estimates of forest biomass from tropical forest.

Supplementary Materials: The supplementary materials are available online at http://www.mdpi.com/1999-4907/10/6/527/s1.

Author Contributions: The authors contributed in the research as follows: conceptualization: A.L., R.C.G., and M.H.; methodology: A.L., R.C.G., H.B., and M.H.; data collection: A.L., R.C.G., H.S., and J.S.; formal analysis: A.L., K.C., R.C.G,. P.R., and M.V.; writing–original draft preparation: A.L. and R.C.G.; writing–review and editing: A.L., R.C.G,. C.M., H.B., and K.C.; visualization: A.L.; K.C., and P.R.; supervision: M.H. and C.M.; funding acquisition: M.H. and C.M. All authors contributed critically to the drafts and gave final approval for publication.

Funding: This research is part of CIFOR's Global Comparative Study on REDD+ with financial support from the donors to the CGIAR Fund. A.L. is supported by the donors to the CGIAR Fund, SilvaCarbon research project 14-IG-11132762-350 and ERA-GAS NWO-3DforMod project 5160957540. KC is funded by BELSPO (Belgian Science Policy Office) in the frame of the STEREO III programme-project 3D-FOREST (SR/02/355). M.V. is funded by CNPq (National Council of Technological and Scientific Development-Brazil) through the programme Science Without Borders (Process number 233849/2014-9).

Acknowledgments: Our field data were acquired through collaboration between Wageningen University and Research and Guyana Forestry Commission. A special thanks to José Gonzalez de Tanago, Jens van der Zee and the forestry team in the Guyana Forestry Commission for all the assistance during the fieldwork and Rong An Inc. concession for allowing us to stay in their camp during fieldwork. We extend a very special thanks to the Guyana Forestry Commission and especially to Pradeepa Bholanath and Nasheta Dewnath for all the assistance before, during, and after the fieldwork.

Conflicts of Interest: The authors declare no conflict of interest. The funders had no role in the design of the study; in the collection, analyses, or interpretation of data; in the writing of the manuscript, or in the decision to publish the results.

Abbreviations

The following abbreviations are used in this manuscript:

AGB	Aboveground biomass
adj-R^2	Adjusted R-square
AICc	Akaike's information criterion
CCC	Concordance correlation coefficient
CD	Crown diameter
CF	Correction factor
Ch05.I.5	Chave et al. [12] Equation I.5
Ch05.II.3	Chave et al. [12] Equation II.3
Ch14.E	Chave et al. [11] Equation (7)
Ch14.H	Chave et al. [11] Equation (4)
CV RMSE	Coefficient of variation of RMSE
D	Diameter at breast height

df	degrees of freedom
dmf	dry mass fraction
GWDD	Global wood density database [8]
H	Height
LiDAR	Light Detection And Ranging
MMRV	Monitoring, measurement, reporting and verification
POM	Point of measurement
Rj17.E	Réjou-Méchain et al. [46] Equation (1)
QSM	Quantitative structure models
TLS	Terrestrial laser scanning
RDVC	Reference Dummy Variable Correction
REDD+	Reducing emissions from deforestation and degradation
RMSE	Root mean square error
RSE	Residual standard error
UAV	Unnamed aerial vehicle
UAV-LS	Unnamed aerial vehicle laser scanning
WD	Wood density

Appendix A

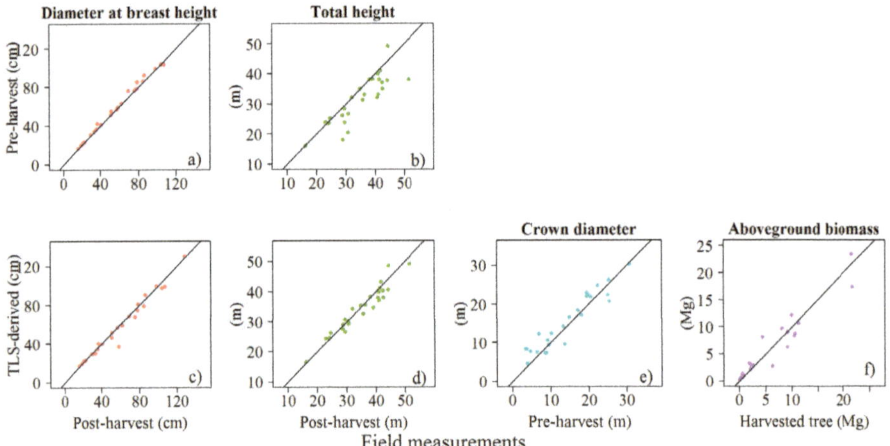

Figure A1. Relationship between pre-harvest against post-harvest values for D (**a**) and H (**b**), TLS-derived against post-harvest values for D (**c**) and H (**d**), TLS-derived against pre-harvest values for CD (**e**), and against harvested tree AGB (**f**), Solid line is 1:1 relationship.

Table A1. Summary of AGB estimates from TLS-derived and pantropical allometric models—R^2, RMSE, CCC, sum of errors (sum, mean, standard deviation (SD)), mean percent error and relative error ($n = 26$) separated in small trees (D $\leq$ 70 cm) and large trees (D > 70 cm). Models are arranged based on the statistical parameters from the best model to the worst model.

Model	Type	R^2		RMSE		CCC		Mean Error (Mg)		Sum Error (Mg)		SD Error (Mg)		Mean rel. Error (%)		SD. rel. Error (%)	
		Small	Large	Small	Large	Small	Large	Small	Large	Small	Large	Small	Large	Small	Large	Small	Large
m5	D.WD.CD	0.83	0.84	1.27	2.69	0.87	0.90	0.03	0.06	0.50	0.53	1.31	2.86	44.08	0.11	70.75	22.73
m4	D.WD.H.CD	0.83	0.81	1.22	2.89	0.89	0.89	0.08	0.01	1.23	0.13	1.26	3.07	44.29	−0.04	65.74	23.30
Ch05.II.3	WD.D.D^2.D^3	0.70	0.78	1.57	3.30	0.79	0.86	−0.21	−0.43	−3.61	−3.88	1.60	3.47	10.71	−4.22	57.51	22.36
Ch05.I.5	D^2.WD.H	0.76	0.67	1.40	4.25	0.84	0.79	−0.11	−0.88	−1.95	−7.91	1.44	4.42	15.26	−7.57	47.45	24.95
Ch14.H	(D^2.WD.H)	0.75	0.67	1.41	4.11	0.84	0.79	−0.09	−1.16	−1.45	−10.44	1.45	4.19	19.54	−9.21	48.42	23.74
m1	D	0.74	0.59	1.55	3.71	0.81	0.76	0.64	0.32	10.81	2.84	1.46	3.92	99.92	7.74	118.63	26.22
Rj17.E	D.D^2.WD.E	0.70	0.77	1.65	3.45	0.75	0.85	−0.40	−1.39	−6.72	−12.55	1.65	3.34	3.61	−11.50	52.33	20.40
Ch14.E	D.D^2.WD.E	0.70	0.77	1.68	3.55	0.74	0.84	−0.44	−1.57	−7.45	−14.11	1.67	3.37	1.39	−13.08	51.60	20.94
m3	D.WD.H	0.67	0.74	1.67	4.39	0.75	0.72	−0.16	−3.27	−2.66	−29.45	1.71	3.11	35.20	−23.72	65.90	19.41
m2	D.WD	0.64	0.77	1.90	4.30	0.76	0.81	0.84	1.99	14.30	17.91	1.75	4.04	105.06	14.74	109.05	28.47

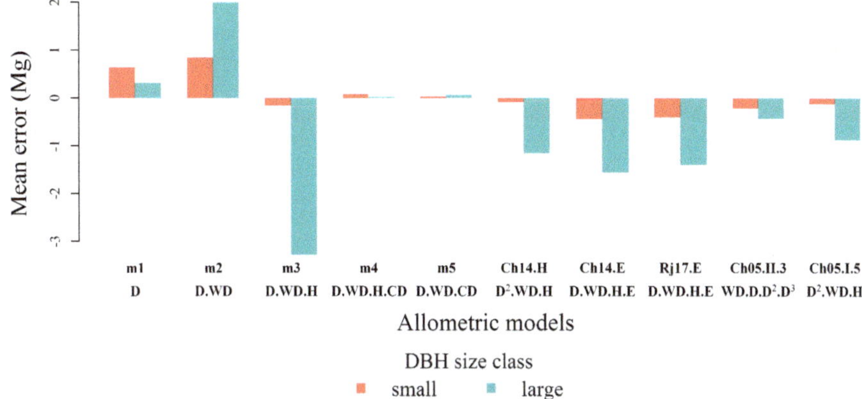

Figure A2. Mean error in estimates (estimated AGB minus reference AGB in Mg) by DBH size class: small trees (D $\leq$ 70 cm; n = 17) and large trees (D $>$ 70 cm; n = 9) for the TLS-derived allometric models and pantropical models.

References

1. Guyana Forestry Commission. *Terms of Reference for Developing Capacities for a National Monitoring, Reporting and Verification System to Support REDD+ Participation of Guyana. Background, Capacity Assessment and Roadmap*; Technical Report; Guyana Foresty Commission: Georgetown, TX, USA, 2009.

2. Butt, N.; Epps, K.; Overman, H.; Iwamura, T.; Fragoso, J.M. Assessing carbon stocks using indigenous peoples' field measurements in Amazonian Guyana. *For. Ecol. Manag.* **2015**, *338*, 191–199. [CrossRef]

3. Henry, M.; Cifuentes Jara, M.; Réjou-Méchain, M.; Piotto, D.; Michel Fuentes, J.M.; Wayson, C.; Alice Guier, F.; Castañeda Lombis, H.; Castellanos López, E.; Cuenca Lara, R.; et al. Recommendations for the use of tree models to estimate national forest biomass and assess their uncertainty. *Ann. For. Sci.* **2015**, *72*, 769–777. [CrossRef]

4. Alvarez, E.; Duque, A.; Saldarriaga, J.; Cabrera, K.; de las Salas, G.; del Valle, I.; Lema, A.; Moreno, F.; Orrego, S.; Rodríguez, L. Tree above-ground biomass allometries for carbon stocks estimation in the natural forests of Colombia. *For. Ecol. Manag.* **2012**, *267*, 297–308. [CrossRef]

5. Goodman, R.C.; Phillips, O.L.; Baker, T.R. The importance of crown dimensions to improve tropical tree biomass estimates. *Ecol. Appl.* **2014**, *24*, 680–698. [CrossRef] [PubMed]

6. Manuri, S.; Brack, C.; Nugroho, N.P.; Hergoualc'h, K.; Novita, N.; Dotzauer, H.; Verchot, L.; Putra, C.A.S.; Widyasari, E.; Hergoualc'h, K.; et al. Tree biomass equations for tropical peat swamp forest ecosystems in Indonesia. *For. Ecol. Manag.* **2014**, *334*, 241–253. [CrossRef]

7. Gibbs, H.K.; Brown, S.; Niles, J.O.; Foley, J.A. Monitoring and estimating tropical forest carbon stocks: Making REDD a reality. *Environ. Res. Lett.* **2007**, *2*, 045023. [CrossRef]

8. Zanne, A.E.; Lopez-Gonzalez, G.; Coomes, D.A.; Ilic, J.; Jansen, S.; Lewis, S.L.S.; Miller, R.B.; Swenson, N.G.; Wiemann, M.C.; Chave, J. Data from: Towards a Worldwide Wood Economics Spectrum. Available online: https://datadryad.org/bitstream/handle/10255/dryad.235/GlobalWoodDensityDatabase.xls?sequence=1 (accessed on 18 June 2019).

9. Chave, J.; Coomes, D.; Jansen, S.; Lewis, S.L.; Swenson, N.G.; Zanne, A.E. Towards a worldwide wood economics spectrum. *Ecol. Lett.* **2009**, *12*, 351–366. [CrossRef]

10. Feldpausch, T.R.; Lloyd, J.; Lewis, S.L.; Brienen, R.J.W.; Gloor, M.; Monteagudo Mendoza, A.; Lopez-Gonzalez, G.; Banin, L.; Abu Salim, K.; Affum-Baffoe, K.; et al. Tree height integrated into pantropical forest biomass estimates. *Biogeosciences* **2012**, *9*, 3381–3403. [CrossRef]

11. Chave, J.; Réjou-Méchain, M.; Búrquez, A.; Chidumayo, E.; Colgan, M.S.; Delitti, W.B.C.; Duque, A.; Eid, T.; Fearnside, P.M.; Goodman, R.C.; et al. Improved allometric models to estimate the aboveground biomass of tropical trees. *Glob. Chang. Biol.* **2014**, *20*, 3177–3190. [CrossRef]

12. Chave, J.; Andalo, C.; Brown, S.; Cairns, M.A.; Chambers, J.Q.; Eamus, D.; Fölster, H.; Fromard, F.; Higuchi, N.; Kira, T.; et al. Tree allometry and improved estimation of carbon stocks and balance in tropical forests. *Oecologia* **2005**, *145*, 87–99. [CrossRef]

13. Larjavaara, M.; Muller-Landau, H.C. Measuring tree height: A quantitative comparison of two common field methods in a moist tropical forest. *Methods Ecol. Evol.* **2013**, *4*, 793–801. [CrossRef]

14. Slik, J.W.F.; Paoli, G.; McGuire, K.; Amaral, I.; Barroso, J.; Bastian, M.; Blanc, L.; Bongers, F.; Boundja, P.; Clark, C.; et al. Large trees drive forest aboveground biomass variation in moist lowland forests across the tropics. *Glob. Ecol. Biogeogr.* **2013**, *22*, 1261–1271. [CrossRef]

15. Ploton, P.; Barbier, N.; Takoudjou Momo, S.; Réjou-Méchain, M.; Boyemba Bosela, F.; Chuyong, G.; Dauby, G.; Droissart, V.; Fayolle, A.; Goodman, R.C.; et al. Closing a gap in tropical forest biomass estimation: Taking crown mass variation into account in pantropical allometries. *Biogeosciences* **2016**, *13*, 1571–1585. [CrossRef]

16. Meyer, V.; Saatchi, S.; Clark, D.B.; Keller, M.; Vincent, G.; Ferraz, A.; Espírito-Santo, F.; Oliveira, M.V.N.; Kaki, D.; Chave, J.; et al. Canopy Area of Large Trees Explains Aboveground Biomass Variations across Nine Neotropical Forest Landscapes. *Biogeosci. Discuss.* **2018**, 1–38. [CrossRef]

17. Calders, K.; Newnham, G.G.; Burt, A.; Murphy, S.; Raumonen, P.; Herold, M.; Culvenor, D.; Avitabile, V.; Disney, M.; Armston, J.; et al. Nondestructive estimates of above-ground biomass using terrestrial laser scanning. *Methods Ecol. Evol.* **2015**, *6*, 198–208. [CrossRef]

18. Gonzalez de Tanago, J.; Lau, A.; Bartholomeus, H.; Herold, M.; Avitabile, V.; Raumonen, P.; Martius, C.; Goodman, R.C.; Disney, M.; Manuri, S.; et al. Estimation of above-ground biomass of large tropical trees with terrestrial LiDAR. *Methods Ecol. Evol.* **2018**, *9*, 223–234. [CrossRef]

19. Clark, D.B.; Kellner, J.R. Tropical forest biomass estimation and the fallacy of misplaced concreteness. *J. Veg. Sci.* **2012**, *23*, 1191–1196. [CrossRef]

20. Goodman, R.C.; Phillips, O.L.; Baker, T.R. Tightening up on tree carbon estimates. *Nature* **2012**, *491*, 527. [CrossRef]

21. Sheil, D.; Eastaugh, C.S.; Vlam, M.; Zuidema, P.A.; Groenendijk, P.; van der Sleen, P.; Jay, A.; Vanclay, J. Does biomass growth increase in the largest trees? Flaws, fallacies and alternative analyses. *Funct. Ecol.* **2017**, *31*, 568–581. [CrossRef]

22. Disney, M.I.; Boni Vicari, M.; Burt, A.; Calders, K.; Lewis, S.L.; Raumonen, P.; Wilkes, P. Weighing trees with lasers: Advances, challenges and opportunities. *Interface Focus* **2018**, *8*, 20170048. [CrossRef]

23. Wilkes, P.; Lau, A.; Disney, M.; Calders, K.; Burt, A.; Gonzalez de Tanago, J.; Bartholomeus, H.; Brede, B.; Herold, M. Data acquisition considerations for Terrestrial Laser Scanning of forest plots. *Remote Sens. Environ.* **2017**, *196*, 140–153. [CrossRef]

24. Malhi, Y.; Jackson, T.; Patrick Bentley, L.; Lau, A.; Shenkin, A.; Herold, M.; Calders, K.; Bartholomeus, H.; Disney, M.I. New perspectives on the ecology of tree structure and tree communities through terrestrial laser scanning. *Interface Focus* **2018**, *8*, 20170052. [CrossRef] [PubMed]

25. Newnham, G.G.; Armston, J.D.; Calders, K.; Disney, M.I.; Lovell, J.L.; Schaaf, C.B.; Strahler, A.H.; Danson, F.M. Terrestrial Laser Scanning for Plot-Scale Forest Measurement. *Curr. For. Rep.* **2015**, *1*, 239–251. [CrossRef]

26. Momo Takoudjou, S.; Ploton, P.; Sonké, B.; Hackenberg, J.; Griffon, S.; de Coligny, F.; Kamdem, N.G.; Libalah, M.; Mofack, G.I.I.; Le Moguédec, G.; et al. Using terrestrial laser scanning data to estimate large tropical trees biomass and calibrate allometric models: A comparison with traditional destructive approach. *Methods Ecol. Evol.* **2018**, *9*, 905–916. [CrossRef]

27. Abd Rahman, M.; Abu Bakar, M.; Razak, K.; Rasib, A.; Kanniah, K.; Wan Kadir, W.; Omar, H.; Faidi, A.; Kassim, A.; Abd Latif, Z. Non-Destructive, Laser-Based Individual Tree Aboveground Biomass Estimation in a Tropical Rainforest. *Forests* **2017**, *8*, 86. [CrossRef]

28. Paynter, I.; Genest, D.; Peri, F.; Schaaf, C. Bounding uncertainty in volumetric geometric models for terrestrial lidar observations of ecosystems. *Interface Focus* **2018**, *8*, 20170043. [CrossRef] [PubMed]

29. Stovall, A.E.; Anderson-Teixeira, K.J.; Shugart, H.H. Assessing terrestrial laser scanning for developing non-destructive biomass allometry. *For. Ecol. Manag.* **2018**, *427*, 217–229. [CrossRef]

30. Burt, A.; Disney, M.; Raumonen, P.; Armston, J.; Calders, K.; Lewis, P. Rapid characterisation of forest structure from TLS and 3D modelling. In Proceedings of the 2013 IEEE International Geoscience and Remote Sensing Symposium—IGARSS, Melbourne, Australia, 21–26 July 2013; pp. 3387–3390. [CrossRef]

31. Krooks, A.; Kaasalainen, S.; Kankare, V.; Joensuu, M.; Raumonen, P.; Kaasalainen, M. Tree structure vs. height from terrestrial laser scanning and quantitative structure models. *Silva Fenn.* **2014**, *48*, 1–11. [CrossRef]

32. Holopainen, M.; Vastaranta, M.; Kankare, V. Biomass estimation of individual trees using stem and crown diameter TLS measurements. *Int. Arch. Photogramm. Remote Sens. Spat. Inf. Sci.* **2011**, *XXXVIII*, 29–31. [CrossRef]

33. Raumonen, P.; Kaasalainen, M.; Åkerblom, M.; Kaasalainen, S.; Kaartinen, H.; Vastaranta, M.; Holopainen, M.; Disney, M.; Lewis, P. Fast Automatic Precision Tree Models from Terrestrial Laser Scanner Data. *Remote Sens.* **2013**, *5*, 491–520. [CrossRef]

34. Hackenberg, J.; Spiecker, H.; Calders, K.; Disney, M.; Raumonen, P. SimpleTree—An Efficient Open Source Tool to Build Tree Models from TLS Clouds. *Forests* **2015**, *6*, 4245–4294. [CrossRef]

35. Kaasalainen, S.; Krooks, A.; Liski, J.; Raumonen, P.; Kaartinen, H.; Kaasalainen, M.; Puttonen, E.; Anttila, K.; Mäkipää, R. Change detection of tree biomass with terrestrial laser scanning and quantitative structure modelling. *Remote Sens.* **2014**, *6*, 3906–3922. [CrossRef]

36. Åkerblom, M.; Raumonen, P.; Mäkipää, R.; Kaasalainen, M. Automatic tree species recognition with quantitative structure models. *Remote Sens. Environ.* **2017**, *191*, 1–12. [CrossRef]

37. Guyana Lands and Surveys Commission. *Guyana National Land Use Plan*; Number June; Guyana Lands and Surveys Commission: Georgetown, TX, USA, 2013; p. 174.

38. Muñoz, G.; Grieser, J. Climwat 2.0 for CROPWAT. Available online: http://www.fao.org/land-water/databases-and-software/climwat-for-cropwat/en/ (accessed on 18 June 2019).

39. Phillips, O.; Baker, T.; Feldpausch, T.; Brienen, R.; Almeida, S.; Arroyo, L.; Aymard, G.; Chave, J.; Cardozo, N.D.; Chao, K.J.; et al. RAINFOR Field Manual for Plot Establishment and Remeasurement. 2009. Available online: http://www.rainfor.org/upload/ManualsEnglish/RAINFOR_field_manual_version_June_2009_ENG.pdf (accessed on 18 June 2019).

40. Kitajima, K.; Mulkey, S.S.; Wright, S.J. Variation in Crown Light Utilization Characteristics among Tropical Canopy Trees. *Ann. Bot.* **2004**, *95*, 535–547. [CrossRef] [PubMed]

41. Williamson, G.B.; Wiemann, M.C. Measuring wood specific gravity...correctly. *Am. J. Bot.* **2010**, *97*, 519–524. [CrossRef] [PubMed]

42. Calders, K.; Burt, A.; Origo, N.; Disney, M.; Nightingale, M.; Raumonen, P.; Akerblom, M.; Lewis, P. Realistic Forest Stand Reconstruction from Terrestrial LiDAR for Radiative Transfer Modelling. *Remote Sens.* **2018**, *10*, 933. [CrossRef]

43. Vicari, M.B.; Disney, M.; Wilkes, P.; Burt, A.; Calders, K.; Woodgate, W. Leaf and wood classification framework for terrestrial LiDAR point clouds. *Methods Ecol. Evol.* **2019**, *10*, 680–694. [CrossRef]

44. Lin, L.I.K. A Concordance Correlation Coefficient to Evaluate Reproducibility. *Biometrics* **1989**, *45*, 255–268. [CrossRef]

45. R Core Team. *R: A Language and Environment for Statistical Computing*; R Foundation for Statistical Computing: Vienna, Austria, 2018.

46. Réjou-Méchain, M.; Tanguy, A.; Piponiot, C.; Chave, J.; Hérault, B. Biomass: An R package for estimating above-ground biomass and its uncertainty in tropical forests. *Method. Ecol. Evol.* **2017**, *8*, 1163–1167. [CrossRef]

47. Baskerville, G.L. Use of Logarithmic Regression in the Estimation of Plant Biomass. *Can. J. For. Res.* **1972**, *2*, 49–53. [CrossRef]

48. Jucker, T.; Caspersen, J.; Chave, J.; Antin, C.; Barbier, N.; Bongers, F.; Dalponte, M.; van Ewijk, K.Y.; Forrester, D.I.; Haeni, M.; et al. Allometric equations for integrating remote sensing imagery into forest monitoring programmes. *Glob. Chang. Biol.* **2017**, *23*, 177–190. [CrossRef] [PubMed]

49. Roșca, S.; Suomalainen, J.; Bartholomeus, H.; Herold, M. Comparing terrestrial laser scanning and unmanned aerial vehicle structure from motion to assess top of canopy structure in tropical forests. *Interface Focus* **2018**, *8*, 20170038. [CrossRef] [PubMed]

50. Hunter, M.; Keller, M.; Victoria, D.; Morton, D.C. Tree height and tropical forest biomass estimation. *Biogeosciences* **2013**, *10*, 8385–8399. [CrossRef]

51. Brede, B.; Lau, A.; Bartholomeus, H.M.; Kooistra, L. Comparing RIEGL RiCOPTER UAV LiDAR derived canopy height and DBH with terrestrial LiDAR. *Sensors* **2017**, *17*, 2371. [CrossRef] [PubMed]

52. Kuyah, S.; Dietz, J.; Muthuri, C.; Jamnadass, R.; Mwangi, P.; Coe, R.; Neufeldt, H. Allometric equations for estimating biomass in agricultural landscapes: I. Aboveground biomass. *Agric. Ecosyst. Environ.* **2012**, *158*, 216–224. [CrossRef]

53. Oliveira, A.A.D.E.; Mori, S.A.; De Oliveira, A.A.; Mori, S.A. A central Amazonian terra firme forest. I. High tree species richness on poor soils. *Biodivers. Conserv.* **1999**, *8*, 1219–1244.:1008908615271. [CrossRef]

54. Lau, A.; Bentley, L.P.; Martius, C.; Shenkin, A.; Bartholomeus, H.; Raumonen, P.; Malhi, Y.; Jackson, T.; Herold, M. Quantifying branch architecture of tropical trees using terrestrial LiDAR and 3D modelling. *Trees Struct. Funct.* **2018**, *32*, 1219–1231. [CrossRef]

55. Raumonen, P.; Casella, E.; Calders, K.; Murphy, S.; Åkerblom, M.; Kaasalainen, M. Massive-scale tree modelling from tls data. *ISPRS Ann. Photogramm. Remote Sens. Spat. Inf. Sci.* **2015**, *II-3/W4*, 189–196. [CrossRef]

56. Burt, A. New 3D Measurements of Forest Structure. Ph.D. Thesis, University College London, London, UK, 2017.

57. Disney, M.I.; Burt, A.; Calders, K.; Schaaf, C.; Stovall, A. Innovations in ground and airborne technologies as reference and for training and validation: Terrestrial Laser Scanning (TLS). *Surv. Geophys.* **2019**. [CrossRef]

© 2019 by the authors. Licensee MDPI, Basel, Switzerland. This article is an open access article distributed under the terms and conditions of the Creative Commons Attribution (CC BY) license (http://creativecommons.org/licenses/by/4.0/).

 *forests*

Article

Estimation of *Pinus massoniana* Leaf Area Using Terrestrial Laser Scanning

Yangbo Deng, Kunyong Yu *, Xiong Yao, Qiaoya Xie, Yita Hsieh and Jian Liu

College of Forestry, Fujian Agriculture and Forestry University, Fujian 350002, China
* Correspondence: yuyky@fafu.edu.cn; Tel.: +86-8639221

Received: 11 June 2019; Accepted: 1 August 2019; Published: 6 August 2019

Abstract: The accurate estimation of leaf area is of great importance for the acquisition of information on the forest canopy structure. Currently, direct harvesting is used to obtain leaf area; however, it is difficult to quickly and effectively extract the leaf area of a forest. Although remote sensing technology can obtain leaf area by using a wide range of leaf area estimates, such technology cannot accurately estimate leaf area at small spatial scales. The purpose of this study is to examine the use of terrestrial laser scanning data to achieve a fast, accurate, and non-destructive estimation of individual tree leaf area. We use terrestrial laser scanning data to obtain 3D point cloud data for individual tree canopies of *Pinus massoniana*. Using voxel conversion, we develop a model for the number of voxels and canopy leaf area and then apply it to the 3D data. The results show significant positive correlations between reference leaf area and mass ($R^2 = 0.8603$; $p < 0.01$). Our findings demonstrate that using terrestrial laser point cloud data with a layer thickness of 0.1 m and voxel size of 0.05 m can effectively improve leaf area estimations. We verify the suitability of the voxel-based method for estimating the leaf area of *P. massoniana* and confirmed the effectiveness of this non-destructive method.

Keywords: *Pinus massoniana*; specific leaf area; leaf area; terrestrial laser scanning; voxelization; forest canopy

---

1. Introduction

As an important organ of trees, leaves play a substantial role in plant photosynthesis, transpiration, and many other physiological activities [1–3]. The leaf area (LA) is an important parameter for expressing the amount of leaves in a tree canopy, and is an important measurement for understanding the growth, development, productivity, and physiology of plants [4,5]. Evaluations of leaf traits at the leaf are receiving more attention in forest ecology and remote sensing studies [6,7], LA directly affects the accumulation of plant dry matter and also directly determines the interception capacity and utilization rate of light energy, as well as changes in transpiration rates. Another indicator related to LA is the specific leaf area (SLA), which refers to the fresh leaf surface area per unit mass of dry matter; its value is directly affected by leaf thickness, shape, and quality. Many ecosystem process models utilize plant SLA as an important input parameter [8]. Therefore, rapid and non-destructive acquisition of parameters is very important for the estimation of the stand structure and the quantification of stand quality.

Currently, commonly used leaf area measurement methods include direct methods and indirect methods [9–11]. Many direct methods require the excision of leaves from plants; this method is labor-intensive and limited in the scope of its application. Indirect measurement methods primarily use a variety of instruments and software to obtain leaf area measurements, such as portable scanning planimeters, hand scanners, laser optic apparatuses, and image analysis software. LA can be obtained by a variety of instruments and software [12]. Most terrestrial laser scanners (TLSs) cannot separate branches and leaves directly. Methods using TLSs are also time consuming, expensive, complex, and only suitable for a few specific species of plants [13]. Especially when studying species with non-flattened blades (e.g., coniferous species, such as *P. thunbergii* and *Pinus massoniana*), such methods result in poor comparability, owing to differences in the understanding of principles of leaf interception [14]. Therefore, new instruments and methods to separate individual trees from the other trees must be explored [6,15].

Recently, many studies have estimated LA by terrestrial light detection and ranging (voxeli). LiDAR is an active remote sensing technology that records the details of three-dimensional information by acquiring the three-dimensional coordinate data and digital images of the research target [16], thereby providing an opportunity to extract the 3D geometry of an individual tree. The estimation of tree characteristics such as height and diameter at breast height (DBH), has been widely used to fully compensate for the limitations of traditional optical remote sensing monitoring in the vertical structure of forest canopy [17,18]. TLSs are also used in measuring vegetation structure information. TLSs are lightweight and portable, have high laser resolution and are safe. Most TLSs use a level 1 laser, which does not harm the human eye. The tree canopy structure is quickly and accurately measured by a pulsed laser in a non-contact manner, thereby obtaining a large-area, complex, irregular forest point cloud data [19,20]. Another advantage of TLSs is their capability to separate a target tree from other trees using its unique distance information. There are three methods for estimating the leaf area of individual trees: regression analysis, gap-based probability, and voxel method.

Regression analysis is mainly based on TLSs to obtain structural parameters of a single tree—such as crown width, breast diameter, tree height—to establish a regression equation of the leaf area [21]. This method is more labor-saving than field measurement and is conducive to data preservation and extraction. There is an error in the extraction process of the forest structure parameters, and this error tends to have an uncertain effect on the subsequent regression equations.

Based on the gap probability and the law of angular distribution, the LAI of the individual tree is derived, and the influence of the relevant parameters on the LAI estimation is analyzed [22], but the trunk effect in the interference probability model is not eliminated.

The voxel-based 3D modeling method has been used for years in the fields of scientific computation and medical imaging. This method is convenient for creating architectural models and for 3D imaging. It is, in fact, a method based on regression models. In forestry research, voxel-based 3D modeling is used to estimate LAI and leaf area density (LAD) by directly counting the contact frequency of each layer of the studied tree [23,24]. However, this method is primarily applied to broad-leaved tree species, and further study is required to adapt this method to conifer species and obtain the leaf areas of conifer needles.

Masson's pine (*Pinus massoniana* Lamb.) is one of the most important tree species in southern China. Because of its high adaptability to drought and barren soils and its capacity to retain water and nutrients, *P. massoniana* has been widely planted in China [25,26]. LA usually represents the quality of a tree [27–29]. However, because of the non-flattened blades of *P. massoniana*, such methods result in poor comparability, owing to differences in the understanding of principles of leaf interception [14].

This study combines different horizontal layers and different voxel sizes to estimate the LA of *P. massoniana* based on the 3D voxel method. Here, 'horizontal layer' indicates the bottom layer of the canopy. The purpose of this study was to: (1) construct a conifer tree LA estimation method based on ground-based remote sensing data; (2) study the optimal voxel size and model in the LA estimation process; and (3) study the LA estimation under the optimal stratification height.

2. Materials and Methods

Figure 1 illustrates the developed LA estimation workflow, including the extraction of the reference LA and the estimation of the LA. The specific steps are as follows:

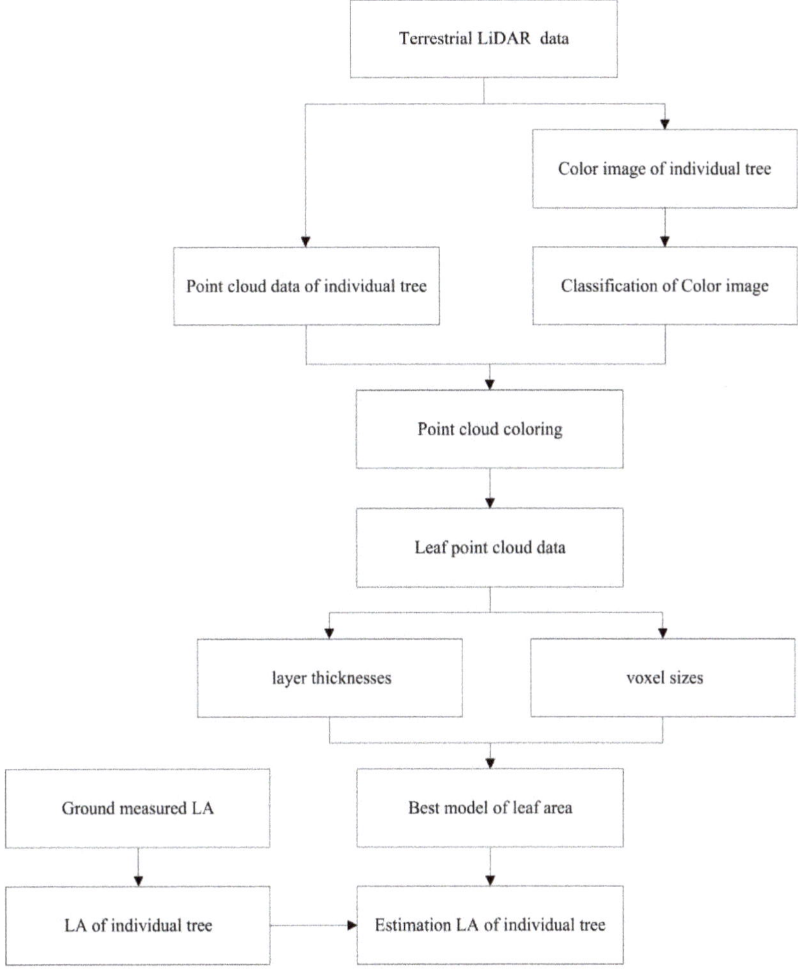

Figure 1. Flowchart of leaf area (LA) estimation.

2.1. Site Characteristics

This study site was located in Hetian Town (25°33′–25°48′ N, 116°18′–116°31′ E), in Fujian Province, China. The site is dominated by *P. massoniana* plantations. The site is characterized by low hills and the average elevation is ~310 m above sea level [30]. The mean annual air temperature is 19 °C, and the annual precipitation is 1621 mm (occurs mainly from April to June) [31].

2.2. Specific Leaf Area Acquisition

A total of 26 canopy samples of *P. massoniana* were selected for sampling. Samples of 10 needles, with no signs of disease and of the same color, were collected from the upper, middle, and lower layers of the canopy. We first used the YMJ-C Digital Leaf Area Meter (HINOTEK, Hangzhou City, China) system to scan the coniferous area, and then loaded the needles into numbered envelopes and brought them back to the laboratory to dry to a constant weight. Then, each sample was weighed and a single mean dry weight was calculated [32]. The specific leaf area of *P. massoniana* was obtained by the least-squares method [4].

$$\text{SLA} = \frac{\sum_{i=1}^{n} LA_i * X_i}{\sum_{i=1}^{n} X_i^2} \tag{1}$$

where n is the number of leaves tested, LA_i is the leaf area of the *ith* leaf, and X_i is the dry mass of the *ith* leaf.

2.3. Point Cloud Data Acquisition

We obtained point cloud data for 26 *P. massoniana* samples using a Stonex X300 laser scanner (Italy), which is a pulsed-static 3D laser scanner. The configuration of the Stonex X300 is shown in Table 1. In order to avoid the influence of light intensity and weather on measurement error, we selected three different angles (Figure 2) [23], corresponding to (1) an instrument shooting angle of 220°, (2) a scanning mode set to fast, and (3) a horizontal field of view of the scanning area with partial station overlap, such that any two stations could be found between multiple points of the same name with a vertical field of view of −25° to 65°. The scan resolution accuracy was <6 mm. Three plastic balls, with a diameter of 20 cm each, were placed next to each sample tree and on the top of a tripod to serve as a fixed target, ensuring that each station could scan at least two targets and match the three stations through the target using the original point cloud data.

Table 1. Configuration of the STONEX X300 laser scanner

Model	STONEX X300
Measuring range	2–300 m (100% Reflectivity)
Visual range	Level 360° (Panoramic view) vertical 90° (−25° to +65°)
Accuracy	<6 mm (50 m) <40 mm (300 m)
Scanning speed	>40,000 points/s
Scan resolution	0.37 mrad
Data storage	32 GB

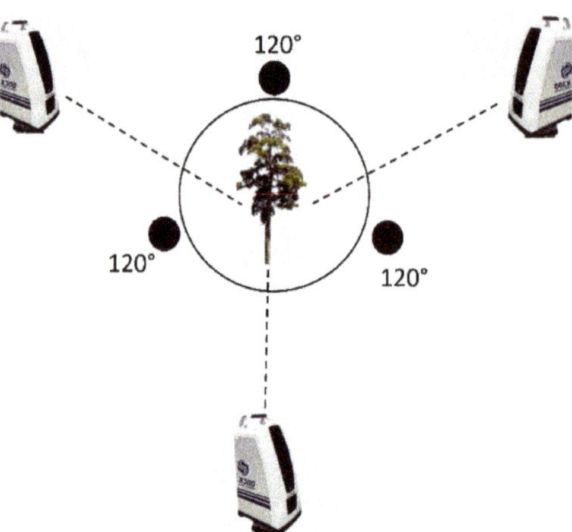

Figure 2. Instrument station locations for individual trees.

2.4. Point Cloud Data Processing

In this study, the area of each layer was calculated according to the edge length, *L*. Choosing a suitable voxel size solved the difference in the number of point clouds in voxels:

$$\begin{cases} a = x_{min} + \left(\frac{int(x-x_{min})}{L} \right) \times L \\ b = y_{min} + \left(\frac{int(y-y_{min})}{L} \right) \times L \\ c = z_{min} + \left(\frac{int(z-z_{min})}{L} \right) \times L \end{cases} \tag{2}$$

where *a*, *b*, and *c* are the coordinates in the voxel array, *int* is a function for rounding off the coordinates to one decimal place to the nearest integer; *x*, *y*, and *z* represent the point coordinates of the registered LiDAR data; x_{min}, y_{min}, and z_{min} are the minimum values of *x*, *y*, and *z*, and *L* represents the voxel element size [33,34]. In this article, the voxel type selected is the filled voxels that contain a group of points [35,36].

First, we layered the point cloud data according to different layer thicknesses and subsequently calculated the number of voxels under different layer thicknesses. Through layering treatment, the overlap between the blades was effectively reduced, so that the number of transformed voxels was more representative of the blade area. The rounding algorithm is shown in Figure 3.

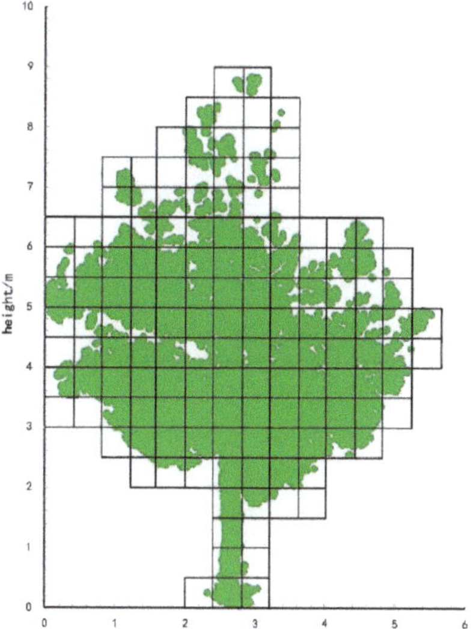

Figure 3. Point cloud data voxelization.

2.5. Point Cloud Data Extraction

Firstly, the original point cloud data obtained by TLS Stonex X300 is used to convert the original point cloud data, and the true color photos are extracted. The leaves, branches and other selections in the original color image are used as training samples to unify other training samples, and then use the maximum likelihood method to classify the image. In order to make the color clear after classification, this article sets the canopy leaves to red, the branches and trunks to green, and the other training samples to white (Figure 4).

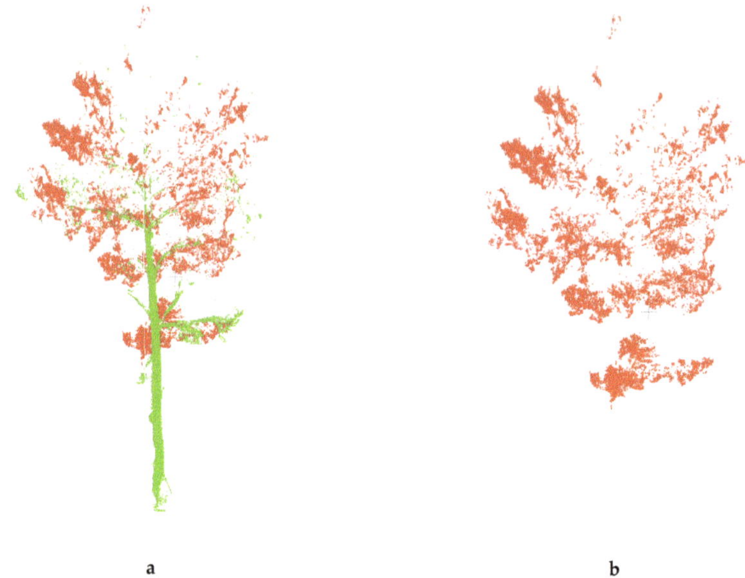

a   b

Figure 4. (**a**) Point cloud data of tree; (**b**) Point cloud data of leaf.

2.6. Model Validation

In order to develop models, the data from 26 individual trees were first classified into two independent subsets for model establishment and model validation. A subset of 18 individual trees was used for model establishment, and the data of five individual trees were used for model validation. With the subset of 18 individual trees used for modelling, the correlations between leaf area and leaf mass were first analyzed, and then single-variable models were established. Single-variable models contained linear, quadratic, and exponential forms, using each of the 10 layer and 10 voxel sizes. The models were validated according to the root mean square error (*RMSE*) and mean relative error (*MRE*) using Equations (3) and (4), respectively:

$$RMSE = \left(\frac{1}{n} \sum_{i=1}^{n} (E_i - M_i)^2 \right)^{\frac{1}{2}} \tag{3}$$

$$MRE = \frac{1}{n} \sum_{i=1}^{n} \frac{\text{abs}(E_i - M_i)}{M_i} \times 100\% \tag{4}$$

Statistical analyses were performed using SPSS version 12.0 (Amos Development Corporation, Chicago, IL, USA).

3. Results

3.1. Specific Leaf Area of Pinus massoniana

The dataset was divided into training data and test data by the random selection of 30% of the total data set as testing data and 70% as training data. The reference leaf mass, reference LA, and SLA of the training data are shown in Table 2. The reference values for leaf mass and LA were obtained as the average of 10 values. The correlations between the reference leaf mass and the reference LA are shown in Figure 5.

Table 2. Leaf parameters

	Sample Number								
	1-1	**1-2**	**1-3**	**1-4**	**1-5**	**1-6**	**1-7**	**1-8**	**1-9**
Reference leaf mass (g)	0.038	0.057	0.032	0.071	0.038	0.046	0.048	0.038	0.052
Reference leaf area (cm^2)	1.528	1.743	1.187	2.002	1.269	1.459	1.592	1.367	1.800
SLA (cm^2/g)	39.705	30.397	37.390	28.142	33.130	31.463	33.394	35.891	34.810
	2-1	**2-2**	**2-3**	**2-4**	**2-5**	**2-6**	**2-7**	**2-8**	**2-9**
Reference leaf mass (g)	0.049	0.035	0.039	0.085	0.063	0.040	0.052	0.036	0.047
Reference leaf area (cm^2)	1.769	1.226	1.535	2.354	1.930	1.653	1.677	1.209	1.774
SLA (cm^2/g)	36.464	35.035	39.196	27.570	30.623	41.159	32.169	33.268	37.365

Abbreviations: Specific leaf area, SLA.

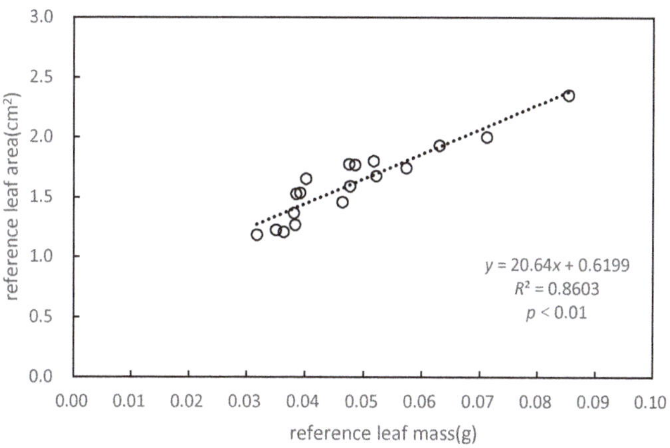

Figure 5. Relationship between reference leaf area (LA) and reference leaf mass.

To obtain a more accurate LAI, we calculated the LA of individual trees based on the SLAs of the upper, middle, and lower parts of *P. massoniana* (Figure 5). Significant correlations were found between reference LA and reference leaf mass, with a coefficient of determination (R^2) of 0.8603. LA was positively correlated to leaf mass ($p < 0.01$). The SLA of *P. massoniana* was obtained by the least-squares method (Table 1).

SLA may be affected by light, temperature, nutritional status, leaf age, and other factors. It can be seen that the average dry weight of *P. massoniana* did not differ much, but the difference in the dry weight of leaves with different leaf ages was larger, indicating that there is a certain difference in dry matter accumulation between new leaves and old leaves. Among these, the maximum LA was 2.354 (sample 2-4), the minimum LA was 1.187 (sample 1-3), the maximum leaf mass was 0.085 (sample 2-4), and the minimum LA was 0.032 (sample 1-3). The maximum SLA was 41.159 (sample 2-6) and the minimum SLA was 27.57 (sample 2-4).

3.2. Leaf Area Estimation at Different Scales

In order to reduce the influence of terrain and scanning distance on the data volume of point clouds in the process of using 3D laser scanners in the field, and to establish a more accurate relationship between point cloud data and LA, a voxel conversion method was adopted in this study based on different voxel size to establish a LA estimation model. Based on the subset of 18 data points, a total of 300 single-variable models were established for each of the 10 needle types, and the R^2 values of these models are shown in Table 3. The largest R^2 was for the quadratic model between 0.1 m and 0.08 m ($R^2 = 0.886$), the lowest was for the exponential model between 0.1 m and 0.01 m ($R^2 = 0.4$), and the average of all single-variable models was $R^2 = 0.757$. The average R^2 values of each of the 75 single-variable models were ranked as: quadratic (0.811) > linear (0.799) > exponential (0.661). As for the layered levels, the average R^2 increased with increasing layer thickness, as 1 m (0.788) > 0.9 m (0.771) > 0.8 m (0.763) > 0.7 m (0.77) > 0.6 m (0.767) > 0.5 m (0.756) > 0.4 m (0.755) > 0.3 m (0.754) > 0.2 m (0.738) > 0.1 m (0.709). It is notable that the average R^2 of the models based on 0.01 m was the lowest. This indicates that the voxel value was too small to reduce the influence of the 3D laser scanning distance on the point cloud density. A 'best' single-variable model was selected for each of the layer types and voxel sizes according to the R^2 values and model stabilities (Table 3). The independent variables of these selected models contained 10 voxel sizes for different layers, and most were quadratic.

Table 3. Coefficient of determination (R^2) for LA estimates.

Layers (m)	0.1	0.2	0.3	0.4	0.5	0.6	0.7	0.8	0.9	1
0.01 (m)										
L	0.532	0.575	0.59	0.594	0.603	0.603	0.606	0.609	0.61	0.63
E	0.4	0.428	0.44	0.444	0.454	0.452	0.457	0.459	0.459	0.481
Q	0.649	0.654	0.659	0.659	0.661	0.662	0.661	0.663	0.662	0.673
0.02 (m)										
L	0.652	0.686	0.707	0.71	0.717	0.72	0.723	0.727	0.729	0.753
E	0.497	0.528	0.548	0.551	0.561	0.561	0.569	0.57	0.571	0.599
Q	0.69	0.71	0.725	0.728	0.732	0.735	0.736	0.74	0.74	0.761
0.03 (m)										
L	0.708	0.746	0.765	0.768	0.773	0.777	0.781	0.782	0.786	0.809
E	0.548	0.586	0.606	0.61	0.62	0.622	0.631	0.629	0.631	0.662
Q	0.729	0.757	0.773	0.776	0.781	0.785	0.787	0.788	0.791	0.812
0.04 (m)										
L	0.749	0.785	0.804	0.806	0.611	0.817	0.821	0.817	0.823	0.844
E	0.587	0.625	0.646	0.652	0.662	0.665	0.676	0.67	0.673	0.703
Q	0.76	0.79	0.807	0.809	0.815	0.82	0.824	0.82	0.825	0.845
0.05 (m)										
L	0.755	0.793	0.81	0.812	0.832	0.838	0.841	0.836	0.844	0.847
E	0.583	0.624	0.645	0.647	0.691	0.693	0.704	0.698	0.701	0.715
Q	0.767	0.798	0.813	0.814	0.835	0.84	0.843	0.839	0.845	0.849
0.06 (m)										
L	0.8	0.831	0.844	0.844	0.847	0.854	0.847	0.845	0.859	0.871
E	0.641	0.68	0.7	0.702	0.712	0.716	0.705	0.706	0.723	0.749
Q	0.804	0.833	0.846	0.846	0.849	0.856	0.848	0.847	0.86	0.87
0.07 (m)										
L	0.815	0.845	0.857	0.856	0.857	0.862	0.865	0.858	0.865	0.877
E	0.659	0.698	0.715	0.718	0.728	0.732	0.742	0.733	0.735	0.763
Q	0.819	0.847	0.858	0.857	0.858	0.864	0.866	0.86	0.866	0.878
0.08 (m)										
L	0.834	0.851	0.869	0.865	0.868	0.873	0.876	0.869	0.877	0.886
E	0.681	0.706	0.732	0.732	0.747	0.749	0.762	0.75	0.75	0.779
Q	0.837	0.852	0.87	0.866	0.869	0.875	0.877	0.871	0.878	0.886
0.09 (m)										
L	0.82	0.862	0.871	0.868	0.868	0.873	0.873	0.868	0.876	0.878
E	0.689	0.722	0.741	0.74	0.75	0.755	0.764	0.755	0.757	0.778
Q	0.829	0.863	0.872	0.869	0.869	0.875	0.874	0.869	0.877	0.878
0.1 (m)										
L	0.859	0.87	0.875	0.873	0.873	0.881	0.878	0.829	0.879	0.881
E	0.71	0.738	0.761	0.755	0.764	0.768	0.775	0.739	0.764	0.788
Q	0.861	0.871	0.877	0.874	0.873	0.882	0.879	0.835	0.88	0.882

Model types: linear, L; exponential, E; quadratic, Q.

3.3. Model Validation

Using the 10 selected models (Equations (5–14) in Table 4), LA were calculated for different voxel sizes with the eight sets of independent validation data. For the 10 voxel sizes, the maximum, minimum, and mean RMSE of the multivariate models were 13.36, 1.94, and 5.75 (Figure 6), respectively. At first, as the scale factor increased, and the scatterplot of the model was more closely distributed along the 1:1 line (Figure 7). When the scale factor was greater than 0.5, the scatterplot of the model was more sparsely distributed along the 1:1 line. Among the 100 models established, the voxel size of 0.05 models for the 10 types performed best (i.e., the models based on a layer thickness of 1 m performed better than those based on a layer thickness from one radiometric correction image). This indicates that the ability to utilize 3D laser point cloud data is well-grounded at a layer thickness of 0.1 m and a voxel size of 0.05 to improve LA estimation [23].

Table 4. Selected single-variable models for estimating LA.

Voxel	Model	R^2
0.01	$y = 1 \times 10^{-8}x^2 - 0.0005x + 35.498$	0.673
0.02	$y = 3 \times 10^{-8}x^2 - 0.0009x + 15.204$	0.761
0.03	$y = 4 \times 10^{-8}x^2 - 0.0024x + 9.6254$	0.812
0.04	$y = 5 \times 10^{-8}x^2 - 0.0042x + 6.4524$	0.845
0.05	$y = 1 \times 10^{-7}x^2 - 0.0055x + 7.1991$	0.849
0.06	$y = 0.0088x + 2.8488$	0.871
0.07	$y = 2 \times 10^{-7}x^2 - 0.0093x + 5.5129$	0.878
0.08	$y = 2 \times 10^{-7}x^2 - 0.0115x + 4.8836$	0.886
0.09	$y = 3 \times 10^{-7}x^2 - 0.0133x + 5.536$	0.878
0.1	$y = 4 \times 10^{-7}x^2 - 0.0156x + 5.2663$	0.882

Note: y = leaf area; x = number of voxels.

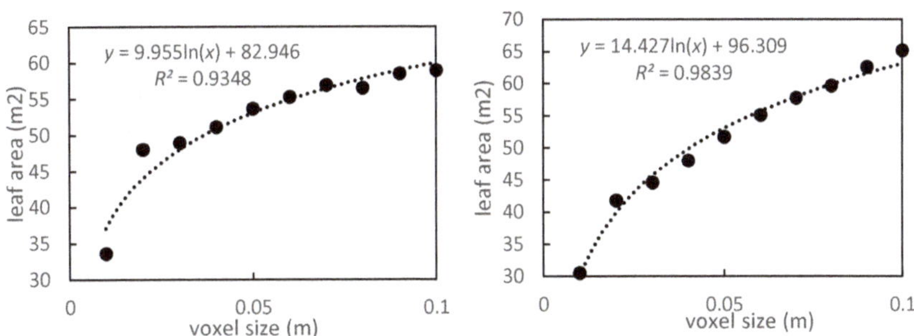

Figure 6. LA of *P. massoniana* trees for different voxel sizes.

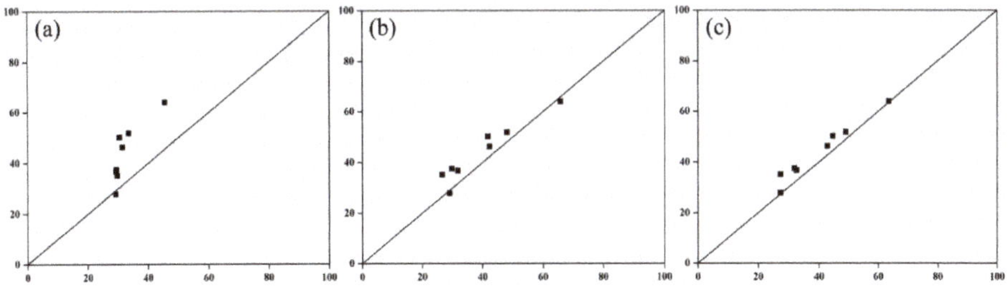

Figure 7. *Cont.*

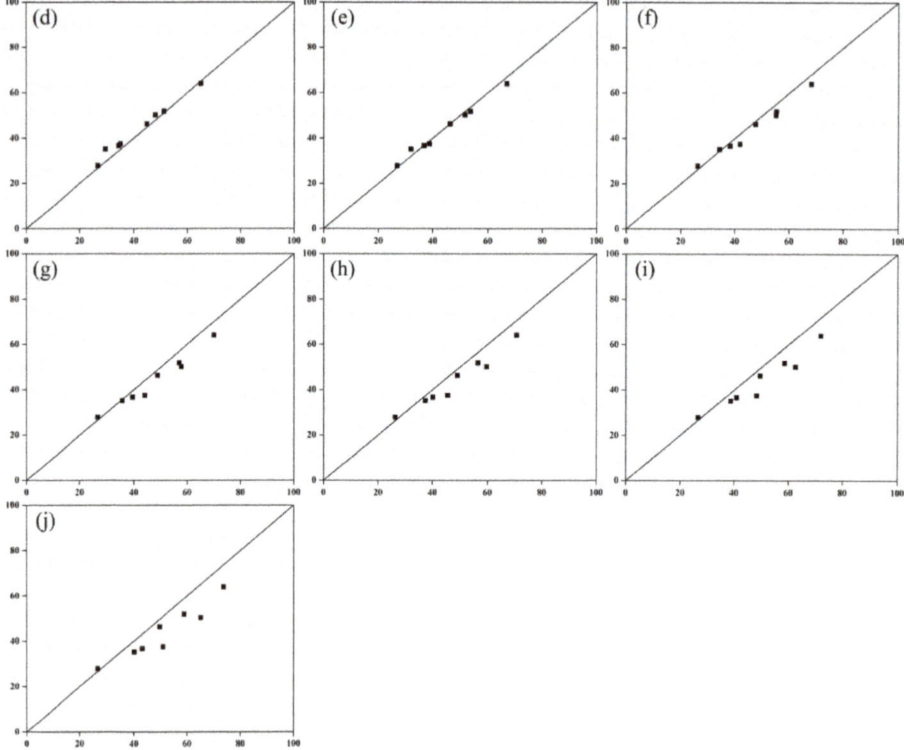

Figure 7. Comparison between field measured leaf area and model estimated leaf area. The voxel sizes are: (**a**) 0.01; (**b**) 0.02; (**c**) 0.03; (**d**) 0.04; (**e**) 0.05; (**f**) 0.06; (**g**) 0.07; (**h**) 0.08; (**i**) 0.09; (**j**) 0.1.

4. Discussion

4.1. Measurement of Specific Leaf Area

Using leaf length and leaf width to establish LA is not feasible, and there are better measurements using leaf weight [37,38]. Furthermore, it is found that the blade quality has a better correlation with the blade area [19,39]. In order to improve the efficiency and obtain a more accurate canopy LA, a method of partially replacing the whole had been used in this study, and the canopy is divided into upper, middle, and lower parts to collect leaves. To ensure the consistency of the canopy structure and reduce the scanning error of the terrestrial LiDAR to the height of different canopy layers of *P. massoniana*, the *P. massoniana* canopy was divided into upper, middle, and lower parts, and the LA of each layer was calculated and subsequently used as the overall LA. However, the obtained leaf mass and LA still contain errors (Figure 5). This is due to the peculiarity of the leaf of the needle of the *P. massoniana*, and it is impossible to accurately measure the LA [12]. Significant correlations were found between the reference LA and reference leaf mass (Figure 1) [39–41]. We found that this rule also applies to *P. massoniana* (R^2 = 0.8603). The SLA of *P. massoniana* was obtained by the least-squares method. Based on the prediction of a linear model, estimates of LA and leaf quality [42] have been obtained, and the correlation is very high, suggesting that the model is still applicable to the LA of *P. massoniana*. Compared with the new leaves, the organic matter of the old leaves is more developed, and old leaves exhibit a higher dry matter content, with smaller LA [39,40].

4.2. Effects of Different Voxel Sizes

There is a clear linear relationship between LA and leaf mass. Many researchers have developed multiple regression models based on leaf width and leaf length [37,43]. However, the LA acquisition of *P. massoniana* remains very difficult. Therefore, in this study, TLSs were selected to estimate the leaf area of *P. massoniana*. In order to improve the spatial coverage and lessen the effects of occlusion, complete point cloud data were obtained by multiple scans. However, moving TLS parts (including supporting bases and reference targets) for multiple scans is labor-intensive and time consuming. Different scanning distances produce different numbers of point cloud data. The closer the scanning distance is, the larger the number of generated point cloud data becomes. Due to the influence of terrain and trees, it is often impossible to ensure the TLS scanning distance in the field. Therefore, we selected the voxel method to convert point cloud into voxel to improve the accuracy of leaf area estimation [44].

Different voxel sizes may affect the estimation of LA. The estimated LA was positively related to the voxel size, which was also related to the algorithm operating efficiency [34]. The smaller the voxel is, the larger the number of data is, and the slower the calculation becomes. If the voxel size is too large, there will be many blanks in the process of converting the point cloud into voxels. These blanks can then be calculated as the leaf area, increasing the estimated LA. The LA of *P. massoniana* increased with increase in voxel size. When the voxel value increases to 0.1 m, the estimated R^2 of the model reaches 0.886 (Table 4). The highest R^2 of the tree was 0.939. Therefore, choosing the appropriate voxel size was beneficial for reducing the leaf area estimation error. Our findings demonstrate that using terrestrial laser point cloud data with a layer thickness of 0.1 m and voxel size of 0.05 m effectively improved leaf area estimations [34,45].

Here, we show that a voxel value of 0.5 is ideal because it can reduce the impact of scanning distance on the establishment of a point cloud-LA model and can also avoid excessive calculations while providing optimal LA estimation results [25]. The purpose of voxelization is to reduce the influence of distance on the density of cloud points in the field during 3D scanning. If the voxel value is too small, it cannot reduce the impact of distance, and if it is too large, the point cloud density is too sparse, reducing the impact of point cloud density too much [46].

5. Conclusions

In this study, we investigated 26 *P. massoniana* trees. We determined the relationship between the number of voxels and LA, we used voxel size and layer thickness as the influencing factors for constructing the leaf area estimation model. The LA of *P. massoniana* was estimated and modeled, and the reference LA was used for model validation.

The results showed that it is feasible to extract *P. massoniana* LA based on terrestrial 3D laser techniques. Larger voxel values result in a greater delamination density and higher estimation accuracy. These findings highlight the feasibility of non-destructive acquisition of single LAs of *P. massoniana* based on TLS data. When the voxel value is 0.05, the optimal layer size is 1 m, and the best estimate model is the quadratic model. Using three stations for scanning, a complete canopy LA can be obtained and fully realized. Methods for estimation of LA are needed, and these findings will help develop and contribute to development of efficient TLS applications for forest inventories. In addition, the voxel-based 3D modeling method involves only a regression model. This study only applied the voxel-based modeling method to the estimation of the leaf area of *P. massoniana*. This method was not applied to other species It may only be effective for this species, and it can be used in future research of other species.

Author Contributions: Y.D and K.Y. contributed equally to the development of the project, J.L. and K.Y.; Methodology, Y.D.; Validation, J.L., K.Y. and Y.D.; Formal analysis, X.Y.; Investigation, Q.X.; Resources, Y.H.; Data curation, X.Y.; Writing—original draft preparation, Y.D.; Writing—review and editing, J.L.; Visualization, Y.D.; Supervision, Y.H.; Project administration, K.Y.; Funding acquisition, K.Y.

Funding: This research was funded by remote sensing integrated monitoring technology for bamboo resources (grant number 2018YFD060010304), Study for remote sensing quantitative simulation of stand vertical structure rainfall reducing cavitation function (grant number 31770760).

Acknowledgments: The authors gratefully acknowledge University Key Lab for Geomatics Technology & Optimize Resources Utilization in Fujian Province.

Conflicts of Interest: The authors declare no conflict of interest.

References

1. Chen, J.M.; Leblanc, S.G. A Four-scale bidirectional reflectance model based on canopy architecture. *IEEE Trans. Geosci. Remote Sens.* **1997**, *35*, 1316–1337. [CrossRef]
2. Sepaskhah, A.R.; Amini-Nejad, M.; Kamgar-Haghighi, A.A. Developing a dynamic yield and growth model for saffron under different irrigation regimes. *Int. J. Plant Prod.* **2013**, *7*, 473–504.
3. Shabani, A.; Sepaskhah, A.R.; Kamkar-Haghighi, A.A. A model to predict the dry matter and yield of rapeseed under salinity and deficit irrigation. *Arch. Agron. Soil Sci.* **2014**, *61*, 525–542. [CrossRef]
4. Diao, J.; Guo, H.; Lu, J.; Lei, X.D.; Tang, S.Z. Leaf area estimation model and specific leaf area of Chinese pine. *For. Res.* **2013**, *26*, 174–180. (In Chinese)
5. Baker, S.C.; Halpern, C.B.; Wardlaw, T.J.; Crawford, R.L.; Bigley, R.E.; Edgar, G.J.; Evans, S.A.; Franklin, J.F.; Jordan, G.J.; Karpievitch, Y.; et al. Short- and long-term benefits for forest biodiversity of retaining unlogged patches in harvested areas. *For. Ecol. Manag.* **2015**, *353*, 187–195. [CrossRef]
6. Weraduwage, S.M.; Chen, J.; Anozie, F.C.; Morales, A.; Weise, S.E.; Sharkey, T.D. The relationship between leaf area growth and biomass accumulation in Arabidopsis thaliana. *Front. Plant Sci.* **2015**, *6*, 167. [CrossRef] [PubMed]
7. Alton, P.B. The sensitivity of models of gross primary productivity to meteorological and leaf area forcing: A comparison between a Penman–Monteith ecophysiological approach and the MODIS Light-Use Efficiency algorithm. *Agric. For. Meteorol.* **2016**, *218*, 11–24. [CrossRef]
8. Landsberg, J.; Gower, S. Canopy Architecture and Microclimate. *Appl. Physiol. Ecol. For. Manag.* **1997**, 51–88.
9. Sepaskhah, A.R. Estimation of individual and total leaf area of safflowers. *Agron. J.* **1977**, *69*, 783–785. [CrossRef]
10. Shabani, A.; Sepaskhah, A.R.; Kamkar-Haghighi, A.A. Growth and physiologic response of rapeseed (*Brassica Napus* L.) to deficit irrigation, water salinity and planting method. *Int. J. Plant Prod.* **2013**, *7*, 569–596.
11. Montero, F.; De Juan, J.; Cuesta, A.; Brasa, A. Nondestructive Methods to Estimate Leaf Area in (*Vitis vinifera*) L. *HortScience* **2000**, *35*, 696–698. [CrossRef]
12. Peksen, E. Non-destructive leaf area estimation model for faba bean (*Vicia Faba* L.). *Sci. Hortic.* **2007**, *113*, 322–328. [CrossRef]
13. Kandiannan, K.; Parthasarathy, U.; Krishnamurthy, K.S.; Thankamani, C.K.; Srinivasan, V. Modeling individual leaf area of ginger (*Zingiber* Officinale Roscoe) using leaf length and width. *Sci. Hortic.* **2009**, *120*, 532–537. [CrossRef]
14. Chen, J.M.; Black, T.A. Defining leaf area index for non-flat leaves. *Plant Cell Environ.* **1992**, *15*, 421–429. [CrossRef]
15. Salazar, J.C.S.; Melgarejo, L.M.; Bautista, E.H.D.; Di Rienzo, J.A.; Casanoves, F. Non-destructive estimation of the leaf weight and leaf area in cacao (*Theobroma Cacao* L.). *Sci. Hortic.* **2018**, *229*, 19–24. [CrossRef]
16. Wehr, A.; Lohr, U. Airborne laser scanning—An introduction and overview. *ISPRS J. Photogramm. Remote Sens.* **1999**, *54*, 68–82. [CrossRef]
17. Hese, S.; Lucht, W.; Schmullius, C.; Barnsley, M.J.; Dubayah, R.C.; Knorr, D.; Neumann, K.; Riedel, T.; Schroeter, K. Global biomass mapping for an improved understanding of the CO_2 balance-the earth observation mission carbon-3D. *Remote Sens. Env.* **2005**, *94*, 94–104. [CrossRef]
18. Bouvier, M.; Durrieu, S.; Fournier, R.A.; Renaud, J.P. Generalizing predictive models of forest inventory attributes using an area-based approach with airborne LiDAR data. *Remote Sens. Environ.* **2015**, *156*, 322–334. [CrossRef]
19. Li, X.R.; Liu, Q.C.; Cai, Z. Specific leaf area and leaf area index of conifer plantations in Qianyanzhou station of subtropical china. *J. Plant Ecol.* **2007**, *31*, 93–101. (In Chinese)

20. Zhili, L.; Yu, Z.; Fengri, L.; Guangze, J. Non-destructively predicting leaf area, leaf mass and specific leaf area based on a linear mixed-effect model for broadleaf species. *Ecol. Indic.* **2017**, *78*, 340–350. [CrossRef]

21. Roberts, S.D.; Dean, T.J.; Evans, D.L.; McCombs, J.W.; Harrington, R.L.; Glass, P.A. Estimating individual tree leaf area in loblolly pine plantations using LiDAR-derived measurements of height and crown dimensions. *For. Ecol. Manag.* **2005**, *213*, 54–70. [CrossRef]

22. Bao, Y.; Ni, W.; Wang, D.; Yue, C.; He, H.; Verbeeck, H. Effects of Tree Trunks on Estimation of Clumping Index and LAI from HemiView and Terrestrial LiDAR. *Forests* **2018**, *9*, 144. [CrossRef]

23. Hosoi, F.; Omasa, K. Voxel-based 3-D modeling of individual trees for estimating leaf area density using high-resolution portable scanning LiDAR. *IEEE Trans. Geosci. Remote Sens.* **2006**, *44*, 3610–3618. [CrossRef]

24. Béland, M.; Baldocchi, D.D.; Widlowski, J.L.; Fournier, R.A.; Verstraete, M.M. On seeing the wood from the leaves and the role of voxel size in determining leaf area distribution of forests with terrestrial LiDAR. *Agric. For. Meteorol.* **2014**, *184*, 82–97. [CrossRef]

25. Yao, X.; Yu, K.; Deng, Y.; Zeng, Q.; Lai, Z.; Liu, J. Spatial distribution of soil organic carbon stocks in Masson pine (*Pinus Massoniana*) forests in subtropical China. *Catena* **2019**, *178*, 189–198. [CrossRef]

26. Ma, Z.; Hartmann, H.; Wang, H.; Li, Q.; Wang, Y.; Li, S. Carbon dynamics and stability between native Masson pine and exotic slash pine plantations in subtropical China. *Eur. J. For. Res.* **2014**, *133*, 307–321. [CrossRef]

27. Schaffer, B.; Schaffer, B.; Whiley, A.W.; Wolstenholme, B.N. *The Avocado Botany, Production and Uses*, 2nd ed.; Schaffer, B.A., Wolstenholme, B.N., Whiley, A.W., Eds.; CABI: Wallingford, UK, 2013.

28. Ghoreishi, M.; Hossini, Y.; Maftoon, M. Simple models for predicting leaf area of mango (*Mangifera indica* L.). *J. Biol. Earth Sci.* **2012**, *2*, 9.

29. McFadyen, L.M.; Morris, S.G.; Oldham, M.A.; Huett, D.O.; Meyers, N.M.; Wood, J.; McConchie, C.A.; Morris, S. The relationship between orchard crowding, light interception, and productivity in macadamia. *Aust. J. Agric. Res.* **2004**, *55*, 1029. [CrossRef]

30. Yu, K.; Yao, X.; Deng, Y.; Lai, Z.; Lin, L.; Liu, J. Effects of stand age on soil respiration in Pinus massoniana plantations in the hilly red soil region of Southern China. *Catena* **2019**, *178*, 313–321. [CrossRef]

31. Liu, J.; Gu, Z.; Shao, H.; Zhou, F.; Peng, S. N–P stoichiometry in soil and leaves of Pinus massoniana, forest at different stand ages in the subtropical soil erosion area of China. *Environ. Earth Sci.* **2016**, *75*, 1091. [CrossRef]

32. Fascella, G.; Darwich, S.; Rouphael, Y. Validation of a leaf area prediction model proposed for rose. *Chil. J. Agric. Res.* **2013**, *73*, 73–76. [CrossRef]

33. Li, S.; Dai, L.; Wang, H.; Wang, Y.; He, Z.; Lin, S. Estimating Leaf Area Density of Individual Trees Using the Point Cloud Segmentation of Terrestrial LiDAR Data and a Voxel-Based Model. *Remote Sens.* **2017**, *9*, 1202. [CrossRef]

34. Guarato, A.Z.; Quinsat, Y.; Mehdi-Souzani, C.; Lartigue, C.; Sura, E. Conversion of 3D scanned point cloud into a voxel-based representation for crankshaft mass balancing. *Int. J. Adv. Manuf. Technol.* **2018**, *95*, 1315–1324. [CrossRef]

35. Nourian, P.; Gonçalves, R.; Zlatanova, S.; Ohori, K.A.; Vu Vo, A. Voxelization algorithms for geospatial applications: Computational methods for voxelating spatial datasets of 3D city models containing 3D surface, curve and point data models. *Methods* **2016**, *3*, 69–86. [CrossRef]

36. Dionne, O.; De Lasa, M. Voxelization Techniques. U.S. Patent Application No. 14/252,399, 1 October 2015.

37. Keramatlou, I.; Sharifani, M.; Sabouri, H.; Alizadeh, M.; Kamkar, B. A simple linear model for leaf area estimation in Persian walnut (*Juglans Regia* L.). *Sci. Hortic.* **2015**, *184*, 36–39. [CrossRef]

38. Tondjo, K.; Brancheriau, L.; Sabatier, S.A.; Kokutsè, A.D.; Akossou, A.Y.J.; Kokou, K.; Fourcaud, T. Non-destructive measurement of leaf area and dry biomass in Tectona grandis. *Trees* **2015**, *29*, 1625–1631. [CrossRef]

39. Wang, Y.J.; Jin, G.Z.; Liu, Z.L. Construction of empirical models for leaf area and leaf dry mass of two broadleaf species in Xiaoxing'an Mountains, China. *Chin. J. Appl. Ecol.* **2018**, *29*, 1745–1752. (In English abstract).

40. Poux, F.; Billen, R. Voxel-based 3D point cloud semantic segmentation: unsupervised geometric and relationship featuring vs deep learning methods. *ISPRS Int. J. Geo-Inf.* **2019**, *8*, 213. [CrossRef]

41. Cai, H.Y.; Di, X.Y.; Jin, G.Z. Allometric models for leaf area and leaf mass predictionsacross different growing periods of elm tree (*Ulmus japonica*). *J. For. Res.* **2017**, *28*, 975–982. [CrossRef]

42. Shipley, B.; Almeida-Cortez, J. Interspecific consistency and intraspecific variability of specific leaf area with respect to irradiance and nutrient availability. *Écoscience* **2003**, *10*, 74–79. [CrossRef]

43. Pompelli, M.; Antunes, W.; Ferreira, D.; Cavalcante, P.; Wanderley-Filho, H.; Endres, L. Allometric models for non-destructive leaf area estimation of Jatropha curcas. *Biomass Bioenergy* **2012**, *36*, 77–85. [CrossRef]
44. Graham, L. Mobile mapping systems overview. *Photogramm. Eng. Remote Sens.* **2010**, *76*, 222–228.
45. Su, W.; Guo, H.; Zhao, D.L.; Zhang, M.Z.; Zhang, L.; Wu, D.Y. Estimation of actual leaf area of maize based on terrestrial laser scanning. *Chin. Soc. Agric. Mach.* **2016**, *7*, 345–353.
46. Cifuentes, R.; Van Der Zande, D.; Farifteh, J.; Salas, C.; Coppin, P. Effects of voxel size and sampling setup on the estimation of forest canopy gap fraction from terrestrial laser scanning data. *Agric. For. Meteorol.* **2014**, *194*, 230–240. [CrossRef]

 © 2019 by the authors. Licensee MDPI, Basel, Switzerland. This article is an open access article distributed under the terms and conditions of the Creative Commons Attribution (CC BY) license (http://creativecommons.org/licenses/by/4.0/).

 *forests*

Article

Determining a Carbon Reference Level for a High-Forest-Low-Deforestation Country

Johannes Pirker [1,2,*], Aline Mosnier [1,3], Tatiana Nana [4], Matthias Dees [5,6], Achille Momo [4], Bart Muys [2], Florian Kraxner [1] and René Siwe [7]

1. International Institute for Applied Systems Analysis (IIASA), 2361 Laxenburg, Austria; aline.mosnier@unsdsn.org (A.M.); kraxner@iiasa.ac.at (F.K.)
2. KU Leuven - Catholic University of Leuven, 3000 Leuven, Belgium; bart.muys@kuleuven.be
3. Sustainable Development Solutions Network (SDSN), 75009 Paris, France
4. Technical REDD+ Secretariat (ST-REDD), Ministry of the Environment (MINEPDED), Yaoundé, Cameroon; tatyng2002@yahoo.fr (T.N.); machibe2003@yahoo.fr (A.M.)
5. UNIQUE forestry and land use GmbH, 79098 Freiburg, Germany; matthias.dees@t-online.de
6. Albert Ludwig University of Freiburg, 79085 Freiburg im Breisgau, Germany
7. US Forest Service (USFS), P.O. Box 14 Brazzaville, Republic of Congo; rene.siwe@gmail.com
* Correspondence: johannes.pirker@unique-landuse.de

Received: 30 September 2019; Accepted: 20 November 2019; Published: 2 December 2019

Abstract: *Research Highlights*: A transparent approach to developing a forest reference emissions level (FREL) adjusted to future local developments in Southern Cameroon is demonstrated. *Background and Objectives*: Countries with low historical deforestation can adjust their forest reference (emission) level (FREL/FRL) upwards for REDD+ to account for likely future developments. Many countries, however, find it difficult to establish a credible adjusted reference level. This article demonstrates the establishment of a FREL for southern Cameroon adjusted to societal megatrends of strong population—and economic growth combined with rapid urbanization. It demonstrates what can be done with available information and data, but most importantly outlines pathways to further improve the quality of future FREL/FRL's in light of possibly accessing performance-based payments. *Materials and Methods*: The virtual FREL encompasses three main elements: Remotely sensed activity data; emission factors derived from the national forest inventory; and the adjustment of the reference level using a land use model of the agriculture sector. Sensitivity analysis is performed on all three elements using Monte Carlo methods. *Results*: Deforestation during the virtual reference period 2000–2015 is dominated by non-industrial agriculture (comprising both smallholders and local elites) and increases over time. The land use model projections are consistent with this trend, resulting in emissions that are on average 47% higher during the virtual performance period 2020–2030 than during the reference period 2000–2015. Monte Carlo analysis points to the adjustment term as the main driver of uncertainty in the FREL calculation. *Conclusions:* The available data is suitable for constructing a FREL for periodic reporting to the UNFCCC. Enhanced coherence of input data notably for activity data and adjustment is needed to apply for a performance-based payment scheme. Expanding the accounting framework to include forest degradation and forest gain are further priorities requiring future research.

Keywords: REDD+; Cameroon; reference level; deforestation; agriculture; forest baseline

---

1. Introduction

The REDD+ mechanism was designed to reward countries financially for reducing emissions from deforestation and forest degradation (DD), which significantly contributes to total greenhouse gas (GHG) emissions worldwide [1]. REDD+ also includes the promotion of sustainable management of forests and increasing and conserving forest carbon stocks [2]. The underlying idea is that avoided deforestation offers a large and very cost-effective potential to curb GHG emissions [3,4]. To measure progress in this regard, a benchmark measure is needed to define how much would be emitted in the absence of REDD+ interventions. This benchmark is called a forest reference level (FRL) or forest reference emission level (FREL). The former refers to an accounting framework where forest gain is considered, the latter refers to forest loss only. FRL/FREL was institutionalized in the Warsaw Framework as one of the four elements required to participate in REDD+ [5].

The establishment of a FRL/FREL is preceded by a number of important policy-related decisions including a national definition of what constitutes a forest; the scale in terms of geographical coverage where FRL/FREL for sub-national jurisdictions or regions can be developed as an interim measure *en route* to developing a national baseline (Decision 1/CP.16 - FCCC/CP/2010/7/Add.1); the scope of the FRL/FREL in terms of the relevant activities causing changes in forest carbon stock, the carbon pools and the gases to be considered; and a reference period in the recent past. Together, these decisions provide the framework for measuring the area extent of change (activity data or AD) and emission factors (EF), which is the difference between carbon stored before and after conversion of forest to another land use. The combination of AD x EF for a given scale, scope, and period results in a historic baseline. Some countries such as Brazil and Indonesia used these historical baselines as their FREL, i.e., as the benchmark to measure their progress towards reducing future emissions from deforestation. Historical deforestation rates are very high for these two countries.

The thus established FRL/FREL can then be submitted to the United Nations Framework Convention on Climate Change (UNFCCC) for technical assessment. This is to assess the degree to which information provided is in accordance with the guidelines for submissions of information on FREL/FRL and offer a facilitative, non-intrusive, technical exchange of information on the construction of the FREL/FRL (FCCC/CP/2013/10/Add.1.). The information provided (including historical information) should be guided by the most recent Intergovernmental Panel on Climate Change (IPCC) guidance and be transparent, complete, consistent, and accurate (FCCC/CP/2011/9/Add.2).

Many countries and jurisdictions such as so-called high-forest-low-deforestation (HFLD) countries, which are characterized by high remaining forest cover and comparatively low rates of deforestation, dispose of atypical starting situations for REDD+ [6]. Historical rates of deforestation alone are inadequate to define an FRL/FREL for HFLD countries, especially in cases where there are clear indications of changes (increase) in drivers of DD. The risk of using a historical baseline in HFLD countries is that the efforts necessary to contain future deforestation could be underestimated. The FRL/FREL will therefore have to take socioeconomic development into consideration, which influence the trajectories of future drivers of change and the effort to be compensated will have to be based on the effectiveness of national/international policies and measures to address these.

However, the potential for inflation of the FRL/FREL and the subsequent creation of "hot air" when historical baselines are adjusted, have been the subject of much thinking about design principles [7–9] and criticism of applied methodologies [10–13]. How to define a benchmark for emissions remains an issue in the era of the Paris Agreement where many countries mention REDD+ in their Nationally Determined Contributions (NDC's) [14] and state that GHG emissions from land use, land-use change and forestry (LULUCF) are likely to increase in the future under the business-as-usual scenario [15,16]. Methodologies used to justify a perceived increase of future emissions vary greatly and often go unreported [15,16].

Engagement in REDD+ will ultimately depend on a country's capacity to demonstrate the level of emissions from forests with and without REDD+ interventions. The lack of data or capacity, or both, to measure, project, and monitor emissions has been put forward as a possible major hindrance, notably for African countries, to effectively participate in REDD+ [17–21], although the stepwise approach of improving reference levels as new and better data become available (See COP decision 12/CP17) recommended by the UNFCCC, largely facilitates the task. Globally, more than 30 countries have already submitted FREL/FRLs to the UNFCCC for technical assessment, some of which have also claimed upwards adjustment [16]. The methodology used to justify the upwards adjustment is often poorly documented.

This paper focuses on the development of a virtual subnational FRL as an interim measure towards the development of a national FRL for Cameroon. Southern Cameroon serves as a case study area since it is facing a bundle of societal megatrends that are poised to have an impact on the still very high forest cover in coming years: The country's economy has recovered from the crisis of the early 2000s [22–24]; the local population is growing rapidly, especially in urban centers where the population tends to have different food consumption patterns; and the area is known as a bread basket for export to neighboring Gabon, Republic of the Congo, and Equatorial Guinea. These trends, combined with continued low agricultural yields, are generally expected to drive agriculture further into forest areas [25–28].

From the perspective of a stepwise approach towards the development of a national FREL, the study uses available datasets to establish an adjusted subnational FREL for Cameroon and critically analyses future steps for improving the FREL as the country aims to access performance-based finance. The working hypothesis underlying the adjustment term of the FREL is that there is a clear set of quantifiable variables related to the development of a society leading to forest conversion that can be projected into the future [29] while aiming at a high degree of transparency. The methodology for FREL development presented in this paper is easily replicable in other HFLD countries.

2. Materials and Methods

The study area is presented in Section 2.1 and a set of policy-related working definitions adopted by the country is presented in Section 2.2. These are the boundary conditions for the establishment of the subnational FREL. The virtual reference level for the reporting period 2020–2030 for the study area is computed based on historical emissions derived from activity data (Section 2.3) combined with emission factors (Section 2.4) and adjusted to national circumstances using a land use model (Section 2.5). The approach to sensitivity analysis of all components of the FREL is presented in Section 2.6.

2.1. The Study Area

The study area encompasses seven (out of a total of 58) administrative divisions located in the humid tropical part of southern Cameroon (Figure 1). It covers a total area of 9.3 million ha, which is equivalent to the land area of Hungary. The climate is dominated by ample rainfall of 1500–4000 mm per year, which allows for the growth of moist evergreen forests that cover close to 90% of the study area [30].

The population density of around 14 people per km^2 is relatively low and concentrated around the coast and in towns and villages along the road to the capital city Yaoundé, located around 50 km north of the boundary of the study region. The main drivers of deforestation in the region are shifting smallholder agriculture, agro-industrial plantations that mainly focus on tree crops and the expansion of transport infrastructure [31].

2.2. Working Definitions: Scale, Scope, Forest Definition, and Virtual Reporting Periods

Cameroon's draft national REDD+ strategy provides the theoretical working definitions that frame the conditions under which an FRL/FREL can be developed. In the scope of REDD+ in Cameroon, the term "forest" is defined by three criteria: Crown cover of 10% or more, an attainable tree height of 3 m or more, and a patch size of 0.5 ha or more. Mono-specific tree crops such as plantations of oil palm, rubber, and full-sun cocoa are explicitly excluded from the forest definition. The country also considers all eligible REDD+ activities (reducing emissions and increasing removals) in their draft strategy. All carbon pools should be considered in the establishment of the FRL, with an emphasis on accounting for significant pools at a Tier 2 level. CO_2 is the most relevant GHG in the forestry sector but in specific cases CH_4 and N_2O may also be considered. The country has not formally fixed a reference period and a performance period.

The establishment of an FRL is further constrained by available data. First, there is no dataset available to reliably trace the degradation or gain of forests, which restricts the scope of this article to establishing an FREL of CO_2 emissions from deforestation only, where other gases potentially emerging from smaller fires are considered negligible. Pools are restricted to above and below-ground biomass due to the uncertainty associated with more liable pools. This study uses the periods 2000–2015 and 2020–2030 as virtual reference and reporting periods, respectively, based on the availability of relevant datasets. "Virtual" reference and reporting periods refer to the working definition made in the context of this article as opposed to the political decision that lies in the sovereignty of the country.

2.3. Remote Sensing of Activity Data

Activity data was derived from a countrywide reference land cover map for a base year (2000) and the assessment of forest loss over the reference period (2000–2015), clipped to the extent of the study area. The reference maps for 2000 and 2015 were developed using a hierarchical supervised decision tree-based wall-to-wall mapping methodology implemented in PCI Geomatica Software. The assessment was run on a composite of 10,517 terrain corrected (L1T) Landsat images (bands 3,4,5,7) with low cloud coverage sourced from the United States Geological Survey (USGS) Earth Explorer and resulted in seven thematic classes: (1) Primary and (2) secondary terra firma and (3) primary and (4) secondary swamp forests, (5) mangroves, (6) agro-industrial plantations of perennial crops, (7) non-forest land, and (8) forest loss since the year 2000. The workflow largely builds on a recent study conducted in the Democratic Republic of the Congo [32]. A minimum mapping unit of 0.5 ha compliant with the national forest definition was applied to forest loss areas to eliminate small-scale degradation and natural disturbances from the resulting deforestation map for 2000–2015. The assessment of accuracy, deforestation dynamics within the 2000–2015 period and identification of drivers of deforestation was performed using a stratified random sampling method [33,34]. To that end, 348 dated reference samples from Landsat and high-resolution images from Google Earth were collected, visually interpreted, and used to validate the reference and the forest change map, respectively [32]. The pre-defined protocol for the assessment of drivers of deforestation distinguishes eleven classes and is presented in Appendix B alongside a link to the online sample visualization.

2.4. Emission Factors from the National Forest Inventory

Emission factors were developed on the basis of the country's first, and thus far only, national forest inventory (NFI) performed during the years 2003–2004 [35]. Tree biomass in Cameroon was inventoried using 204 valid census tracts distributed according to a systematic stratified sample design across the country - 45 of which are located inside the study region. Tree biomass in both forest and non-forest (such as agroforests, fallows, tree plantations, etc.) areas was considered, although non-forest classes are only sparsely present in the final sample.

Re-analyzed for the purpose of REDD+, the NFI data allows for distinguishing carbon stocks in five land cover types in the study area: (1) Forest, (2) settlements, (3) grassland, (4) fallow land, (5) annual crops, and (6) perennial crops. Carbon stocks were calculated by combining data and information on tree diameter and tree height from the NFI with pan-tropical allometric equations [36] and proxy shoot-to-root ratios for moist tropical forests [37]. Emission factors were computed as the difference between initial forest carbon stocks of the living biomass and the carbon stocks left in the vegetation after forest conversion and presented in the results section. Further details on NFI data and assessments can be found in the background report [38].

2.5. Adjustment of the Reference Level to National Circumstances

Given that non-industrial agriculture is by far the main driver of deforestation in the region [39], the main adjustment of the reference level was calculated for non-industrial agriculture using a land use model. The model's rationale builds on demand for agricultural products that needs to be satisfied by a matching supply. The model is implemented in MS Excel and is available for download from http://dare.iiasa.ac.at/56. It provides results in five-year intervals from 2000 until 2030 for each of the five agro-ecological zones of Cameroon [40] and the study area, and in terms of the contributions of the 15 most prevalent agricultural crops in the country (see a list of crops in the annex). The model comprises six computation steps, each of which is parameterized using available data for Cameroon: (1) Population, (2) food and feed consumption, (3) trade within Cameroon and with the rest of the world, (4) agricultural production, (5) cultivated area, and (6) resulting forest cover change.

2.5.1. Population

Demographic development plays a key role in the land use model. It is fed by national data and projections by division (third administrative level) and separately for urban and rural areas. In this study, towns with more than 50,000 inhabitants are considered urban areas. Population data for the years 1987 and 2005 is available from the national population censuses conducted by the Central Bureau of Census and Population Studies of Cameroon (BUCREP).

The population for reference year 2000 of the FREL is calculated based on 1987 census data and the average growth rate between 1987 and 2005. Existing population projections according to the shared socioeconomic pathway (SSP) [41,42] scenario number 3 of the SSP Framework were used to project the population growth available from the censuses to the future for the years after 2005 until the end of the virtual performance period in 2030. Rather than the middle-of-the road scenario SSP2, SSP3 is used as it is very close to the official national population projections [43]. SSP scenarios about future population growth for Cameroon are only available at the national level and were therefore used to project national data in the future only.

The data shows that the population in the study area has increased from 706,000 people at the time of the first census in 1987 to 1.05 million in 2005 with the share of the urban population increasing from 16% in 1987 to 22% in 2015. According to the SSP3 scenario, the population will increase further to 1.33 million by 2030, with 38% of the population living in urban areas.

2.5.2. Consumption

Living conditions and relative wealth in Cameroon are improving with an average of 4.2% GDP growth per annum over the last 10 years [44]. This has direct repercussions on people's dietary habits. Income growth and urbanization, for instance, lead to changes in consumption such as a more diverse diet that includes a larger share of animal protein, fats, and oils (a phenomenon known as Bennett's Law [45]).

Data on diets for the different regions of Cameroon comes from the UN World Food Program [46] and projected to the future using GDP projections from the World Bank and income elasticities for food consumption [47] that translate into future diets specific to Cameroon. These socioeconomic changes result in a projected increased per-capita consumption of beans and groundnuts (+35% each until 2030), bananas and plantains (+27%), as well as pork and poultry (+45%).

The resulting per capita consumption, combined with population projections result in an estimate of future food consumption that can be met either by local production, imports from other regions of the country, or imports from other countries.

2.5.3. Trade

The model considers flows of agricultural goods within Cameroon, as well as to and from neighboring countries and the international market. Trade of crops and foodstuff with third countries is documented by national publications [48] and the Food and Agriculture Organization Corporate Statistical Database (FAOSTAT). Statistics on internal trade within Cameroon are not available and therefore estimated as the complement of local agricultural production needed to feed the local population. The share of the consumption in each of the five agro-ecological zones (AEZ) and rural/urban area that is satisfied by each source, that is local, each other AEZ or the rest of the world. This estimation is made according to three guiding principles: (1) The (over) supply status of a certain crop; (2) the proximity of the other regions with the importing region; and (3) it is assumed that imported goods from other countries mostly go to the urban areas. Further, the shares of imported versus locally produced food are assumed to be constant over time.

2.5.4. Agricultural Production

Production in each region is computed based on the local demand times the share that is domestically produced in case the region is a net importer, plus the sum of the shares of the region in the consumption of other regions times their projected level of consumption if the region is a net exporter. The local demand for crops is the sum of the demand for human consumption and for animal feed such as poultry.

A significant share of production is lost or wasted at different stages in the value chain [49,50]. This leads to a higher computed production compared to the computed consumption requirement. According to the Food and Agriculture Organization (FAO) [51], post-harvest losses reached 32% of the cassava production in 2010. For periods beyond 2010, constant post-harvest losses shares are assumed.

2.5.5. Harvested Area and Arable Land

Once the production level is computed, the harvested area results from dividing production by agricultural yields. Yields in the model are specified per administrative region based on data from the ministry of Agriculture's (MINADER), the AGRISTAT statistics report series, and counter-checked with country-level production data from FAOSTAT. Yields are kept constant at reported levels for future projections.

In order to compute the impact on arable land, two other parameters are used: (1) The average number of harvests per year and (2) the share of fallow land in total arable land. The average number of harvests per year is crop and region specific and is taken from the AGRISTAT reports over 2000–2004. The area that lies fallow is mainly driven by the fallow period, which varies with the population density, that is to say, fallow periods are shorter in densely populated areas [52]. Furthermore, fallow periods are longer in humid tropical regions than in drier savannah areas. In the model these are capped to two years (fallow multiplier: 1). A typology of fallow periods is presented in Table 1, where the resulting fallow coefficient is applied in the model. No fallow period is assumed for perennial crops such as oil palms, cocoa, rubber, and banana plantations.

Table 1. Typology of fallow duration as a function of population density in Cameroon. Source: Modified after Gillet et al. (2014).

Population Density (inhab./km^2)	Cultivation vs. Fallow Duration	Fallow Multiplier
<20	2y cultivation, 7y fallow	3.5
20–30	2y cultivation, 5y fallow	2.5
>30	2y cultivation, 3y fallow	1.5

At this stage, the model predicts the amount of arable land required to satisfy the demand for food and feed, which are the key drivers of deforestation, in intervals of five years.

2.5.6. Deforestation and Emissions

Forest clearing is a direct result of cropland expansion into forest areas. However, the share of cropland claimed from forests as opposed to other, non-forested lands varies from one region to another and is determined by the availability of non-forest land. For Cameroon, the share of new cropland claimed from forest ranges from 6% in the Far North region, which is dominated by open vegetation, to 90% for the tropical humid zone [53]. For the study area in southern Cameroon, the same study finds a share of 85% of new cropland claimed from forest.

Next, projected deforestation is combined with carbon stocks for the land cover types presented in Section 3.2. Carbon stock values are derived from the first and thus far only NFI performed in 2003 [35] (see details in Section 2.4). In this context, the results of the NFI are preferred over those of other studies, since reporting for REDD+ requires a long-term monitoring framework where carbon stocks can be traced over time using one coherent methodology.

2.5.7. Model Validation—Comparison of Model Results with Observed Variables

Obtaining a good match of model results with independently observed data is key in terms of making a credible argument for an adjustment to national circumstances. To this end, checks that allow the comparison of intermediary model outputs at each calculation step with independent statistics—such as agricultural production computed by the model as described in 2.5—is compared with statistical data from the ministry of agriculture (MINADER) or the FAO.

The deforestation dynamics computed by the model are also compared to data from available remote sensing products such as that used to define activity data as described in Section 2.3 and independent global-scale remote sensing products [54,55] clipped to the extent of the study area.

2.5.8. Projecting Other Drivers—Industrial Agriculture and Infrastructure

The expansion of industrial agriculture and infrastructure in the context of Cameroon are based on discrete political decisions, and their future impact is estimated in terms of legal claims to land clearing. Legal claims to clearing land takes the form of sales of standing volumes (SSVs) that allow for unsustainable wood harvesting, and typically precedes the establishment of agro-industrial plantations and infrastructure projects [56,57]. That being said about land allocation, future land use in these areas is uncertain as only a small fraction of planned agricultural development projects in the country are actually implemented [58].

A total of 66,971 ha of SSVs that are almost entirely (97%) covered by forests are currently located inside concession areas flagged for the development of agro-industrial oil palm and rubber plantations, while another 17,980 ha are allocated around the Kribi deep water port to make way for port infrastructure. To account for the uncertainty associated with the development of these areas, it is conservatively assumed that only 10% of SSVs will be cleared and replaced by perennial crops and infrastructure for agro-industrial concessions and the deepwater port, respectively, by 2035.

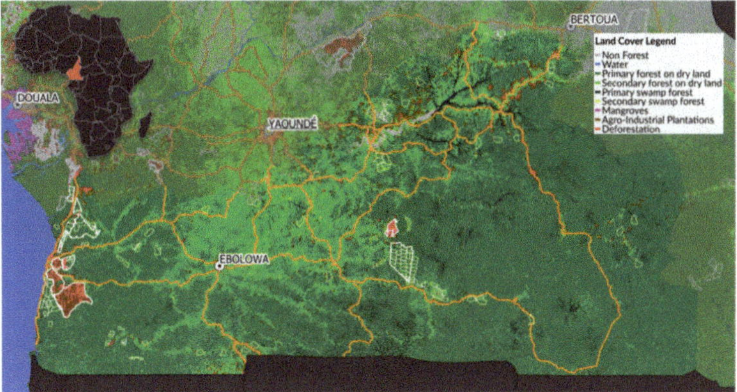

Figure 1. Overview of land cover and land allocation in the study area. Land allocation in the form of agro-concessions (thick white outline) and unsustainable logging concessions (Sales of Standing Volumes - SSVs; thin white outline) are located at the coast and in the center of the area.

2.6. Sensitivity Analysis of Input Data and Methods

Monte Carlo methods iterate greenhouse gas calculations many times where input variables randomly take different values from the variables AD*EF*Adjustment (Adj hereafter) in each iteration according to a pre-defined probabilistic distribution. The resulting solution space for the FRL emerges from a random combination of input variables and therefore gives a complete picture of the uncertainties associated with the FRL calculation.

For the southern Cameroon case study, a Monte Carlo analysis with 1000 iterations was defined based on the arithmetic mean and the standard deviation at a 90% confidence interval for a normal distribution of the AD, EF, and Adj variables listed in Table 2. For the adjustment term, the variation comes from the deployment of alternative SSP scenarios. From the optimistic SSP 1 "Sustainability" with moderate population and GDP growth to the most pessimistic SSP 5 scenario "Conventional development", a gap of 9% for population, and 21% for GDP, respectively, can be observed for the year 2030. It should be noted that this is a somewhat simplified assessment for demonstration purposes: For AD it only considers smallholder deforestation, for EF two biomass pools (before and after conversion as opposed to land use specific ones) and various socioeconomic development scenarios leading to varying adjustment factors (as opposed to variation of each input variable).

Data describing the mean and the shape of AD is sourced from the remote sensing exercise described in Section 2.3, which is backed by 127 validation points. EF is composed of biomass in a forest before conversion (50 sample plots) and biomass after conversion (26 sample plots), where the latter is an average over annual and perennial cropland (see Section 3.2). The variation of the adjustment term is calculated from the standard deviation across the six scenarios of socioeconomic pathways for Cameroon (Table 2, Line 4 - Adj).

Table 2. Activity data, emission factor, and adjustment variables drive the uncertainty of a forest reference level.

Domain	Potential Source of Error to Analyze	Unit	Number of Samples	Mean	SD (CV)
AD	Deforestation observed 2000–2015	ha/year	127	10,602	1837 (17%)
EF	Forest biomass sampling	tCO_2/ha	45	490.13	35.39 (7%)
EF	Non-Forest biomass sampling	tCO_2/ha	26	174.99	48 (27%)
Adj	Potential development trajectories	Adjustment multiplier(dimensionless)	6	1.44	0.12 (42%)

SD = Standard Deviation; CV = Coefficient of Variation (SD/mean); Mean = Arithmetic mean; Adjustment multiplier: Applied to historical smallholder deforestation.

3. Results

3.1. Forest Loss during the Reference Period 2000–2015

Remotely sensed deforestation accounts for 219,948 ha (14,633 ha/yr or 0.16% per annum of the initial forest cover) over the period 2000–2015 with a standard error of 10.45%. The analysis of drivers of deforestation revealed that the expansion of industrial agriculture, notably palm oil and rubber plantations, contributed 30,128 ha (13.7%), while the expansion of infrastructure—notably the deep water port of Kribi and various projects involving the construction or upgrading of roads, contributed 10,260 ha or 4.7%. Non-industrial agriculture with 159,037 ha or more than 72%, contributed the most by far to forest loss during the reference period 2000–2015. Almost 10% of forest clearings cannot be clearly attributed to an anthropogenic driver and are therefore not further considered as relevant for REDD+ (Table 3).

Trend analysis further shows a strong increase from 36,807 ha of forest loss from 2000–2005, to 48,737 during 2005–2010 and 134,403 ha during 2010–2015. The relative standard error of remotely sensed forest loss at a 90% confidence interval is 60. These results are in the range of other remote sensing products available for Cameroon (Figure 2, different colored dots).

Table 3. Per-driver presentation of the area and proportion of remotely-sensed deforestation observed during the reference period.

Driver of Deforestation (2000–2015)	Area		Standard Error (%)
	(ha)	%	
Non-industrial agriculture	159,037	72.3	
Infrastructure	10,260	4.7	
Industrial agriculture	30,128	13.7	
Other	20,521	9.3	
Total	**219,947**	**100**	**10.45**

3.2. Emission Factors from the NFI

Emission factors are developed from the difference in above- and below-ground biomass between forest and the land use after clearing, as derived from the NFI. The highest EF is associated with the conversion of forest to built-up areas where all the forest carbon (490 tCO_2/ha) is lost (Table 4). Transitions of forest to grassland, annual crops, and fallow/wasteland are also emissions intensive. Perennial cropland houses more than half of the carbon stored in forests and therefore have a relatively low EF of 226.8 tCO_2/ha. Uncertainties are smallest in the forest class with a standard error of 4.3% but are significantly higher for the agricultural land cover classes that relate to the lower number of NFI sample plots located in these land cover classes.

Table 4. Emission factors and associated uncertainty for five land use transitions developed from the NFI.

Transition of Forest to	Emission Factor (tCO_2/ha) *	Standard Error ($\pm$%)
Annual crops **	347.7	14.1
Perennial crops	226.8	31.2
Fallow land	332.1	11.2
Grassland	462.3	5.9
Built-up areas	490.1	4.3

* Comprised of above ground biomass (AGB) and belowground biomass (BGB). ** Comprising 15 crops listed in Table A2 in the annex.

3.3. Adjustment of the Reference Level to National Circumstances

The rate of forest clearing is projected to increase in the future. This results in projected cleared areas of 15,900 ha/yr on average through 2020–2030 across all anthropogenic drivers of deforestation, which is 20% above the remotely sensed deforestation rate during the 2000–2015.

Non-industrial agriculture is responsible for the majority of the projected increase. Fueled by increases in population and consumption rates, the land use model predicts an increasing demand for land for the virtual performance period 2020–2030. This leads to projected forest loss from non-industrial agriculture of 14,600 ha/yr, which is 3900 ha/yr (+76%) higher than during the virtual reference period of 2000–2015 (Figure 2, left; light blue bar). This increase occurs gradually where the model for the period 2000-2015 estimates an average cleared area of 8300 ha/yr which then increases to 13,100 ha/yr during 2020–2025 and 16,100 ha/yr in the following period 2025–2030.

This expansion of non-industrial agriculture is mainly fueled by staple crops such as groundnuts, corn, cassava, plantains and bananas, and to a lesser extent by smallholder oil palm plantations and beans (Figure 2, right). Further drivers of deforestation include the expansion of infrastructure (almost constantly at 2500 ha/yr accounting for 16% of total deforestation), and industrial plantations that contribute substantially (12,300 ha or 14% of total deforestation) but are assumed to slow down to 780 ha or 5% during the performance period. As a result, the overall projected deforestation for the virtual performance period is 3250 ha/yr (26%) higher than during the virtual reference period.

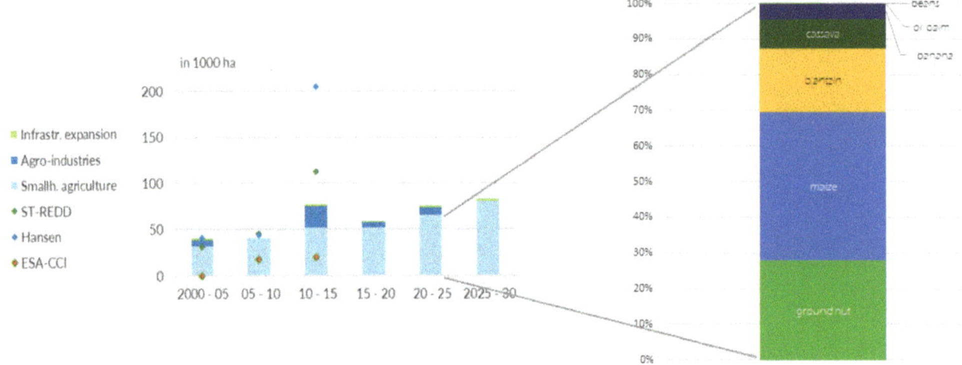

Figure 2. Reconstructed and projected deforestation from the land use model (stacked bars) as observed by three independent remote sensing products (diamond symbols) on the left and decomposition of modeled non-industrial agricultural drivers per crop (stacked % chart on the right).

Both historical and projected deforestation, as presented in Figure 2 are translated into emissions by applying the relevant emission factors derived from carbon stocks presented in Table 4.

The resulting forest reference level (Figure 3) for the virtual performance period 2020–2030 with 5.63 $MtCO_2$/yr is 1.26 $MtCO_2$/yr or 29% higher than the emissions for the reference period. This adjustment is driven by the expansion of smallholder agriculture, from which the associated annual emissions are projected to increase by 48% from 2020–2030, as compared to the virtual reference period 2000–2015.

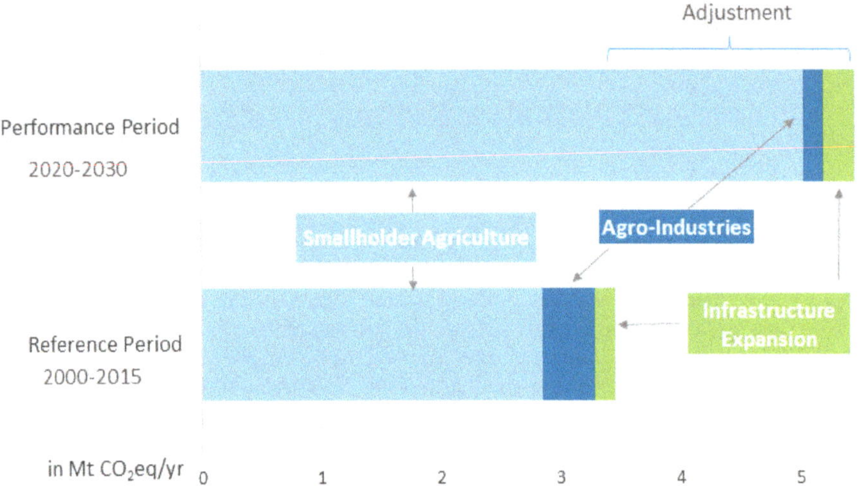

Figure 3. Projected emissions during the performance period (top bar) are 29% higher than emissions during the reference period (left bar); the increase is driven by expanding smallholder agriculture for which emissions are projected to increase by 48%.

3.4. Sensitivity of the Reference Level to Input Data

The spread of the results of the Monte Carlo analysis as shown in Figure 4. gives an indication of the sensitivity of the reference level calculation (AD*EF*Adj) where the adjustment term varies across five SSP scenarios. The resulting distribution is skewed to the right (with a skew factor of around 0.40) as can be expected due to the multiplication of normally distributed variables. The assumption of a normal distribution of the data can however be maintained since the skew factor is predominantly < 0.5 [59]. The mean FREL is 4.96 $MtCO_2$/year and the confidence interval is $\pm$ 2.19 $MtCO_2$/yr at a 90% confidence level (the z-score multiplier for the SD is 1.64). This means that the true mean value for the FREL lies between 2.73 and 6.78 $MtCO_2$/yr in 90% of the iterations. The aggregated variation of the reference level expressed as the coefficient of the variation (SD divided by the mean) is 27%.

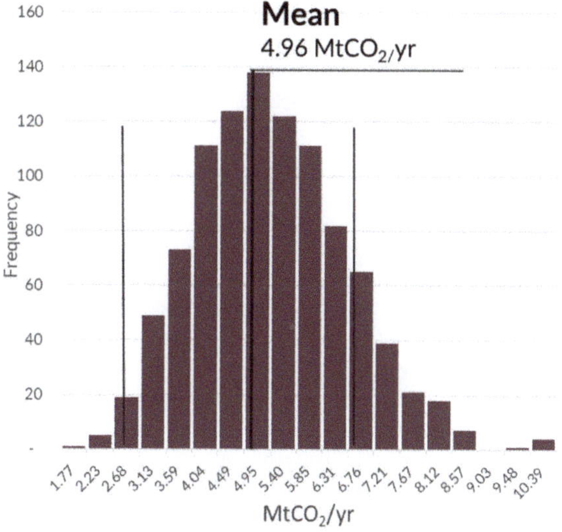

Figure 4. The results of 1000 Monte Carlo iterations of the reference level calculation; the mean (bold black line) lies within the 2-tailed standard deviation ($\alpha = 0.1$; thin black lines).

4. Discussion

This article demonstrates the establishment of a forest reference emission level (FREL) adjusted to national circumstances using southern Cameroon as a case study. The results show that during the virtual reference period of 2000–2015, deforestation was mostly driven by the expansion of smallholder agriculture. Using a land use model based on food and feed consumption, deforestation is projected to the future using a virtual performance period spanning 2015–2030, thus leading to an adjusted FREL that lies 26% above historical emission levels over the virtual reference period. There is a great number of parameters in the land use model—population growth, agricultural yields, food losses, to name a few—that influence the future demand for land. A sensitivity analysis of activity data, emission factors, and the adjustment factor computed with the land use model using Monte Carlo techniques demonstrates that at a 90% confidence level calculated from 1000 calculation iterations, the FREL's coefficient of variation (CV) is $\pm$ 27%, while the CV of its single components varies between 7% and 43%. The remainder of this section is structured into two blocks: The first reviews the limitations of the approach and the uncertainties of the results, whereas the second block discusses the policy implications derived from the findings.

4.1. Uncertainties and Limitations of Approach and Results

Increasing domestic food consumption is the main driver of deforestation. This finding complements and confirms the conclusions of other pieces of research showing that the expansion of non-industrial agriculture is the principle driver of land conversion. This fact has been confirmed through various methodologies including analyses through the political economy lens [60–62], expert knowledge [25], analyses of the spatial drivers of deforestation [39,63], scenario trend projection [64] and, last but not least, a review of the Congo Basin countries' submissions to the UNFCCC [65]. In this study the non-industrial agriculture sector comprises smallholders and medium-sized "elites" landowners without distinction. Local or urban elites are gaining importance in the land use sector in Cameroon [63,66,67]. Not distinguishing between both agent groups is nevertheless justifiable because both groups respond to the same immediate market signals, which are subject to increasing demand for food and feed, as depicted in the land use model (Section 2.5). The results of this study show a relatively modest and even decreasing proportion of agro-industrial plantations. This is due to the conservative methodology adopted: Only clearings with a legal basis, provided by the allocation of sales of standing volumes, are projected to be cleared within the ten years timeframe of the study. This is in line with approved proposals for performance-based jurisdictional REDD+ funding provided by other countries [68,69]. Moreover, the allocation of new concessions is a political endeavor [70] with discretionary elements unknown to the public and therefore neither scale nor location can be predicted in sufficient detail to contribute to an FREL.

Consistency of the FREL with a GHG monitoring system is a challenge for many REDD+ countries [71]. A GHG monitoring system should allow the consistent tracking of AD, EF, and adjustments over time, meaning that each land cover transition should find its respective emission and adjustment factors. The currently available data does not allow for this consistency. This has to do with the aimed-for granularity of the analysis (Tier 3), which generally increases the complexity of monitoring [72] and in this case, over-stretches the degree of granularity of the available data.

For example, the national forest definition (Section 2.2) considers oil palm plantations as non-forest. Activity data (Section 2.3) map agro-industrial oil palm plantations but fail to map out the smallholder oil palm plantations that supply up to two-thirds of the national palm oil harvest [40,73] but are nonetheless classified as forest (if established before 2000) or deforestation in the currently available land cover map. This is mainly due to the technical difficulty of the task and the fact that technical advances in reliably mapping oil palm plantations at different scales have only recently become available [74,75]. Acknowledging this lack, the land use model used for the adjustment of the FREL (Section 2.5) relies on national statistics of oil palm supply rather than on the AD and uses emission factors (Section 2.4) developed for all perennial crops. However, the emission factor for perennial crops is partly derived from sample plots located within cocoa agro-forests. These are, however, considered forests according to the national forest definition (Section 2.2) and should therefore be accounted for as forest degradation. More generally, currently available AD data maps deforestation with unknown post-forest land use that inhibits the application of land use specific EF derived from the NFI, in other words no EF can be associated with remotely sensed deforestation but the land use following the clearing would need to be determined. Further, a definition of land use after conversion, rather than land cover, generally results in lower estimated deforested area as temporarily unstocked areas such as fallow are not considered [76]. Assessment of the land use should potentially be done a few years after clearing, to ensure the correct post-forest land use (or forest regrowth) is identified.

The uncertainty of the FREL calculation is considerable but within the range of results of comparable studies. The coefficient of variation as a measure of uncertainty in biomass estimates ranges from 7% for remaining forest to 27% for cropland. To put this in context, the IPCC recommends applying a default uncertainty of ± 75% for its default (Tier 1) emission factors. Global biomass maps—when aggregated to the national level for Cameroon—and the FAO's forest resources assessment (FAO FRA) yield relatively consistent results with differences across sources being in the range of 7% to 16% [77]. The authors however also note that on a local scale, differences across global maps are significantly higher. The land use modeling approach to adjusting the FRL certainly increases the transparency of the adjustment term as compared to other approaches. It is an approach that requires a wide range of input data, which raises questions about data reliability beyond the scale of the sensitivity analysis performed in Section 2.6. The literature does notably point out potential problems with national datasets relating to population [78,79] and international trade [80]. Further, challenges relate to data on domestic trade, consumption rates and the characterization of complex multi-crop and multi cropping cycle systems, and the general political situation and stability that could evolve in the next 15 years.

Pertinent datasets and information are not available at present. The UNFCCC reporting guidelines stipulate that all significant gases, pools, and activities shall be covered [16] where the threshold for "significant" is often defined as 10% of total emissions [81]. Forest degradation is a significant source of emissions, as suggested by both qualitative national analyses [25,31] and global studies [25,82,83], the latter placing forest degradation in the range of 25% of total land-based emissions. The main drivers of forest degradation are—listed in decreasing order of available documentation—industrial logging aimed at the international timber market [57,84,85], cocoa encroaching into forests [86–88], informal logging for the domestic and regional timber market [89–96], and fuel wood collection [97,98]. None of these drivers of degradation is considered in this study because they are not assessed by activity data mapping and only the industrial logging and cocoa is tracked by national export statistics. In addition, it is not possible to retrieve emission factors for degradation from the one-time national forest inventory performed in 2003/2004. Moreover, not included is forest gain, which is the main element absent from many countries' submissions to the UNFCCC [65,71]. In the case of southern Cameroon, the forest definition (threshold ≥ 10% canopy cover), available AD and EF from the one-time NFI, as well as the dynamic nature of clearing and regrowth in the shifting cultivation landscape [95,99] make it challenging to effectively report on forest gain.

GHG reporting should be fit for the specific purpose. The virtual FREL described in this paper would qualify for the comparatively lenient requirements of jurisdictional REDD+ results reporting to the UNFCCC. This process is a facilitative, non-intrusive, technical exchange of information rather than a critical assessment of data and approach. However, this FREL requires further improvement, if the aim is to obtain performance-based payments such as those offered by the FCPF Carbon Fund [81]. This is notably due to the lack of coherence between AD, EF, and adjustment as listed above in this section and the resulting use of Tier 1 proxies in the absence of alternative national data. Further, the crediting period should start no longer than two years after the end of the reference period [81], not five years as is the case in this virtual reference level.

4.2. Policy Implications

As outlined in Section 2.1, a number of policy-related choices preceded the virtual FREL developed in this article. Most are straightforward as they adopted the widest possible definition. The forest definition and the scale of the project, however, deserve further discussion.

Forest definitions should be tailored to the specific policy question they address [100]. The current forest definition for Cameroon as stated in the forest code does not have any quantitative parameters. The definition used for the FAO forest resources assessment (FAO FRA) set a high bar of 30% canopy cover to what is considered forest. In contrast, the forest working definition in place for REDD+ and other climate change-related processes in Cameroon aims to cover the widest possible ranges of forest with a minimum canopy cover of 10%. The choice of either forest definition has minimal effects on the forest area in the dense evergreen forests of southern Cameroon which easily fulfill both canopy cover thresholds. It will, however, make a difference in the open forests of the savannah-like northern regions of the country [31,101]. More importantly, picking the appropriate forest definition is crucial for Cameroon as a whole to be or not be considered an HFLD country and is expected to provide preferential access to climate finance [81,102]. While the working definition of forests for REDD+ has not yet been officially adopted in political circles, the entire issue of multiple and conflicting forest definitions raises questions of legitimacy given that each definition serves only the specific needs of the policy process that is en vogue at a given time [103,104].

High-level political risks are pending in the current project area that might drive future deforestation beyond the regular development trajectory. The development of the Heveasud plantation (Figure 1a,b), for instance, is not completed yet and information about the area to be cleared or logged unsustainably [57] in the future is not available. There have also been other development projects initiated but then stalled, such as the Mballam iron ore mine and the associated Kribi-Mballam railway stretching 500 km through thus far densely forested zones of the virtual project area [105,106]. On the other hand, considering these projects for the FREL is difficult to justify, given the low implementation rate of large-scale projects in the past [58]. The ongoing conflict in the Northwest and Southwest administrative regions of the country has had negative impacts on agricultural production and expansion in the conflict region, which might lead to some forest regrowth there. On the other hand, recent agricultural statistics point to a leakage of agricultural production away from the conflict zone to other parts of the country. Implicitly, the socioeconomic pathways underlying the Monte Carlo analysis likely cover the effects of this regional conflict.

To date, 39 countries have submitted FREL's, some of which have also claimed adjustment. The most commonly used approach in doing so is projecting trends of past deforestation into the future, which is generally considered a robust and conservative approach. Does the modeling approach presented here give rise to *"formulating incredibly high deforestation scenarios"* [13], as has been reported for other HFLD countries? Not for our case of Southern Cameroon. The average modeled deforestation for the virtual performance period 2020–2030 is 15,900 ha/yr, which is +50% above the reference period 2000–2015 but is exactly equal to deforestation observed in the period 2005–2015. Hence, shortening the reference period from 15 to the last 10 years (which is often the recommended duration of a reference period suggested by performance-based payment schemes [81]) would have the same effect on a reference level as deploying the mechanistic land use model, but without the benefits of having the breakdown per supply chain of the drivers of deforestation.

The scenarios underlying the most contested reference levels proposed for HFLD countries or regions (see for instance [10,13]) follow an ad-hoc narrative, i.e., "what would happen if … ?". This approach differs significantly from the bottom-up, data driven approach presented here, which stands in the scientific tradition of mechanistic land use modeling (for instance [29,107]). For other countries and jurisdictions to capitalize on this methodology, solid data on population, food, and feed consumption and agriculture and wood production are needed. For countries with a low population and low level of agricultural activity, the methodology presented here will yield in low adjustments of the reference level (see [108]).

5. Conclusions and Recommendations

This article aims at determining a transparent FREL adjusted to likely future developments and highlights significant uncertainties associated with this process. Adjustment of reference levels, such as other outcomes of international climate negotiations [109], are sometimes perceived as loopholes putting actual GHG reductions at risk [10]. In that sense, this article contributes to narrowing these loopholes by proposing a transparent approach to determining a FREL adjusted to future circumstances.

This section, consequently, focuses on making concrete suggestions for improving each element of the FREL with the ultimate objective of developing a FREL for a performance-based payment program.

To that end, the overall objective is to create a GHG monitoring system that consistently spans across all components (AD, EF, and adjustment) of a FREL, subject to TCCA principles and policy-related decisions described in Section 2.2.

Creating and improving the necessary data, information and infrastructure is costly and funding is limited. This section therefore aims to define priorities for future working directions.

5.1. Priorities for Improving Activity Data

The alignment of forest definition and AD mapping would also require differentiating natural forest from mono-specific plantations such as rubber and oil palms in the activity data. The technical feasibility of doing so has been demonstrated for other regions in Cameroon [27,110]. Further, the proof of concept is also made for distinguishing cocoa agroforests from full sun cocoa [111], which will be a requirement for aligning the forest definition and AD for REDD+ in the country.

The reference map of all six IPCC classes should place more emphasis on mapping agricultural land uses, given its role as a primary driver of deforestation. The reference map should split agriculture at least by annual and perennial crops with very different carbon densities (see Table 4). Radar-based systems with the capacity to penetrate clouds and detect canopy texture features (such as Sentinel 1) enable the differentiation of different crops and have been applied both in research [111–113] and in an operational context [111,114].

Forest loss mapping of all REDD+ activities: AD mapping should encompass all D's in REDD: Deforestation and forest degradation, as well as carbon stock enhancement.

Deforestation mapping should allow to trace the fate of cleared forest patches, as having data about land use after forest clearing is critical in terms of choosing the relevant emission factor. This will require a frequent update of the reference land cover map, which appears possible using radar sensors such as Sentinel 1 [111,112,115] and the definition of cut-off dates [116,117] given the often rapid consequential transitions between various land uses. For example, perennial plantains are often planted immediately after clearing [118] since they seem to benefit from the nutrient cocktail remaining in the wood ashes but after around three years they are replaced by other (often annual) crops. In short, spatially explicit tracking of drivers of deforestation and degradation across all relevant sectors will be needed.

Forest gain mapping will require long-term time series analysis to retrieve stand age data [119] combined with long-term forest inventory plots [120]. The inclusion of forest gain and/or forest stock enhancement would allow Cameroon to be one of the few REDD+ countries to report an FRL to the UNFCCC, which would testify to a significant improvement in reporting capacity [71]. It should also be noted that defining forest by a minimum crown cover of ≥ 10% might be an impediment to tracking forest gain in the dense forest areas of southern Cameroon [100].

Mapping of deforestation is mandatory. Activity data other than deforestation can be assessed according to two methods: remote sensing, which gives direct evidence of the extent and intensity of an activity [121–123], or through the use of proxy data. Using proxy data means that emissions from degradation are inferred from land use activity intensities, typically national statistics on forest use, such as national wood harvest. To assess emissions from selective logging, for instance, national wood harvesting data is combined with biomass expansion factors plus a logging damage factor to the remaining forest stand [57,124,125]. Proxy data are only as reliable as the statistics, which underpin them. Another, issue using proxy data is the overlap of different land uses in space and time. For instance, logging for timber and fuel wood for the local market occurs primarily in fallow areas [95], which have already been cleared once and accounted for as such and therefore pose the risk of double-counting.

The combination of both methods as demonstrated on the local scale [123] might improve results. Acknowledging the technical difficulty of detecting degradation, using forest fragmentation and the associated decrease in tree height and biomass in degraded forest edges has been proposed as a robust spatially explicit workaround [126].

5.2. Priorities for Improving Emission Factors and the Next NFI

A long-term network of forest inventory plots will be a requirement for measuring progress towards sustainable forest management and the enhancement of carbon stocks [120], which necessitates a rapid repetition of the NFI dating from 2003/2004.

Spatial misalignment errors of EF and AD should be avoided by using a stratified approach focused on moist and dry forest regions and systematic sampling inside the strata to ensure adequate coverage of forest types proportional to their area coverage with a higher density compared to the NFI from 2003/2004.

Implementation of a cost-effective NFI repetition cycle of 5-10 years building on the clusters and plots from the NFI 2003/3004 will be key to allow assessment of the change in carbon storage. A too long time-lapse since the last assessment should be avoided to not diminish the value of the NFI 2003/2004.

Densifying the sample grid is important in order to represent (1) all five agro-ecological zones of Cameroon, (2) all land use classes, and (3) drivers of deforestation and forest degradation with due statistical representation.

Forest degradation and enhancement of forest carbon stocks, as well as changes in the carbon stocks in remaining forests need to be, and can be, assessed by repeated measurements of the sample plots.

Balancing density and frequency: There should be a reasonable balance of the number of inventory plots and a short (ideally five-year) repetition cycle to enable a timely assessment of forest degradation and enhancement.

Technical and financial resources to the forest inventory unit in the ministry in charge of forestry should be made available to enable it carry out forest inventory in a repetitive manner 5.3. Adjustment to National Circumstances

Demographic dynamics are expected to be better represented by including data from the third national census of 2015. Alternatively, the possibility of using remotely sensed population data should be exploited [127].

Reliable agricultural statistics at the best possible granular level of detail should be collected. Previously, agricultural statistics were available at the department level in Cameroon but in recent years statistics are only available at the regional level—this is a major degradation of data availability.

Drivers of forest degradation should be represented in the model once relevant AD and EF data become available. This should encompass industrial and smallholder logging, cocoa, and fuel wood consumption.

5.3. From UNFCCC Reporting to Performance-Based Payments

The enhancement of coherence both at the institutional- and technical level are major priorities. Moreover, a content analysis of approved applications for performance-based payments from Central- and West Africa [128] and associated technical assessment reports have revealed simplification as an overarching strategy.

This is true for activity data mapping where a re-aggregated binary forest-non-forest land cover map, for example, yields higher accuracies than a thematic one [128]. This is also true for adjusting reference levels: Declaring deforestation for agro-industrial concessions in the most remote parts of the country as "planned" [68] might indeed appear more intuitive than going to lengths in elucidating the socioeconomic drivers underlying deforestation.

When it comes to ex-ante estimation of emissions reductions (ER's), the focus will probably need to be put on two to three well-organized supply chains (such as cocoa, oil palm, or industrial logging) with high potential to reduce emissions. The land use modelling approach offers a convenient framework for quantification potential ER's at the design stage of an ER program, notably for supply chains pursuing a land-sparing strategy. Thereby, projected yield increases resulting from REDD+ interventions translate into potential land sparing, moderated by a discount factors for imperfect translation of yield improvements to spared land. It should be noted, however, that the short-term policy framework of performance-based payment schemes (for instance, the FCPF prescribes a performance period of five years) will make effective implementation and results delivery by REDD+ a major challenge.

Author Contributions: Conceptualization, J.P., B.M. and R.S.; methodology, J.P.; software, A.M. (Aline Mosnier) (land use model), J.P. (Monte Carlo analysis), T.N. (Remote Sensing) and M.D. (NFI assessment); validation, A.M. (Aline Mosnier) (land use model), J.P. (Monte Carlo analysis), T.N. and A.M. (Achille Momo) (Remote Sensing); resources, R.S.; writing—original draft preparation, J.P.; writing—review and editing, B.M., F.K.; visualization, J.P.; supervision, B.M; project administration, F.K.; funding acquisition, F.K.

Funding: This work was supported by the RESTORE+ Project (www.restoreplus.org) which is part of the International Climate Initiative (IKI), supported by the Federal Ministry for the Environment, Nature Conservation, Building and Nuclear Safety (BMU) based on a decision adopted by the German Bundestag. JP and AM received funding under service contracts of MINEPDED and the FCPF. The re-analysis of the national forest inventory data performed by MD was funded by the SilvaCarbon Program (egsc.usgs.gov/silvacarbon).

Acknowledgments: This work was supported by the RESTORE+ Project (www.restoreplus.org) which is part of the International Climate Initiative (IKI), supported by the Federal Ministry for the Environment, Nature Conservation, Building and Nuclear Safety (BMU) based on a decision adopted by the German Bundestag. J.P. and A.M. received supplementary funding under service contracts of MINEPDED and the FCPF. The re-analysis of the national forest inventory data performed by M.D. was funded by the SilvaCarbon Program (egsc.usgs.gov/silvacarbon).

Conflicts of Interest: The authors declare no conflict of interest.

Appendix A

Table A1. Results of the land use model in terms of cropland expansion 2000–2035 for the Southern Cameroon Area (in 1000 ha per 5-year period).

	2000–2005	2005–2010	2010–2015	2015–2020	2020–2025	2025–2030	2030–2035
Cassava	4.84	9.59	8.52	5.68	5.99	7.22	7.99
Mais	5.13	6.64	9.55	8.75	11.63	15.57	19.47
Beans	0.02	0.02	0.03	0.02	0.01	0.01	0.00
Millet/Sorghum	0.00	0.00	0.00	0.00	0.00	0.00	0.00
Oil palm	0.34	0.33	0.52	0.51	0.60	0.67	0.73
Plantain	4.82	5.16	6.68	8.43	11.23	14.22	17.67
Ground nuts	4.57	4.61	6.31	7.63	9.24	10.22	11.66
Banana	1.44	1.21	2.07	2.10	2.68	3.37	4.10
Cacao	19.84	23.08	47.57	15.55	33.56	33.56	33.56
Cotton	0.00	0.00	0.00	0.00	0.00	0.00	0.00

Appendix B

Table A2. List of crops represented in the model and classification as annual or perennial (only with regard to the Emission factor to apply).

Crop Name	Annual/Perennial (for EF Calculation)
Banana	Annual
Beans	Annual
Cassava	Annual
Cocoa	Perennial
Ground nut	Annual
Maize	Annual
Oil palm	Perennial
Plantain	Perennial
Rubber	Perennial

Appendix C

Table A3. Protocol used to drivers assessment.

Disturbance Class	Description of Disturbance
Infrastructure	Geometric areas with very high reflectance value
Croplands	Permanent small and medium-scale agriculture
Logging (Road, selective) – Industrial	Located inside allocated logging concessions; signs of logging infrastructure visible
Mining	Permanent openings with high, stable reflectance
Natural (Wildfires, windfalls, river meandering and other natural disturbances)	Immediate proximity to rivers; fires database
Non-industrial logging	Very short (annual) openings
Road construction	Linear shapes with high reflectance values
Smallholder clearing	Openings for smallholder agriculture (≤ca 1ha) visible for 2–3 years; remainder of all above

The collection of images is available from this website: http://glad.geog.umd.edu/Potapov/Cameroon/Cameroon_index_part2.html.

References

1. IPCC. *Climate Change and Land: An IPCC Special Report on Climate Change, Desertification, Land Degradation, Sustainable land Management, Food Security, and Greenhouse Gas Fluxes in Terrestrial Ecosystems*; IPCC: Geneva, Switzerland, 2019.
2. Pachauri, R.K.; Reisinger, A. *Climate Change 2007: Synthesis Report. Contribution of Working Groups I, II and III to the Fourth Assessment Report of the Intergovernmental Panel on Climate Change*; IPCC: Geneva, Switzerland, 2007.
3. Griscom, B.W.; Adams, J.; Ellis, P.W.; Houghton, R.A.; Lomax, G.; Miteva, D.A.; Schlesinger, W.H.; Shoch, D.; Siikamäki, J.V.; Smith, P.; et al. Natural climate solutions. *Proc. Natl. Acad. Sci. USA* **2017**. [CrossRef] [PubMed]
4. Rockström, J.; Gaffney, O.; Rogelj, J.; Meinshausen, M.; Nakicenovic, N.; Schellnhuber, H.J. A roadmap for rapid decarbonization. *Science* **2017**, *355*, 1269–1271. [CrossRef] [PubMed]
5. Fischer, R.; Hargita, Y.; Günter, S. Insights from the ground level? A content analysis review of multi-national REDD+ studies since 2010. *Forest Policy Econ.* **2016**, *66*, 47–58. [CrossRef]
6. Da Fonseca, G.A.B.; Rodriguez, C.M.; Midgley, G.; Busch, J.; Hannah, L.; Mittermeier, R.A. No Forest Left Behind. *PLoS Biol.* **2007**, *5*, e216. [CrossRef] [PubMed]
7. Obersteiner, M.; Huettner, M.; Kraxner, F.; McCallum, I.; Aoki, K.; Böttcher, H.; Fritz, S.; Gusti, M.; Havlik, P.; Kindermann, G.; et al. On fair, effective and efficient REDD mechanism design. *Carbon Balance Manag.* **2009**, *4*, 11. [CrossRef]
8. Griscom, B.; Shoch, D.; Stanley, B.; Cortez, R.; Virgilio, N. Sensitivity of amounts and distribution of tropical forest carbon credits depending on baseline rules. *Environ. Sci. Policy* **2009**, *12*, 897–911. [CrossRef]
9. Pirard, R.; Karsenty, A. Climate Change Mitigation: Should "Avoided Deforestation" Be Rewarded? *J. Sustain. For.* **2009**, *28*, 434–455. [CrossRef]
10. Seyller, C.; Desbureaux, S.; Ongolo, S.; Karsenty, A.; Simonet, G.; Faure, J.; Brimont, L. The "virtual economy" of REDD+ projects: Does private certification of REDD+ projects ensure their environmental integrity? *Int. For. Rev.* **2016**, *18*, 231–246. [CrossRef]
11. Karsenty, A.; Aubert, S.; Brimont, L.; Dutilly, C.; Desbureaux, S.; Ezzine de Blas, D.; Le Velly, G. The Economic and Legal Sides of Additionality in Payments for Environmental Services. *Environ. Policy Gov.* **2017**, *27*, 422–435. [CrossRef]
12. Sax, S. Fears of a Dire Precedent as Brazil Seeks Results-based REDD+ Payment. 2019. Available online: https://news.mongabay.com/2019/02/fears-of-a-dire-precedent-as-brazil-seeks-results-based-redd-payment/ (accessed on 16 May 2019).
13. Dezécache, C.; Salles, J.-M.; Hérault, B. Questioning emissions-based approaches for the definition of REDD+ deforestation baselines in high forest cover/low deforestation countries. *Carbon Balance Manag.* **2018**, *13*, 21. [CrossRef]
14. Hein, J.; Guarin, A.; Frommé, E.; Pauw, P. Deforestation and the Paris climate agreement: An assessment of REDD+ in the national climate action plans. *Forest Policy Econ.* **2018**, *90*, 7–11. [CrossRef]
15. Forsell, N.; Turkovska, O.; Gusti, M.; Obersteiner, M.; den Elzen, M.; Havlik, P. Assessing the INDCs' land use, land use change, and forest emission projections. *Carbon Balance Manag.* **2016**, *11*, 26. [CrossRef] [PubMed]
16. FAO. *From Reference Levels to Results Reporting: REDD+ under the UNFCCC*; 2018 update; FAO: Rome, Italy, 2018.
17. Gizachew, B.; Duguma, L.A. Forest Carbon Monitoring and Reporting for REDD+: What Future for Africa? *Environ. Manag.* **2016**, *58*, 922–930. [CrossRef] [PubMed]
18. Bucki, M.; Cuypers, D.; Mayaux, P.; Achard, F.; Estreguil, C.; Grassi, G. Assessing REDD+ performance of countries with low monitoring capacities: The matrix approach. *Environ. Res. Lett.* **2012**, *7*, 014031. [CrossRef]
19. Pelletier, J.; Martin, D.; Potvin, C. REDD+ emissions estimation and reporting: Dealing with uncertainty. *Environ. Res. Lett.* **2013**, *8*, 034009. [CrossRef]
20. Bos, A.B.; De Sy, V.; Duchelle, A.E.; Herold, M.; Martius, C.; Tsendbazar, N.-E. Global data and tools for local forest cover loss and REDD+ performance assessment: Accuracy, uncertainty, complementarity and impact. *Int. J. Appl. Earth Obs. Geoinf.* **2019**, *80*, 295–311. [CrossRef]

21. Romijn, E.; De Sy, V.; Herold, M.; Böttcher, H.; Roman-Cuesta, R.M.; Fritz, S.; Schepaschenko, D.; Avitabile, V.; Gaveau, D.; Verchot, L.; et al. Independent data for transparent monitoring of greenhouse gas emissions from the land use sector – What do stakeholders think and need? *Environ. Sci. Policy* **2018**, *85*, 101–112. [CrossRef]

22. Mertens, B.; Sunderlin, W.D.; Ndoye, O.; Lambin, E.F. Impact of Macroeconomic Change on Deforestation in South Cameroon: Integration of Household Survey and Remotely-Sensed Data. *World Dev.* **2000**, *28*, 983–999. [CrossRef]

23. Bikie, H.; Ndoye, O.; Sunderlin, W.D. *L'Impact de la Crise Economique sur les Systemes Agricoles et la Changement du Couvert Forestier Dans la Zone Forestiere Humide du Cameroun*; CIFOR Occasional Paper; CIFOR: Bogor Indonesia, 2000. [CrossRef]

24. Ndoye, O.; Kaimowitz, D. Macro-economics, markets and the humid forests of Cameroon, 1967–1997. *J. Mod. Afr. Stud.* **2000**, *38*, 225–253. [CrossRef]

25. Tegegne, Y.T.; Lindner, M.; Fobissie, K.; Kanninen, M. Evolution of drivers of deforestation and forest degradation in the Congo Basin forests: Exploring possible policy options to address forest loss. *Land Use Policy* **2016**, *51*, 312–324. [CrossRef]

26. Ordway, E.M.; Asner, G.P.; Lambin, E.F. Deforestation risk due to commodity crop expansion in sub- Saharan Africa. *Environ. Res. Lett.* **2017**, *12*, 044015. [CrossRef]

27. Ordway, E.M.; Naylor, R.L.; Nkongho, R.N.; Lambin, E.F. Oil palm expansion at the expense of forests in Southwest Cameroon associated with proliferation of informal mills. *Nat. Commun.* **2019**, *10*, 114. [CrossRef] [PubMed]

28. Carodenuto, S.; Merger, E.; Essomba, E.; Panev, M.; Pistorius, T.; Amougou, J. A Methodological Framework for Assessing Agents, Proximate Drivers and Underlying Causes of Deforestation: Field Test Results from Southern Cameroon. *Forests* **2015**, *6*, 203–224. [CrossRef]

29. Stehfest, E.; van Zeist, W.-J.; Valin, H.; Havlik, P.; Popp, A.; Kyle, P.; Tabeau, A.; Mason-D'Croz, D.; Hasegawa, T.; Bodirsky, B.L.; et al. Key determinants of global land-use projections. *Nat. Commun.* **2019**, *10*, 2166. [CrossRef]

30. WWF. Africa: Cameroon, Central African Republic, Gabon, and Republic of the Congo. Available online: https://www.worldwildlife.org/ecoregions/at0126 (accessed on 17 May 2019).

31. République du Cameroun. *Analyse Approfondie Des Options Strategiques Susceptibles De Regler La Deforestation Et La Degradation Dans Rapport Final Analyse Approfondie Des Options Strate- Giques Susceptibles De Regler La Defores- Rapport Final*; Republic of Cameroon: Yaoundé, Cameroon, 2017.

32. Potapov, P.V.; Turubanova, S.A.; Hansen, M.C.; Adusei, B.; Broich, M.; Altstatt, A.; Mane, L.; Justice, C.O. Quantifying forest cover loss in Democratic Republic of the Congo, 2000-2010, with Landsat ETM+ data. *Remote Sens. Environ.* **2012**, *122*, 106–116. [CrossRef]

33. Olofsson, P.; Foody, G.M.; Stehman, S.V.; Woodcock, C.E. Making better use of accuracy data in land change studies: Estimating accuracy and area and quantifying uncertainty using stratified estimation. *Remote Sens. Environ.* **2013**, *129*, 122–131. [CrossRef]

34. Olofsson, P.; Foody, G.M.; Herold, M.; Stehman, S.V.; Woodcock, C.E.; Wulder, M.A. Good practices for estimating area and assessing accuracy of land change. *Remote Sens. Environ.* **2014**, *148*, 42–57. [CrossRef]

35. Abena, J.C.; Medjo, R.; Blaise, J.; Salomon, P.; Méfé, N.; Salomon, B.; Fonweban, J.; Lekealem, J. *Évaluation des Ressources Forestières Nationales du Cameroun*; FAO: Rome, Italy, 2005.

36. Chave, J.; Condit, R.; Aguilar, S.; Hernandez, A.; Lao, S.; Perez, R. Error propagation and scaling for tropical forest biomass estimates. *Philos. Trans. R. Soc. B Biol. Sci.* **2004**, *359*, 409–420. [CrossRef]

37. Mokany, K.; Raison, J.; Prokushkin, A. Critical analysis of root: Shoot ratios in terrestrial biomes. *Glob. Chang. Biol.* **2006**, *12*, 84–96. [CrossRef]

38. Dees, M. *Analysis of Cameroon NFI 2003–2004 for REDD+ Reporting*; [Unpublished]; Republic of Cameroon: Yaoundé, Cameroon, 2018.

39. Tyukavina, A.; Hansen, M.C.; Potapov, P.; Parker, D.; Okpa, C.; Stehman, S.V.; Kommareddy, I.; Turubanova, S. Congo Basin forest loss dominated by increasing smallholder clearing. *Sci. Adv.* **2018**, *4*, eaat2993. [CrossRef]

40. Ngom, E.; Ndjogui, E.; Nkongho, R.N.; Iyabano, A.; Levang, P.; Miaro, L., III; Feintrenie, L. *Diagnostic du Secteur élæicole au Cameroun*; CIRAD: Yaoundé Cameroun, 2014. [CrossRef]

41. UN-DESA. World Population Prospects - Population Division-United Nations. 2019. Available online: https://population.un.org/wpp/Graphs/ (accessed on 12 July 2019).

42. KC, S.; Lutz, W. The human core of the shared socioeconomic pathways: Population scenarios by age, sex and level of education for all countries to 2100. *Glob. Environ. Chang.* **2017**, *42*, 181–192. [CrossRef] [PubMed]

43. Dayang, R.; Minya, J. *Caracteristiques de l'habitat et Cadre de vie des Populations*; BUCREP: Yaoundé, Cameroon, 2015.

44. The World Bank DataBank. 2019. Available online: https://databank.worldbank.org/home.aspx (accessed on 12 July 2019).

45. Bennett, M.K. Wheat in National Diets. *Wheat Stud.* **1941**, *18*. [CrossRef]

46. PAM Cameroon. *Analyse Globale de la Sécurité Alimentaire et de la Vulnerabilité (2007)*; WFP: Yaoundé, Cameroon, 2007.

47. Muhammad, A.; D'Souza, A.; Meade, B.; Micha, R.; Mozaffarian, D. *The Influence of Income and Prices on Global Dietary Patterns by Country, Age, and Gender*; The World Bank: Washington, DC, USA, 2017.

48. Nkendah, R.; Ako, E.; Tamokwe, B.; Nzouessin, C.; Njoupouognigni, M.; Melingui, E.; Azeufouet, A. *Le Commerce Transfrontalier Informel des Produits Agricoles et Horticoles Entre le Cameroun et ses Voisins de la CEMAC: Implications sur la Sécurité Alimentaire Sous régionale*; Économie rurale Investment Climate and Business Environment Research Fund (ICBE-RF): Dakar, Sénégal, 2011; p. 324.

49. Gromko, D.; Abdurasulova, G. *Climate Change Mitigation and Food Loss and Waste Reduction: Exploring the Business Case*; CCAFS: Wageningen, The Netherlands, 2018; p. 54.

50. HLPE-CFS. *Food Losses and Waste in the Context of Metropolitan Food and Nutrition Security*; HLPE-CFS: Rome, Italy, 2014.

51. FAOSTAT FAOSTAT Gateway. 2019. Available online: http://www.fao.org/faostat/en/#home (accessed on 12 July 2019).

52. Raintree, J.B.; Warner, K. Agroforestry pathways for the intensification of shifting cultivation. *Agrofor. Syst.* **1986**, *4*, 39–54. [CrossRef]

53. IGN. *France Cartographies Forestières Historiques et Détaillées du Cameroun Résultats Statistiques Finaux*; IGN: Paris, France, 2015.

54. Defourny, P.; Bontemps, S.; Lamarche, S.; Brockmann, C.D.; Wevers, J.; Boettcher, M.; Santoro, M.; Kirches, G.; Moreau, I. *Land Cover CCI Product User Guide Version 2.0*; ESA: Louvain-la-Neuve, Belgium, 2017.

55. Hansen, M.C.; Potapov, P.V.; Moore, R.; Hancher, M.; Turubanova, S.A.; Tyukavina, A.; Thau, D.; Stehman, S.V.; Goetz, S.J.; Loveland, T.R.; et al. High-resolution global maps of 21st-century forest cover change. *Science* **2013**, *342*, 850–853. [CrossRef]

56. Sartoretto, E.; Henriot, C.; Bassalang, M.; Nguiffo, S. *How Existing Legal Frameworks Shape Forest Conversion to Agriculture*; FAO: Rome, Italy, 2017.

57. Cerutti, P.O.; Poufoun, J.N.; Karsenty, A.; Eba'a, R.; Atyi, A.; Nasi, R.; Fomete Nembot, T. The technical and political challenges of the industrial forest sector in Cameroon. *Int. For. Rev.* **2016**, *18* (Suppl. 1), 25–39. [CrossRef]

58. Feintrenie, L. Agro-industrial plantations in Central Africa, risks and opportunities. *Biodivers. Conserv.* **2014**, *23*, 1577–1589. [CrossRef]

59. Bulmer, M.G. *Principles of Statistics*; Dover Publications: New York, NY, USA, 1979.

60. Tchatchou, B.; Sonwa, D.; Ifo, S.; Tiani, A. *Deforestation and Forest Degradation in the Congo Basin: State of Knowledge, Current Causes and Perspectives*; CIFOR: Bogor, Indonesia, 2015.

61. Dkamela, G.P.; Brockhaus, M.; Djiegni, F.K.; Schure, J.; Mvondo, S.A. Lessons for REDD + from Cameroon's past forestry law reform: A political economy analysis. *Ecol. Soc.* **2014**, *19*, 30. [CrossRef]

62. Megevand, C.; Dulal, H.; Braune, L.; Wekhamp, J. *Deforestation Trends in the Congo Basin*; Open Konwledge Repository: Washington, DC, USA, 2013.

63. Austin, K.G.; González-Roglich, M.; Schaffer-Smith, D.; Schwantes, A.M.; Swenson, J.J. Trends in size of tropical deforestation events signal increasing dominance of industrial-scale drivers This. *Environ. Res. Lett.* **2017**, *12*. [CrossRef]

64. Galford, G.L.; Soares-Filho, B.S.; Sonter, L.J.; Laporte, N. Will passive protection save Congo forests? *PLoS ONE* **2015**, *10*, e0128473. [CrossRef]

65. Sandker, M.; Lee, D.; Crete, P.; Sanz-Sanchez, M. *Emerging Approaches to Forest Reference Emission Levels and/or Forest Reference Levels for REDD+*; FAO: Rome, Italy, 2014.

66. Njogui, T.E.; Levang, P. Elites urbaines, élaiculture et question foncière au Cameroun. *Territoires d'Afrique* **2013**, *5*, 35–46.

67. Levang, P. *Le Développement du Palmier à Huile au Cameroun: Entre Accaparements Massifs, Agro-industries, élites et Petits Planteurs*; CIRAD, IRD: Montpellier, France, 2012.

68. République du Congo. *Document de Programme de Réductions des Émissions (ER-PD). Programme de Réduction des Émissions dans la Sangha et la Likouala, République du Congo Date*; Republic of Congo: Brazzaville, Republic of Congo, 2017.

69. Forest Carbon Partnership Facility (FCPF). *Emission Reductions Program Document (ER-PD) East Kalimantan Jurisdictional Emission Reductions Program, Indonesia*; The World Bank: Washington, DC, USA, 2018.

70. Sonkoue, M.; Nguiffo, S. *Apes, Crops and Communities: Land Concessions and Conservation in Cameroon*; IIED Briefing: London, UK, 2019.

71. FAO. *From Reference Levels to Results Reporting: REDD+ under the UNFCCC*; FAO: Rome, Italy, 2017.

72. GOFC-GOLD. *A Sourcebook of Methods and Procedures for Monitoring and Reporting Anthropogenic Greenhouse Gas Emissions and Removals Associated with Deforestation, Gains and Losses of Carbon Stocks in Forests Remaining Forests, and Forestation*, COP22-1st ed.; Achard, F., Boschetti, L., Brown, S., Brady, M., DeFries, R., Grassi, G., Herold, M., Mollicone, D., Mora, B., Pandey, D., et al., Eds.; Wageningen University: Wageningen, The Netherlands, 2016.

73. Nkongho, R.; Nchanji, Y.; Tataw, O.; Levang, P. Less oil but more money! Artisanal palm oil milling in Cameroon. *Afr. J. Agric. Res.* **2014**, *9*, 1586–1596.

74. Chong, K.L.; Kanniah, K.D.; Pohl, C.; Tan, K.P. A review of remote sensing applications for oil palm studies. *Geo-Spat. Inf. Sci.* **2017**, *20*, 184–200. [CrossRef]

75. Miettinen, J.; Gaveau, D.L.A.; Liew, S.C. Comparison of visual and automated oil palm mapping in Borneo. *Int. J. Remote Sens.* **2019**, *40*, 8174–8185. [CrossRef]

76. Molinario, G.; Hansen, M.C.; Potapov, P.V.; Tyukavina, A.; Stehman, S.; Barker, B.; Humber, M. Quantification of land cover and land use within the rural complex of the Democratic Republic of Congo. *Environ. Res. Lett.* **2017**, *12*, 104001. [CrossRef]

77. Mitchard, E.T.; Saatchi, S.S.; Baccini, A.; Asner, G.P.; Goetz, S.J.; Harris, N.L.; Brown, S. Uncertainty in the spatial distribution of tropical forest biomass: A comparison of pan-tropical maps. *Carbon Balance Manag.* **2013**, *8*, 10. [CrossRef]

78. Randall, S.; Coast, E. The quality of demographic data on older Africans. *Demogr. Res.* **2016**, *34*, 143–174. [CrossRef]

79. Randall, S. Visibilité et invisibilité statistique en Afrique. *Afr. Contemp.* **2016**, *258*, 41. [CrossRef]

80. Richer, E. *Trade Data Overview: Cameroon*; Forest Trends: Prague, Czech Republic, 2016.

81. FCPF. Carbon Fund Methodological Framework. The Forest Carbon Partnership Facility. 2016. Available online: https://www.forestcarbonpartnership.org/carbon-fund-methodological-framework (accessed on 3 January 2019).

82. Pearson, T.R.H.; Brown, S.; Murray, L.; Sidman, G. Greenhouse gas emissions from tropical forest degradation: An underestimated source. *Carbon Balance Manag.* **2017**, *12*, 3. [CrossRef]

83. Maria Roman-Cuesta, R.; Rufino, M.; Herold, M.; Butterbach-Bahl, K.; Rosenstock, T.S.; Herrero, M.; Ogle, S.; Li, C.; Poulter, B.; Verchot, L.; et al. Hotspots of gross emissions from the land use sector: Patterns, uncertainties, and leading emission sources for the period 2000–2005 in the tropics. *Biogeosciences* **2016**, *13*, 4253–4269. [CrossRef]

84. Tsanga, R.; Lescuyer, G.; Cerutti, P.O. What is the role for forest certification in improving relationships between logging companies and communities? Lessons from FSC in Cameroon. *Int. For. Rev.* **2014**, *16*, 14–22. [CrossRef]

85. Cerutti, P.O.; Suryadarma, D.; Nasi, R.; Forni, E.; Medjibe, V.; Delion, S.; Bastin, D. The impact of forest management plans on trees and carbon: Modeling a decade of harvesting data in Cameroon. *J. For. Econ.* **2017**, *27*, 1–9. [CrossRef]

86. Alemagi, D.; Duguma, L.; Minang, P.A.; Nkeumoe, F.; Feudjio, M.; Tchoundjeu, Z. Intensification of cocoa agroforestry systems as a REDD plus strategy in Cameroon: Hurdles, motivations, and challenges. *Int. J. Agric. Sustain.* **2015**, *13*, 187–203. [CrossRef]

87. Nijmeijer, A.; Lauri, P.-É.; Harmand, J.-M.; Saj, S. Carbon dynamics in cocoa agroforestry systems in Central Cameroon: Afforestation of savannah as a sequestration opportunity. *Agrofor. Syst.* **2019**, *93*, 851–868. [CrossRef]

88. Nijmeijer, A.; Lauri, P.E.; Harmand, J.M.; Freschet, G.T.; Essobo Nieboukaho, J.D.; Fogang, P.K.; Enock, S.; Saj, S. Long-term dynamics of cocoa agroforestry systems established on lands previously occupied by savannah or forests. *Agric. Ecosyst. Environ.* **2019**, *275*, 100–111. [CrossRef]

89. Lescuyer, G.; Cerutti, P.O.; Mendoula, E.E.; Eba, R.; Nasi, R. Chainsaw milling in the Congo Basin. In *European Tropical Forest Research Network (ETFRN) News*, 52nd ed.; Wit, M., van Dam, J., Cerutti, P.O., Lescuyer, G., Kerrett, R., Mckeown, J.P., Eds.; Tropenbos International: Wageningen, The Netherlands, 2010; pp. 121–128.

90. Cerutti, P.O.; Eba'a Atyi, R.; Mendoula, E.E.; Gumbo, D.; Lescuyer, G.; Moombe, K.; Tsanga, R.; Walker, J. Sub-Saharan Africa's invisible timber markets. *ITTO Trop. For. Update* **2017**, *26*, 3–5.

91. Cerutti, P.O.; Lescuyer, G. *The Domestic Market for Small-Scale Chainsaw Milling in Cameroon Present Situation, Opportunities and Challenges*; CIFOR: Bogor, Indonesia, 2011.

92. Bayol, N.; Anquetil, F.; Bile, C.; Bollen, A.; Bousquet, M.; Castadot, B.; Cerruti, P.; Kongape, J.A.; Leblanc, M.; Lescuyer, G.; et al. Filière bois d'oeuvre et gestion des forêts naturelles: Les bois tropicaux et les forêts d'Afrique centrale face aux évolutions des marchés. In *Les Forêts du Bassin du Congo–État des Forêts 2013*; de Wasseig. Neufchâteau: Weyrich, Belgique, 2014.

93. Cerutti, P.O.; Tacconi, L.; Lescuyer, G.; Nasi, R. Cameroon's Hidden Harvest: Commercial Chainsaw Logging, Corruption, and Livelihoods. *Soc. Nat. Resour.* **2013**, *26*, 539–553. [CrossRef]

94. Cerutti, P.O.; Mbongo, M.; Vandenhaute, M. *State of the timber sector in Cameroon (2015)*; FAO and CIFOR: Rome, Italy; Bogor, Indonesia, 2016.

95. Robiglio, V.; Lescuyer, G.; Cerutti, P.O. From Farmers to Loggers: The Role of Shifting Cultivation Landscapes in Timber Production in Cameroon. *Small-Scale For.* **2013**, *12*, 67–85. [CrossRef]

96. CIFOR, I. *Domestic Markets, Cross-Border Trade and the Role of the Informal Sector in Côte d'Ivoire, Cameroon and the Democratic Republic of the Congo*; CIFOR Report for ITTO: Yokohama, Japan, 2016.

97. Eba'a Atyi, R.; Ngouhouo Poufoun, J.; Mvondo Awono, J.-P.; Ngoungoure Manjeli, A.; Sufo Kankeu, R. Economic and social importance of fuelwood in Cameroon. *Int. For. Rev.* **2016**, *18*, 52–65. [CrossRef]

98. Schure, J.; Marien, J.-N.; De Wasseige, C.; Drigo, R.; Salbitano, F.; Dirou, S.; Nkoua, M. Contribution of woodfuel to meet the energy needs of the population of Central Africa: Prospects for sustainable management of available resources. In *The Forests of the Congo Basin-State of the Forest 2010*; Publications Office of the European Union: Luxembourg, 2012; pp. 109–122. [CrossRef]

99. Kanmegne, J. *Slash and Burn Agriculture in the Humid Forest Zone of Southern Cameroon: Soil Quality Dynamics, Improved Fallow Management and Farmers' Perceptions*; Wageningen University: Wageningen, The Netherlands, 2004.

100. Chazdon, R.L.; Brancalion, P.H.S.; Laestadius, L.; Bennett-Curry, A.; Buckingham, K.; Kumar, C.; Moll-Rocek, J.; Vieira, I.C.G.; Wilson, S.J. When is a forest a forest? Forest concepts and definitions in the era of forest and landscape restoration. *Ambio* **2016**, *45*, 538–550. [CrossRef]

101. Bouvet, A.; Mermoz, S.; Le Toan, T.; Villard, L.; Mathieu, R.; Naidoo, L.; Asner, G.P. An above-ground biomass map of African savannahs and woodlands at 25 m resolution derived from ALOS PALSAR. *Remote Sens. Environ.* **2018**. [CrossRef]

102. Heads of Delegation. Krutu of Paramaribo Joint Declaration on HFLD Climate Finance Mobilization. In *Heads of Delegation and Representatives of High Forest Cover and Low Deforestation*; Heads of Delegation: Paramaribo, Surinam, 2019.

103. Verchot, L.V.; Zomer, R.; Van Straaten, O.; Muys, B. Implications of country-level decisions on the specification of crown cover in the definition of forests for land area eligible for afforestation and reforestation activities in the CDM. *Clim. Chang.* **2007**, *81*, 415–430. [CrossRef]

104. Romijn, E.; Ainembabazi, J.H.; Wijaya, A.; Herold, M.; Angelsen, A.; Verchot, L.; Murdiyarso, D. Exploring different forest definitions and their impact on developing REDD+ reference emission levels: A case study for Indonesia. *Environ. Sci. Policy* **2013**, *33*, 246–259. [CrossRef]

105. Business in Cameroon. Portuguese Company Mota-Engil to Build Mbalam-Kribi Railway. Business in Cameroon. 2014. Available online: https://www.businessincameroon.com/mining/0706-4873-portuguese-company-mota-engil-to-build-mbalam-kribi-railway (accessed on 17 July 2019).

106. Business in Cameroon. Mbalam Mining Agreement between Australian Sundance Resources and Cameroon Lapsed. Business in Cameroon. 2019. Available online: https://www.businessincameroon.com/finance/3101-8808-mbalam-mining-agreement-between-australian-sundance-resources-and-cameroon-lapsed (accessed on 17 July 2019).

107. Valin, H.; Sands, R.D.; van der Mensbrugghe, D.; Nelson, G.C.; Ahammad, H.; Blanc, E.; Bodirsky, B.; Fujimori, S.; Hasegawa, T.; Havlik, P.; et al. The future of food demand: Understanding differences in global economic models. *Agric. Econ.* **2014**, *45*, 51–67. [CrossRef]

108. Mosnier, A.; Mant, R.; Pirker, J.; Bodin, P.; Ndinga, R.; Tonga, P.; Havlik, P.; Bocqueho, G.; Maukonen, P.; Obersteiner, M.; et al. *Modelling Land Use Changes in the Republic of the Congo 2000–2030. A report by the REDD-PAC Project*; IIASA, COMIFAC, UNEP-WCMC: Laxenburg, Austria; Yaounde, Cameroon; Cambridge, UK, 2015.

109. Maraseni, T.; Reardon-Smith, K. Meeting National Emissions Reduction Obligations: A Case Study of Australia. *Energies* **2019**, *12*, 438. [CrossRef]

110. Ordway, E.M.; Naylor, R.L.; Nkongho, R.N.; Lambin, E.F. Oil palm expansion in Cameroon: Insights into sustainability opportunities and challenges in Africa. *Glob. Environ. Chang.* **2017**, *47*, 190–200. [CrossRef]

111. Wielaard, N. Satellite Imaging Making Deforestation-Free Cocoa Possible—Satelligence. 2019. Available online: https://satelligence.com/news/satellite-imaging-makingdeforestation-free-cocoa (accessed on 19 June 2019).

112. Bégué, A.; Arvor, D.; Bellon, B.; Betbeder, J.; de Abelleyra, D.; PD Ferraz, R.; Lebourgeois, V.; Lelong, C.; Simões, M.; R Verón, S. Remote Sensing and Cropping Practices: A Review. *Remote Sens.* **2018**, *10*, 99. [CrossRef]

113. Reiche, J.; Verhoeven, R.; Verbesselt, J.; Hamunyela, E.; Wielaard, N.; Herold, M. Characterizing Tropical Forest Cover Loss Using Dense Sentinel-1 Data and Active Fire Alerts. *Remote Sens.* **2018**, *10*, 777. [CrossRef]

114. Starling Verify your Forest Impact. 2019. Available online: https://www.starling-verification.com/#ourTestimonies (accessed on 27 September 2019).

115. Reiche, J.; Hamunyela, E.; Verbesselt, J.; Hoekman, D.; Herold, M. Improving near-real time deforestation monitoring in tropical dry forests by combining dense Sentinel-1 time series with Landsat and ALOS-2 PALSAR-2. *Remote Sens.* **2018**, *204*, 147–161. [CrossRef]

116. Gaveau, D.; Sloan, S.; Molidena, E.; Yaen, H.; Sheil, D.; Abram, N.K.; Ancrenaz, M.; Nasi, R.; Quinones, M.; Wielaard, N.; et al. Four Decades of Forest Persistence, Clearance and Logging on Borneo. *PLoS ONE* **2014**, *9*, e101654. [CrossRef]

117. Austin, K.G.; Mosnier, A.; Pirker, J.; McCallum, I.; Fritz, S.; Kasibhatla, P.S. Shifting patterns of oil palm driven deforestation in Indonesia and implications for zero-deforestation commitments. *Land Use Policy* **2017**, *69*, 41–48. [CrossRef]

118. Kotto-Same, J.; Moukam, A.; Njomgang, R.; Tiki-Manga, T.; Tonye, J.; Diaw, C.; Gockowski, J.; Hauser, S.; Weise, S.; Nwaga, D.; et al. *Alternatives to Slash-and-Burn: Summary Report and Synthesis of Phase II in Cameroon*; ICRAF: Nairobi, Kenya, 2002.

119. Chazdon, R.; Broadbent, E.N.; Rozendaal, D.M.A.; Bongers, F.; Zambrano, A.M.A.; Aide, T.M.; Balvanera, P.; Becknell, J.M.; Boukili, V.; Brancalion, P.H.; et al. Carbon sequestration potential of second-growth forest regeneration in the Latin American tropics. *Sci. Adv.* **2016**, *2*. [CrossRef] [PubMed]

120. Poorter, L.; Bongers, F.; Aide, T.M.; Almeyda Zambrano, A.M.; Balvanera, P.; Becknell, J.M.; Boukili, V.; Brancalion, P.H.; Broadbent, E.N.; Chazdon, R.L.; et al. Biomass resilience of Neotropical secondary forests. *Nature* **2016**, *530*, 211–214. [CrossRef] [PubMed]

121. Hirschmugl, M.; Steinegger, M.; Gallaun, H.; Schardt, M. Mapping Forest Degradation due to Selective Logging by Means of Time Series Analysis: Case Studies in Central Africa. *Remote Sens.* **2014**, *6*, 756–775. [CrossRef]

122. Kleinschroth, F.; Laporte, N.; Laurance, W.F.; Goetz, S.; Ghazoul, J. Road expansion and persistence in forests of the Congo Basin. *Nat. Sustain.* **2019**. [CrossRef]

123. Sufo Kankeu, R.; Sonwa, D.J.; Eba'a Atyi, R.; Moankang Nkal, N.M. Quantifying post logging biomass loss using satellite images and ground measurements in Southeast Cameroon. *J. For. Res.* **2016**, *27*, 1415–1426. [CrossRef]

124. République du Congo. *Niveau des Emissions de Référence pour les Forêts-Soumission au Secrétariat CCNUCC*; République du Congo: Brazzaville, Republic of Congo, 2016.

125. Mosnier, A.; Mant, R.; Pirker, J.; Makoudjou, A.; Awono, E.; Bodin, B.; Gillet, P.; Havlik, P.; Obersteiner, M.; Kapos, V.; et al. *Modelling Land Use Changes in Cameroon 2000–2030*; IIASA: Laxenburg, Austria, 2016.

126. Shapiro, A.C.; Aguilar-Amuchastegui, N.; Hostert, P.; Bastin, J.-F. Using fragmentation to assess degradation of forest edges in Democratic Republic of Congo. *Carbon Balance Manag.* **2016**, *11*, 11. [CrossRef] [PubMed]

127. Wardrop, N.A.; Jochem, W.C.; Bird, T.J.; Chamberlain, H.R.; Clarke, D.; Kerr, D.; Bengtsson, L.; Juran, S.; Seaman, V.; Tatem, A.J. Spatially disaggregated population estimates in the absence of national population and housing census data. *Proc. Natl. Acad. Sci. USA* **2018**, *115*, 3529–3537. [CrossRef]
128. Republic of Côte d'Ivoire. *Emissions Reduction Program Document for Taï National Park*; Republic of Côte d'Ivoire: Abidjan, Côte d'Ivoire, 2019.

© 2019 by the authors. Licensee MDPI, Basel, Switzerland. This article is an open access article distributed under the terms and conditions of the Creative Commons Attribution (CC BY) license (http://creativecommons.org/licenses/by/4.0/).

MDPI

St. Alban-Anlage 66

4052 Basel

Switzerland

Tel. +41 61 683 77 34

Fax +41 61 302 89 18

www.mdpi.com

Forests Editorial Office

E-mail: forests@mdpi.com

www.mdpi.com/journal/forests

www.ingramcontent.com/pod-product-compliance
Lightning Source LLC
LaVergne TN
LVHW071513180726
843512LV00014B/1074